# Monitoring Animal Populations
## and Their Habitats
### A Practitioner's Guide

Brenda McComb
Benjamin Zuckerberg
David Vesely
Christopher Jordan

CRC Press
Taylor & Francis Group
Boca Raton  London  New York

CRC Press is an imprint of the
Taylor & Francis Group, an **informa** business

CRC Press
Taylor & Francis Group
6000 Broken Sound Parkway NW, Suite 300
Boca Raton, FL 33487-2742

© 2010 by Taylor and Francis Group, LLC
CRC Press is an imprint of Taylor & Francis Group, an Informa business

No claim to original U.S. Government works

Printed in the United States of America on acid-free paper
10 9 8 7 6 5 4 3 2 1

International Standard Book Number: 978-1-4200-7055-2 (Hardback)

This book contains information obtained from authentic and highly regarded sources. Reasonable efforts have been made to publish reliable data and information, but the author and publisher cannot assume responsibility for the validity of all materials or the consequences of their use. The authors and publishers have attempted to trace the copyright holders of all material reproduced in this publication and apologize to copyright holders if permission to publish in this form has not been obtained. If any copyright material has not been acknowledged please write and let us know so we may rectify in any future reprint.

Except as permitted under U.S. Copyright Law, no part of this book may be reprinted, reproduced, transmitted, or utilized in any form by any electronic, mechanical, or other means, now known or hereafter invented, including photocopying, microfilming, and recording, or in any information storage or retrieval system, without written permission from the publishers.

For permission to photocopy or use material electronically from this work, please access www.copyright.com (http://www.copyright.com/) or contact the Copyright Clearance Center, Inc. (CCC), 222 Rosewood Drive, Danvers, MA 01923, 978-750-8400. CCC is a not-for-profit organization that provides licenses and registration for a variety of users. For organizations that have been granted a photocopy license by the CCC, a separate system of payment has been arranged.

**Trademark Notice:** Product or corporate names may be trademarks or registered trademarks, and are used only for identification and explanation without intent to infringe.

**Library of Congress Cataloging-in-Publication Data**

Monitoring animal populations and their habitats : a practitioner's guide / authors, Brenda McComb ... [et al.]. -- 1st ed.
    p. cm.
Includes bibliographical references and index.
ISBN 978-1-4200-7055-2 (alk. paper)
1. Wildlife monitoring. 2. Habitat (Ecology) 3. Wildlife management. I. McComb, Brenda C.

QL83.17.M66 2010
639.9--dc22                                                                2009044284

Visit the Taylor & Francis Web site at
http://www.taylorandfrancis.com

and the CRC Press Web site at
http://www.crcpress.com

# Monitoring
Animal
Populations
and Their Habitats

# Contents

Preface .................................................................................................. xiii
The Authors ........................................................................................ xv

**Chapter 1**  Introduction ................................................................... 1
　　　　　　Monitoring Resources of High Value ....................................... 2
　　　　　　　　Economic Value ................................................................. 2
　　　　　　　　Social, Cultural, and Educational Value ............................ 3
　　　　　　　　Economic Accountability ................................................... 3
　　　　　　Monitoring as a Part of Resource Planning ............................. 5
　　　　　　Monitoring in Response to a Crisis .......................................... 7
　　　　　　Monitoring in Response to Legal Challenges ........................ 10
　　　　　　Adaptive Management ............................................................ 11
　　　　　　An Example of Monitoring and Use of Adaptive Management ........ 12
　　　　　　Summary ................................................................................. 13
　　　　　　References .............................................................................. 14

**Chapter 2**  Lessons Learned from Current Monitoring Programs ....... 17
　　　　　　Federal Monitoring Programs ................................................ 18
　　　　　　　　The Biomonitoring of Environmental Status and Trends (BEST) .... 18
　　　　　　　　　　What Is the Goal of the Monitoring Program and
　　　　　　　　　　How Is It to Be Achieved? ......................................... 19
　　　　　　　　　　Where to Monitor? ...................................................... 19
　　　　　　　　　　What to Monitor? ........................................................ 19
　　　　　　　　The North American Breeding Bird Survey (BBS) ......... 20
　　　　　　　　　　What Is the Goal of the Monitoring Program? .......... 20
　　　　　　　　　　Where and How to Monitor? ...................................... 21
　　　　　　　　　　What to Monitor? ........................................................ 21
　　　　　　　　Environmental Monitoring and Assessment Program (EMAP) .... 23
　　　　　　　　　　What Is the Goal of the Monitoring Program? .......... 24
　　　　　　　　　　Where and How to Monitor? ...................................... 24
　　　　　　　　　　What to Monitor? ........................................................ 25
　　　　　　Nongovernmental Organizations and Initiatives ................... 26
　　　　　　　　Monitoring the Illegal Killing of Elephants (MIKE) ....... 26
　　　　　　　　　　What Is the Goal of the Monitoring Program? .......... 27
　　　　　　　　　　Where and How to Monitor? ...................................... 27
　　　　　　　　　　What to Monitor? ........................................................ 28
　　　　　　　　Learning from Citizen-Based Monitoring ........................ 30
　　　　　　　　　　What Is the Goal of the Monitoring Program? .......... 30
　　　　　　　　　　Where and How to Monitor? ...................................... 31

|  |  |  |
|---|---|---|
| | What to Monitor? | 32 |
| | Summary | 33 |
| | References | 34 |
| **Chapter 3** | Community-Based Monitoring | 37 |
| | A Conflict Over Benefits | 38 |
| |    Economic | 38 |
| |    Ethical | 39 |
| |    Education and Community-Enrichment | 40 |
| |    Effectiveness | 43 |
| | Designing and Implementing a Community-Based Monitoring Program | 44 |
| |    The Prescriptive Approach | 45 |
| |       In What Context Does It Work? | 47 |
| |    The Collaborative Approach | 48 |
| |       In What Context Does It Work? | 49 |
| | Suggestions for Scientists | 50 |
| |    Resolving the Underlying Conflict | 51 |
| |    Participatory Action Research | 51 |
| |    Systems Thinking | 52 |
| | Summary | 54 |
| | References | 55 |
| **Chapter 4** | Goals and Objectives Now and Into the Future | 59 |
| | Targeted Versus Surveillance Monitoring | 59 |
| | Incorporating Stakeholder Objectives | 61 |
| |    Participants | 62 |
| |    Data | 62 |
| |    Analysis | 62 |
| |    Results | 62 |
| |    No Surprise Management | 63 |
| | Identifying Information Needs | 63 |
| | The Anatomy of an Effective Monitoring Objective | 64 |
| |    Scientific Objectives | 64 |
| |    Management Objectives | 65 |
| |    Sampling Objectives | 65 |
| |    What? | 65 |
| |    Where? | 66 |
| |    When? | 66 |
| |    Who? | 66 |
| | Articulating the Scales of Population Monitoring | 67 |
| |    Project or Site Scale | 67 |
| |    Landscape Scale | 67 |
| |    Rangewide Scale | 68 |

Contents

|  | Organism-Centered Perspective | 70 |
|---|---|---|
|  | Data Collected to Meet the Objectives | 70 |
|  | Which Species Should Be Monitored? | 74 |
|  | Intended Users of Monitoring Plans | 75 |
|  | Summary | 76 |
|  | References | 76 |

**Chapter 5** Designing a Monitoring Plan .............................................................. 79

    Articulating Questions to Be Answered ........................................ 80
    Inventory, Monitoring, and Research ............................................ 82
    Are Data Already Available? ......................................................... 82
    Types of Monitoring Designs ........................................................ 87
        Incidental Observations ...................................................... 88
        Inventory Designs .............................................................. 88
        Status and Trend Monitoring Designs ............................... 90
        Cause and Effect Monitoring Designs ............................... 94
            Retrospective Analyses and ANOVA Designs ........ 94
            Before–After Control Versus Impact (BACI) Designs ............ 95
            EDAM: Experimental Design for Adaptive Management ........ 97
    Beginning the Monitoring Plan ..................................................... 97
        Sample Design .................................................................... 98
        Selection of Specific Indicators ......................................... 98
        Selection of Sample Sites ................................................ 100
            Detecting the Desired Effect Size ........................... 100
            The Proposed Statistical Analyses ........................... 100
            The Scope of Inference ........................................... 101
    Summary ...................................................................................... 101
    References .................................................................................... 101

**Chapter 6** Factors to Consider When Designing the Monitoring Plan ............ 103

    Use of Existing Data to Inform Sampling Design ...................... 103
        Detectability ...................................................................... 104
        Estimating Detection Distances ....................................... 105
        Estimating Variance Associated with Indicators ............. 105
        Estimating Sample Size ................................................... 105
        Logistical Tradeoff Scenarios .......................................... 106
        Variance Stabilization ...................................................... 106
        Spatial Patterns ................................................................. 107
        Temporal Variation .......................................................... 109
    Cost .............................................................................................. 110
    Stratification of Samples ............................................................. 111
    Adaptive Sampling ...................................................................... 111
    Peer Review ................................................................................. 112
    Summary ...................................................................................... 112

References .................................................................................................. 113

**Chapter 7** Putting Monitoring to Work on the Ground ........................................... 115

Creating a Standardized Sampling Scheme .................................................. 115
    Selecting Sampling Units ........................................................................ 115
    Size and Shape of Sampling Units .......................................................... 115
Selection of Sample Sites ................................................................................ 117
    Simple Random Sampling ....................................................................... 117
    Systematic Sampling ............................................................................... 117
    Stratified Random Sampling ................................................................... 118
Logistics ............................................................................................................ 120
    Safety Plan ............................................................................................... 120
    Resources Needed ................................................................................... 121
    Permits .................................................................................................... 121
Biological Study Ethics ................................................................................... 122
Voucher Specimens ......................................................................................... 122
Schedule and Coordination Plan .................................................................... 122
Qualifications for Personnel ........................................................................... 123
Sampling Unit Marking and Monuments ...................................................... 123
Documenting Field Monitoring Plans ............................................................ 125
    Quality Control and Quality Assurance ................................................. 126
Critical Areas for Standardization .................................................................. 126
    Season and Elevation .............................................................................. 126
    Diurnal Variability .................................................................................. 127
    Clothing Observers Wear While Monitoring ........................................ 127
Budgets ............................................................................................................. 127
Summary ........................................................................................................... 128
References ........................................................................................................ 129

**Chapter 8** Field Techniques for Population Sampling and Estimation ................. 131

Data Requirements .......................................................................................... 131
    Occurrence and Distribution Data .......................................................... 131
    Population Size and Density ................................................................... 132
    Abundance Indices .................................................................................. 133
    Fitness Data ............................................................................................. 133
    Research Studies ..................................................................................... 134
Spatial Extent ................................................................................................... 134
Frequently Used Techniques for Sampling Animals ..................................... 134
    Aquatic Organisms ................................................................................. 135
    Terrestrial and Semi-Aquatic Organisms .............................................. 136
Life History and Population Characteristics ................................................. 141
    Amphibians and Reptiles ........................................................................ 141
    Birds ........................................................................................................ 142
    Mammals ................................................................................................. 142

## Contents

|   |   |   |
|---|---|---|
| | Effects of Terrain and Vegetation | 143 |
| | Merits and Limitations of Indices Compared to Estimators | 144 |
| | Estimating Community Structure | 145 |
| |     Estimating Biotic Integrity | 147 |
| | Standardization and Protocol Review | 147 |
| | Budget Constraints | 147 |
| | Summary | 148 |
| | References | 148 |
| **Chapter 9** | Techniques for Sampling Habitat | 155 |
| | Selecting an Appropriate Scale | 156 |
| |     Hierarchical Selection | 157 |
| | Remotely Sensed Data | 159 |
| |     Aerial Photography | 160 |
| |     Satellite Imagery | 161 |
| |     Accuracy Assessment and Ground-Truthing | 162 |
| |     Vegetation Classification Schemes | 162 |
| | Consistent Documentation of Sample Sites | 163 |
| | Ground Measurements of Habitat Elements | 163 |
| | Methods for Ground-Based Sampling of Habitat Elements | 165 |
| |     Random Sampling | 165 |
| |     Vegetation Sampling | 166 |
| |         Measuring Density | 166 |
| |         Estimating Percent Cover | 166 |
| |         Estimating Tree Height | 167 |
| |         Estimating Basal Area | 168 |
| |         Sampling Dead Wood | 169 |
| |         Estimating Biomass | 169 |
| | Using Estimates of Habitat Elements to Assess Habitat Availability | 170 |
| | Using Estimates of Habitat Elements to Assess Habitat Suitability | 170 |
| | Assessing the Distribution of Habitat Across the Landscape | 172 |
| | Linking Inventory Data to Satellite Imagery and GIS | 172 |
| | Measuring Landscape Structure and Change | 174 |
| | Summary | 175 |
| | References | 176 |
| **Chapter 10** | Database Management | 179 |
| | The Basics of Database Management | 179 |
| | The General Structure of a Monitoring Database | 180 |
| | Digital Databases | 180 |
| | Data Forms | 182 |
| | Data Storage | 184 |
| | Metadata | 184 |

Consider a Database Manager ................................................................ 186
An Example of a Database Management System: FAUNA ............ 187
Summary ..................................................................................................... 188
References .................................................................................................. 188

**Chapter 11** Data Analysis in Monitoring .......................................................... 189

Data Visualization I: Getting to Know Your Data ......................... 190
Data Visualization II: Getting to Know Your Model ..................... 193
    Independence of Data Points ..................................................... 193
    Homogeneity of Variances ......................................................... 193
    Normality ..................................................................................... 194
Possible Remedies if Parametric Assumptions Are Violated .......... 195
    Data Transformation ................................................................... 196
    Nonparametric Alternatives ....................................................... 196
Statistical Distribution of the Data .................................................. 197
    Poisson Distribution ................................................................... 197
    Negative Binomial Distribution ................................................. 198
Abundance and Counts ..................................................................... 198
    Absolute Density or Population Size ........................................ 198
    Relative Abundance Indices ...................................................... 199
    Generalized Linear Models and Mixed Effects ....................... 200
Analysis of Species Occurrences and Distribution ........................ 202
    Possible Analysis Models for Occurrence Data ..................... 203
        Species Diversity ................................................................ 203
        Binary Analyses .................................................................. 203
        Prediction of Species Density ........................................... 204
        Occupancy Modeling ......................................................... 204
    Assumptions, Data Interpretation, and Limitations ................. 204
Analysis of Trend Data ..................................................................... 205
    Possible Analysis Models .......................................................... 206
    Assumptions, Data Interpretation, and Limitations ................. 207
Analysis of Cause-and-Effect Monitoring Data ............................. 207
    Possible Analysis Models .......................................................... 208
    Assumptions, Data Interpretation, and Limitations ................. 209
Paradigms of Inference: Saying Something with Your
Data and Models ................................................................................ 209
    Randomization Tests .................................................................. 209
    Information Theoretic Approaches: Akaike's Information
    Criterion ....................................................................................... 209
    Bayesian Inference ..................................................................... 210
Retrospective Power Analysis ......................................................... 210
Summary ............................................................................................. 212
References ........................................................................................... 212

Contents

**Chapter 12** Reporting ........................................................................................... 219

    Format of a Monitoring Report .................................................... 220
        Title ............................................................................................ 220
        Abstract or Executive Summary ............................................... 220
        Introduction .............................................................................. 221
        Study Area ................................................................................ 221
        Methods .................................................................................... 223
        Results ...................................................................................... 225
        Discussion ................................................................................ 226
        Management Recommendations ............................................. 228
        List of Preparers ...................................................................... 229
        References ................................................................................ 229
        Appendices .............................................................................. 229
        Summary .................................................................................. 230
    Summary ....................................................................................... 230
    References .................................................................................... 230

**Chapter 13** Uses of the Data: Synthesis, Risk Assessment, and
Decision Making ............................................................................ 233

    Thresholds and Trigger Points ...................................................... 233
    Forecasting Trends ........................................................................ 235
    Predicting Patterns Over Space and Time ................................... 236
        Geographic Range Changes ..................................................... 237
        Home Range Sizes ................................................................... 238
        Phenological Changes .............................................................. 238
        Habitat Structure and Composition ......................................... 239
    Synthesis of Monitoring Data ....................................................... 239
    Risk Analysis ................................................................................ 241
    Decision Making .......................................................................... 242
    Summary ....................................................................................... 243
    References .................................................................................... 243

**Chapter 14** Changing the Monitoring Approach ......................................... 247

    General Precautions to Changing Methodology ......................... 247
    When to Make a Change .............................................................. 248
        Changing the Design ............................................................... 248
        Changing the Variables That Are Measured ............................ 249
        Changing the Sampling Techniques ........................................ 250
        Changing the Sampling Locations ........................................... 251
        Changing the Precision of the Samples ................................... 252
        Changing the Frequency of Sampling ..................................... 253

Logistical Issues with Altering Monitoring Programs......................254
Economic Issues with Altering Monitoring Programs ....................254
Terminating the Monitoring Program.............................................255
Summary........................................................................................255
References .....................................................................................256

**Chapter 15** The Future of Monitoring ................................................................257

Emerging Technologies..................................................................258
  Genetic Monitoring ..................................................................258
  Monitoring Environmental Change with Remote Sensing .........259
  Advances in Community Monitoring and the Internet ..............260
A New Conceptual Framework for Monitoring..............................261
  A Reflection on Ecological Thinking.........................................262
  Dealing with Complexity and Uncertainty ...............................263
Summary........................................................................................264
References .....................................................................................264

**Appendix** Scientific Names of Species Mentioned in the Text........................267

**Index**.................................................................................................................271

# Preface

In the face of so many unprecedented changes occurring in our lives, our ecosystems, and our globe, society is more often expecting scientists to provide information that can help guide communities toward a more sustainable future. This book is our attempt to provide a framework that managers of natural resources can use to design monitoring programs that will benefit future generations by providing the information needed to make informed decisions. In addition, we offer tools and approaches that engage individuals in our society in monitoring programs. We firmly believe that people and communities who are empowered in the design and implementation of monitoring programs are more likely to use the information that results from the program, and support it over time.

There are several excellent books on monitoring animal populations, and so what does this book add to the literature? We designed this book to offer a comprehensive overview of the monitoring process, from start to finish. Although there are books that deal with sampling design and the quantitative analysis of population data, there are few that provide practical advice covering the entire evolution of a monitoring plan from incorporating stakeholder input to data collection to data management and analysis to reporting. This book strives to present an overview of this process. We also acknowledge that any such effort tends to reflect the interests and expertise of the authors, and as such, there is a distinct emphasis on monitoring vertebrate populations and upland habitats. Although many of our examples tend to focus on bird populations and forested habitats, we have made an attempt to cover other taxa and habitat types as well, and many of the recommendations and suggestions that we present are applicable to a diversity of monitoring programs.

This book was written to fill a practical need and also to embrace a set of values that we hold dear. We wanted a book that could be used in a classroom because we feel that students in natural resources programs need to know how to design a monitoring program when they enter the workforce. We also realize that many former students now in the workforce did not have that training and may find this book of value to them.

The values that we hold are for a world in which biodiversity is allowed to be maintained, to evolve, to adapt, and to flourish in the face of such uncertainties as climate change, invasive species proliferation, land use expansion, and population growth. These are huge challenges and the information needed to address them must not only be reliable but also available to all affected parties involved in decision-making processes. The stakes, at least to us, are high. The loss of biodiversity robs future generations of opportunities to experience as rich a diversity of life as the world is capable of offering them. Through the proper monitoring and management of our natural resources we would hope that a foundation is laid for this generation to do the same for the ones that follow.

Many people contributed to the material in this book. The impetus for writing this book came from an early effort proposed by Lowell Suring, Richard Holthausen, and

Christina Hargis; they and Juraj Halaj made many contributions to an early monitoring protocol development guide that was the basis for this book. Reviewers of individual chapters made excellent suggestions: Todd Fuller, Brett Butler, Marty Roberts, David Barton Bray, Daniel Kramer, Daniel Fink, Wesley M. Hochachka, and James P. Gibbs, as well as the students in a graduate level course at the University of Massachusetts who provided excellent feedback on early drafts of the chapters: Dennis Babaasa, Laurel Carpenter, Paul Ekness, Jennifer Fill, Michelle Labbe, Rachel Levine, Maili Page, Theresa Portante, Jennifer Strules, and Rebecca Weaver.

Photos were generously provided by Cheron Ferland, Laura Navarrette, Nancy McGarigal, Mike Jones, Katharine Perry, and Laura Erickson (www.lauraerickson.com/), as well as federal agencies including U.S. Geological Survey, U.S. Forest Service, National Park Service, and U.S. Fish and Wildlife Service. Numerous publishers and journals generously allowed us to use previously published figures and text, and they are cited herein.

The College of Natural Resources and the Environment at the University of Massachusetts allowed the senior author the opportunity to continue working on this book while covering administrative duties. Randy Brehm from the Taylor & Francis Group of publishers provided outstanding editorial support and assistance throughout the project.

Finally, we thank our families and the friends who have supported us through the development of this book, all our many other projects, and the various trials and tribulations of life that led us to this place and time in our lives. B. McComb is thankful for all of the support expressed by Kevin, Michael, and Gina throughout many years of projects and life challenges. B. Zuckerberg is eternally grateful for the support of his wife, Frieda, and two daughters, Isabel and Leila, for their unwavering support and patience, and for the confidence and guidance of his parents, Richard and Joan. Thank you all for your encouragement and good humor throughout the years. D. Vesely wishes to thank his wife, Joan C. Hagar, for introducing him to peregrine falcons on a cliff above Lake Superior more than 20 years ago, and for providing inspiration by her own efforts toward wildlife conservation ever since. C. Jordan would like to thank his family for their unwavering support, the community of La Reserva El Patrocinio for its many life lessons, and Brenda McComb for her guidance and respect, and for encouraging him to pursue an academic career with passion, humanity, and honesty.

# The Authors

**Brenda C. McComb** is a professor and head of the Department of Forest Ecosystems and Society, Oregon State University. She is author of over 130 technical papers dealing with forest and wildlife ecology, habitat relationships, and habitat management. She was born and raised in Connecticut at a time and place when the rural setting provided opportunities to roam forests and fields. She received a B.S. degree in natural resources conservation from the University of Connecticut, an M.S. degree in wildlife management from the University of Connecticut, and a Ph.D. in forestry from Louisiana State University. She has served on the faculty at the University of Kentucky and Oregon State University. She was head of the Department of Natural Resources Conservation at the University of Massachusetts–Amherst for 7 years, served as the chief of the Watershed Ecology Branch in Corvallis for U.S. Environmental Protection Agency for 1 year, and was associate dean for research and outreach in the College of Naural Resources and the Environment at the University of Massachusetts for 1 year. She has been a member of The Wildlife Society and the Society of American Foresters for over 25 years. Her current work addresses interdisciplinary approaches to management of multiownership landscapes in Pacific Northwest forests and agricultural areas.

**Benjamin Zuckerberg** is a postdoctoral research associate in the citizen science program at the Cornell Lab of Ornithology in Ithaca, New York. He was born and raised in Brooklyn, and despite his urban surroundings, became a nature enthusiast and conservationist. He received his B.A. in zoology from Connecticut College, an M.S. in wildlife and fisheries conservation from the University of Massachusetts–Amherst, and a Ph.D. in ecology at the State University of New York College of Environmental Science and Forestry. His research has focused on the management and conservation of early successional birds and the use of breeding bird atlases for documenting the effects of climate change and land use practices on bird populations at regional scales. His current research focuses on using spatial statistics and citizen science for addressing the effects of climate change and habitat loss on bird populations throughout the United States.

**David G. Vesely** is the executive director of the Oregon Wildlife Institute in Corvallis, Oregon. He received a B.A. degree in psychology from the University of Minnesota, a B.F.A. degree in illustration from Oregon State University, and an M.S. degree in forest science also from Oregon State University. He was president of Pacific Wildlife Research Inc., a consulting firm that assisted government agencies and private companies conduct special wildlife studies and prepare environmental analyses. He is a member of The Wildlife Society and Northwest Scientific Association. The current focus of his work involves the conservation of threatened wildlife in the Pacific Northwest and investigating the use of conservation detector dogs for surveys of rare plants and animals.

**Christopher A. Jordan** is a Ph.D. candidate in the Department of Fisheries and Wildlife at Michigan State University. He received a B.S. degree in wildlife and fisheries conservation, a B.A. degree in Spanish, and a certificate in Latin American studies from the University of Massachusetts–Amherst. He has worked with a number of universities and organizations to engage communities in the United States, Guatemala, and Nicaragua in the monitoring of their local ecosystems and has extensive experience playing with motion sensor cameras in all of these countries. His proposed research aims to assess the impact of globalization on resource use and biodiversity along the Caribbean coast of Nicaragua by collaborating closely with local assistants and a team of researchers from multiple disciplines.

# 1 Introduction

There are many reasons to enter into a monitoring program, but the reasons must be well considered before doing so. Long-term monitoring takes time, money, and effort that could be spent in other endeavors such as management, research, and outreach. Monitoring is conducted most often when the resources of concern are of high economic or social value, part of a legally-mandated planning process, the result of a judicial decision, or the result of a crisis. Many times it is a combination of these factors that gives rise to a monitoring program. In addition, monitoring may be conducted as part of a formal research program where long-term trends in an ecologically or socially important response variable are the most important outcome. Also if a population or natural community seems rare or there is the perception by scientists or stakeholders that a decline is evident, then monitoring may be called for to clarify the perception. Most often monitoring programs are designed to help managers and policy makers make more informed decisions. Monitoring allows decisions to be based less on beliefs and more on facts. We may believe that grasshopper sparrows in New England are decreasing in abundance because grasslands are being converted to housing (Figure 1.1). Only after a rigorous, unbiased monitoring program has been in place can we say that yes, indeed, the population seems to be declining (Figure 1.2) and that the decline is associated with the loss of grasslands. However, we cannot ascribe the cause of the decline to grassland loss unless a more rigorous research program is put into place. Monitoring provides the hypothesis for the decline; research is often implemented in a structured before–after control–impact design to assess cause-and-effect relationships.

**FIGURE 1.1** Grasslands in the northeast have declined in abundance due to ecological succession to forests, and to housing developments. Does this lead to a decline in species associated with grasslands? (Photo by Laura Erickson. With permission.)

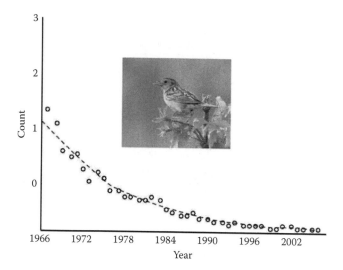

**FIGURE 1.2** Monitoring data provides evidence that at least one species of grassland bird is declining markedly in New York State. (Redrafted from Sauer, J.R., J.E. Hines, and J. Fallon. 2007. The North American Breeding Bird Survey, Results and Analysis 1966–2006. Version 10.13.2007. U.S. Geological Survey Patuxent Wildlife Research Center, Laurel, MD; photo inset by Laura Erickson. With permission.)

## MONITORING RESOURCES OF HIGH VALUE

Typically, we think of monitoring as something that is done because we value a resource and we do not want to lose it, or we wish to maximize it. For goods and services, we often will use monitoring so that we can maximize profit margins while minimizing adverse effects. But economics are not the only values placed on resources. Monitoring the haze present over the Grand Canyon is in response to aesthetic values as well as human health concerns. Ottke et al. (2000) described the importance of considering cultural values in natural resources monitoring and provided examples from 13 case studies around the world. As an example, rarity, in of itself, is often used to initiate a monitoring program. Rare species, populations, or gene pools may be valued sufficiently to initiate and maintain a monitoring program to ensure that these rare organisms persist. Regardless of the motivation for initiating a monitoring program, all require a monitoring approach that allows unbiased sampling, assessment of trends over time, the potential for extrapolation to unsampled areas, and (in some cases) comparisons between areas managed in different ways.

### Economic Value

Monitoring the antler size of white-tailed deer killed on a property leased for hunting, crop loss from Canada goose foraging, or monitoring tree growth on a timber industry landholding all represent examples of specific resources that the owner or manager may wish to manage for economic gain. Economic value commonly drives monitoring programs on a variety of scales. The U.S. Forest Service Forest Inventory and Analysis program is a good example of a well-structured national monitoring program that was initiated to assess the timber value, primarily economic, on non-federal lands in the United States (Sheffield et al. 1985). Over time, however, the program evolved into a multiresource monitoring program, and has since been used to assess other natural resource values on non-federal forest lands (McComb et al. 1986). Similarly, but on a local scale, farmers may monitor the effects of birds on corn seed depredation, or deer on soybean production. Communities in Africa are directly involved in monitoring programs to assess the potential for crop damage and then work with authorities to find ways of minimizing the adverse effects of having African elephants in their fields (Songorwa 1999).

## SOCIAL, CULTURAL, AND EDUCATIONAL VALUE

Monitoring systems that provide the basis for resource management decisions are often initiated and maintained to support resources held in the public trust. Yet not all resources held in the public trust are monitored. While the selection of resources that are monitored is partially driven by economics, the perceptions, concerns, and cultural values of society also play a role. Programs such as the Breeding Bird Survey Program (Sauer et al. 2007), the North American Amphibian Monitoring Program run by the U.S. Geological Survey, the Monitoring Avian Productivity and Survivorship (MAPS) Program created by the Institute for Bird Populations in 1989, and the Environmental Protection Agency's (EPA) program of monitoring and assessing water quality all represent organized, large-scale efforts to acquire data to make more informed resource management decisions. Each has indicators chosen due to a variety of social, cultural, and economic values.

Programs such as these that are supported by federal agencies also have a long-standing reputation for monitoring various biophysical components of ecosystems. The Long Term Ecological Research (LTER) program maintains sites throughout the United States that provide long-term information on ecosystem structure and function. More recently, the National Ecological Observatory Network (NEON) was initiated as a continental-wide program to help understand the impacts of climate change, land-use change, and invasive species on ecosystems and ecosystem services. The importance of these data may not be apparent for years or decades, but the educational benefit that accrues over time may be invaluable. Consider the impact of having monitored carbon dioxide in the atmosphere, ice cover, and plant phenology (the timing of flowering and fruiting) that collectively provided evidence for climate change and insights into likely changes in biota.

## ECONOMIC ACCOUNTABILITY

When push comes to shove, however, economics is almost always the horse that pulls the cart in natural resource programs. When an instructor wants to monitor the progress of a student in learning material she gives a test, or asks for a paper or report. When members of Congress on an Appropriations Committee allocate millions of dollars to the U.S. Fish and Wildlife Service to ensure protection of endangered species, they want to know that the money is being spent wisely and that the actions being taken are effective. Indeed, the Government Accountability Office (GAO) has as its primary responsibility monitoring of appropriations to ensure that the taxpayers' dollars are being spent wisely by our federal agencies (GAO 2007). In both these cases, whether a teacher or an appropriations committee, the supervisory power is asking for a sense of accountability that can only be ascertained through careful monitoring.

In a 2007 report developed jointly by the GAO and the National Academy of Sciences, they state, "One of the greatest challenges facing the United States in the 21st century is sustaining our natural resources and safeguarding our environmental assets for future generations while promoting economic growth and maintaining our quality of life," if that is even possible (Czech 2006). "To manage natural resources effectively and efficiently, policymakers need information and methods to analyze the dynamic interplay between the economy and the environment. Enhancing the information used to make sound decisions can be facilitated by developing national environmental assessments. These assessments provide a framework for organizing information on the status, use, and value of natural resources and environmental assets, as well as on expenditures on environmental protection and resource management" (GAO 2007). Forums such as this one (GAO 2007) provide a strategic and economic framework for the integration of monitoring efforts that span agencies and resources. Whether it is a student taking a quiz or a researcher managing a multi-million dollar monitoring program, the goal is to find the answer to a simple question: "How are we doing?"

So who cares about monitoring and the millions spent on it? You should. This is because public funds often drive monitoring programs, and resources held in the public trust are frequently the targets being monitored! Those who represent you in Congress and in state legislatures, local planning boards, and nongovernmental organizations (NGOs) boards of directors should also care about monitoring. Government agencies and NGOs have issued "state of the environment" reports for countries such as the United States, Australia, and Canada (Environment Canada 1996; Heinz Center 2002; Beeton et al. 2006). A compilation of over 50 such reports has been assembled by the National Council on Science, Policy, and the Environment. These reports are based on whatever monitoring data are available to directly report changes over time in important resources or indicators of those resources. Similar reports, although less common, are also beginning to emerge from scientists working in developing nations (Guarderas et al. 2008).

In addition to these broad "state of the environment" reports, policy makers and elected officials often demand that agencies provide periodic updates on the effectiveness of their work. Is the U.S. taxpayer getting the "biggest bang for the buck?" Is our management effective? Why should we continue to pay for collecting data year after year? According to managers and regulators employed by state and federal agencies, NGOs, and industries, accountability has become a key component of their work. Industry often is most concerned about the economic efficiency of certain management actions. If management actions are not as effective as planned and the monitoring influences the bottom line (it will), then industry will demand a change to more efficient and effective management and monitoring. A timber company may wish to ensure that the goals of leaving a riparian buffer strip are met to the extent that it was worth foregoing the profits, or an NGO may wish to ensure that their limited funds are being effective in restoring a prairie ecosystem. Hence, from purely a practical standpoint, monitoring questions are often of utmost importance to a manager because they are designed to assess how far her expenses go toward meeting her goals. At the end of the day, the results of such assessments will determine whether or not a management action is viable.

But monitoring is not free. It costs money to do it correctly. Hence, monitoring efforts are also driven by the money available to spend on monitoring. Indeed, whether we like it or not, budgets determine our options in resource management, and funds for monitoring are always among the first parts of the budget to be critically reviewed. The system tends to encourage short sightedness: in many budget planning processes it is easier to acquire funding for innovative projects than to continue ongoing efforts. Getting funding to build a new visitor center at a refuge may be easier than maintaining it. Getting a monitoring program initiated may be easier than finding the funding to continue it for a long enough period of time to ensure that the results are used. The implications associated with continuing a commitment to a monitoring program must be accounted for in the design of monitoring programs.

## MONITORING AS A PART OF RESOURCE PLANNING

Monitoring is also a key part of the planning process used by federal agencies, many NGOs, and some industries. People make plans. You have plans for the weekend, for your next vacation, or for your retirement. Plans are based on assumptions, some of which may turn out to not be correct, and despite the best plans, there are often uncertainties that arise to disrupt plans. If you get a flat tire on your car then your plans change for the weekend. Monitoring the function of your car by regularly checking the tire pressure may have prevented that flat. A U.S. Fish and Wildlife Service (USFWS) Refuge may have a refuge management plan, but if an invasive species should establish itself unexpectedly, then the plan may have to change. Monitoring the changes in the primary structures and functions of a refuge (plant communities, distributions of species, erosion, sedimentation, rates of change in species dominance) may allow quick response and rapid removal of the invasive species that may not be possible if one must wait for the next planning cycle. Hence, monitoring is almost always included as a key component of natural resource (and other) plans.

Certainly there are specific guidelines regarding monitoring resources on federal landholdings such as USFWS National Wildlife Refuges (Schroeder 2006). Yet the specifics of the monitoring goals, strategies, and interpretation are often left somewhat vague in Comprehensive Conservation Plans (CCPs), National Forest Management Plans, and many others. Clearly there are exceptions to this (see the Northwest Forest Plan example below), but quite often the development of a detailed monitoring plan comes after the management plan has been developed and approved and not developed as an integral component of the management plan. If we truly

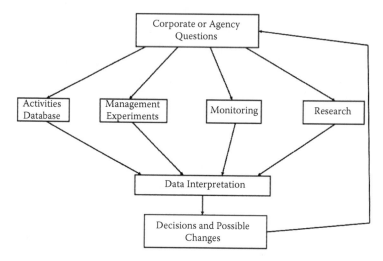

**FIGURE 1.3** In order to make wise management decisions, monitoring is one important avenue for gaining new information. But it is not the only avenue. Formal research and management experiments also contribute to the information. (Redrafted from Haynes, R.W., B.T. Bormann, D.C. Lee, and J.R. Martin, Jon R., tech. eds. 2006. Northwest Forest Plan—the first 10 years (1994–2003): synthesis of monitoring and research results. U.S. Department of Agriculture, Forest Service, Pacific Northwest Research Station. Gen. Tech. Rep. PNW-GTR-651. Portland, OR. 292 pp.)

do care if we are being effective in our management and if we are spending money wisely to achieve goals, then the monitoring plan should be an integral component of a management plan (Figure 1.3).

From the standpoint of achieving planning goals that relate to wildlife species and their habitats, a properly designed monitoring effort allows managers and biologists to understand the long-term dependency of selected species on various habitat elements. Habitat is defined as the set of resources that support a viable population over space and through time (McComb 2007). Identifying those key resources, or reliable indicators of them, can provide information on how a species may respond to changes. The challenge when developing a monitoring plan is to assess the impacts of the dynamic nature of resource availability on a species. In other words, we must assess if changes in occurrence, abundance, or fitness in a population are independent from or related to changes in the availability of resources assumed to contribute to the species' habitat (Cody 1985).

Even with timely planning, implementation of any natural resources plan is done with some uncertainty that the actions will achieve the desired results. Nothing in life is certain (except death!). But by incorporating uncertainty into a project we can reduce many of the risks associated with not knowing. Managers should expect to change plans following implementation based on measurements taken to see if the implemented plan is meeting their needs. If not, then mid-course corrections will be necessary. Many natural resource management organizations in North America use some form of adaptive management (Figure 1.4) as a way of anticipating changes to plans and continually improving plans (Walters 1986).

# Introduction

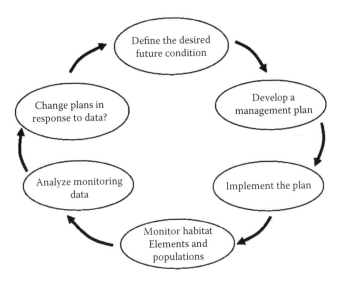

**FIGURE 1.4** The adaptive management cycle is designed to improve information used to make better management decisions.

## MONITORING IN RESPONSE TO A CRISIS

Monitoring to address a perceived crisis has been repeated with many species: northern spotted owls, red-cockaded woodpeckers, and nearly every other species that has been listed as threatened or endangered in the United States under the Endangered Species Act or similar state legislation. Although many of these monitoring programs were developed during a period of social and ecological turmoil, many are also remarkably well structured because the stakes are so high. For instance, in the case of the northern spotted owl, the butting heads of high economic stakes and the palpable risk of loss of a species culminated in a crisis that spawned one of the most comprehensive and costly wildlife monitoring programs in U.S. history: the Northwest Forest Plan (NWFP). The NWFP was designed to fulfill the mandate of the Endangered Species Act by enabling recovery of the federally endangered northern spotted owls and also addressed other species associated with late successional forests over 10 million hectares of federal forest land in the Pacific Northwest of the United States. In his record of decision regarding the plan, Judge Dwyer emphasized the importance of effectiveness monitoring to the NFWP, and monitoring has been an integral part of it since its implementation: "Monitoring is central to the [plan's] validity. If it is not funded or done for any reason, the plan will have to be reconsidered" (Dwyer 1994; USDA, USDI 1994).

One component of NWFP effectiveness monitoring was a plan for the northern spotted owl. The northern spotted owl monitoring program is one of the most intensive avian population monitoring efforts in North America. The purpose of the plan is to record data that reveal trends in spotted owl populations and habitat to assess the success of the NFWP at reversing the population decline for this species (Lint 2005). To this end, the specific objectives of the monitoring program are to (1) assess

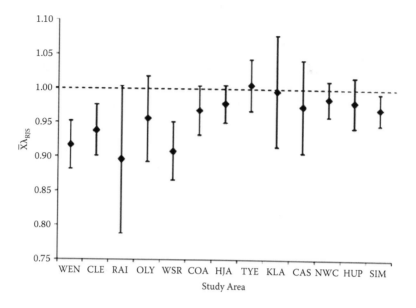

**FIGURE 1.5** Estimates of mean annual rate of population change, $\lambda_{RJS}$, with 95% confidence intervals for northern spotted owls in 13 study areas in Washington, Oregon, and California, based on random effects modeling and with model $\{\varphi(t)\ p(t)\ \lambda(t)\}$, where t represents annual time changes. (Adapted from Anthony, R.G. et al. 2004. Status and trends in demography of northern spotted owls, 1985–2003. Final Report to the Regional Interagency Executive Committee, Portland, OR. With permission from R.G. Anthony.)

changes in population trends and demographic performance of spotted owls on federally administered forest land within the range of the owl, and (2) assess changes in the amount and distribution of nesting, roosting, foraging, and dispersal habitat for spotted owls on federally administered forest land (Lint 2005).

Population monitoring for northern spotted owls encompasses 13 demographic study areas from northern Washington to northern California. Three parameters are estimated from the data to assess trends: survival, fecundity, and lambda (population rate of change). As you can see from Figure 1.5 the trends in population change varied quite widely among the demographic study areas, lending support for use of these study areas as strata within the monitoring framework.

Populations seem to be declining on the Wenatchee (WEN) site in the eastern Washington Cascades, but remaining somewhat stable on the Tyee (TYE) site in the Oregon Coastal Ranges (Figure 1.6). In a case such as this, with such wide differences in trends, where does that leave managers regarding use of these data? The magnitude of population declines on the Wenatchee study site raises significant concerns and the first reaction is that the plan has failed. But the Tyee data indicate that the plan is succeeding. So which is it? Lint (2005) concluded that it is too early to say if the plan has failed or succeeded because restoration of habitat for the species takes longer than the 10 years that monitoring had occurred. But monitoring also revealed other stressors on the population such as competition with barred owls and the potential for increased mortality from West Nile virus, further complicating the

# Introduction

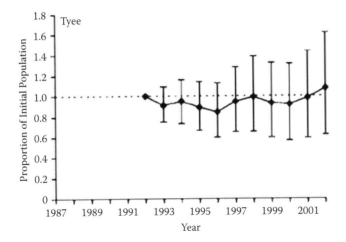

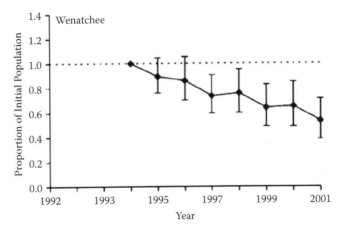

**FIGURE 1.6** Estimates of realized population change, Dt, with 95% confidence intervals for northern spotted owls in the Tyee (Oregon Coast Range) and Wenatchee (Washington Cascades). (Adapted from Anthony, R.G. et al. 2004. Status and trends in demography of northern spotted owls, 1985–2003. Final Report to the Regional Interagency Executive Committee, Portland, OR. With permission from R.G. Anthony.)

interpretation of the monitoring data. Indeed, even with the most rigorous design, uncertainty is inevitable.

Given the importance of economic considerations, the question begs to be asked: Was the monitoring worth over $2 million spent per year (Lint 2001)? Consider the price that taxpayers would pay for not monitoring. First, we could easily lose a species due to plan failure or from other more contemporary stressors. Second, the NWFP would likely be challenged in court again, costing the taxpayers a considerable amount in legal fees. Third, we learned much more about the species and drivers of populations by having collected these data so that the results can (and do) influence how managers make decisions. We also have research-quality data to address future issues with this species and others like it. But the answer to the question "Was the monitoring worth it?" is, clearly, "It depends." It depends on who is doing the evaluating. Some segments of society will answer "Of course, it was worth it." Others will say that "It had to be done legally." Still others would say that "the money would have been better spent on addressing the needs of the displaced forest workers." All are valid points. And the data collected provide each segment of society information on which to base their arguments. Thus, although the cost is the deciding factor, societal values can never be ignored; it is, after all, society who grants us a social license to manage animals and their habitats.

## MONITORING IN RESPONSE TO LEGAL CHALLENGES

Effectiveness monitoring was strongly suggested by Judge Dwyer in his Record of Decision on the implementation of the NFWP over 10 million hectares of federal lands in the Pacific Northwest. His decision emphasized the importance of monitoring as a component of this multiagency plan and influenced plan design. But legal decisions can not only influence the structure of a plan but sometimes determine if the plan and all of its principal components, including monitoring, are implemented.

If a resource is valued highly enough, litigation may be enacted that results in judicial decisions that influence the likelihood that monitoring will be conducted. For instance, there are times when monitoring is an integral part of a written plan, but agencies and managers do not have the funding to initiate or maintain a monitoring program. Concerned citizens may file a lawsuit that results in the reappropriation of public funds to provide for monitoring. A slightly different example of this involves the McLean Game Refuge, a 1,700-ha private tract located in north-central Connecticut in the towns of Granby and Simsbury. The refuge was established in 1932 by bequest of former Connecticut Governor and U.S. Senator George McLean. Decisions about refuge use, maintenance, and management are made by a manager under the oversight of a board of trustees. A proposal to use partial cutting approaches (thinning, group selection, and shelterwood methods) in the McLean Game Refuge in 2001 met with significant opposition by local residents.

The decision to manage the forest was based on suggestions from natural resources professionals that active management could diversify forest structure and composition and hence could lead to more diverse animal communities. Following a public meeting and a series of hearings in civil court, this opposition culminated in a judicial ruling that allowed the refuge manager to proceed with the harvest. However, the judge also encouraged the manager to monitor the changes in animal species composition and habitat so that any future harvests could be informed by the information gained from the monitoring effort. The judicial decision not only stipulated that monitoring ought to be carried out but also influenced how managers monitor animals and habitat. Monitoring of bird communities and habitat structure and composition was initiated prior to the timber harvest and again after the harvests (Figure 1.7). Monitoring indicated that following one growing season after harvests, detections of seven species were higher in the thinned stands while detections of wood thrushes were higher in uncut controls. Was the cutting the correct thing to do? That depends on who is asking the question, but now the debate can be more informed than it was in 2001.

**FIGURE 1.7** A forest stand harvested in 2003 on the McLean Game Refuge in Granby and Simsbury, Connecticut, following considerable debate regarding social concerns for this ecosystem. A judicial ruling allowed the harvest to proceed, but monitoring the effects of the harvest was encouraged by the judge.

## ADAPTIVE MANAGEMENT

Although adaptive management has already been introduced above, it deserves to be addressed at greater length because it is central to successful monitoring and management practices. Adaptive management is a process to find better ways of meeting natural resource management goals by treating management as a hypothesis (Figure 1.4). The results of the process also identify gaps in our understanding of ecosystem responses to management activities. The adaptive management process incorporates learning into the management planning process, and the data collected from the monitoring conducted within this framework provides feedback about the effectiveness of preferred or alternative management practices. The information gained can help reduce the uncertainty associated with ecosystem and human system responses to management. Adaptive management has been classified as both active and passive (Walters and Holling 1990).

Passive adaptive management is a process where the best management options and associated actions are identified, implemented, and monitored. The monitoring may or may not include unmanaged reference areas as points of comparison to the managed areas. The changes observed over time in the managed and reference areas are documented, and the information is used to alter future plans. Hence, the manager learns by managing and monitoring, but the information that is gained from the process is limited, especially if reference areas are not used. Without reference areas we do not know if changes over time are due to management or some other exogenous factors.

Active adaptive management treats the process of management much more like a scientific experiment than passive adaptive management. Under active adaptive management, management approaches are treated as hypotheses to be tested. The hypotheses are developed specifically to identify knowledge gaps and management actions are designed to fill those gaps. Typically, hypotheses are developed based on modeling the responses of the system to management (e.g., using forest growth models, or landscape dynamics models). Management is then conducted and key states and processes are monitored to see if the system responded as it was predicted. Reference areas are also monitored and the data from these areas are used as controls to compare responses of ecosystems and human systems to management. By collecting monitoring data in a more structured hypothesis-testing framework, system responses can be quantified and used to identify probabilities associated with achieving desired outcomes in the future. Whereas passive adaptive management is somewhat reactive in approach (reacting to monitoring data), active adaptive management is proactive and follows a formal experimental design.

Adaptive management generally consists of six major steps (Figure 1.4):

- Set goals (define the desired future condition)
- Develop a plan to meet the goals based on best current information
- Implement the plan
- Monitor the responses of key states and processes to the plan
- Analyze the monitoring data
- Adjust the plan based on results from analyzing the monitoring data.

Before anything is implemented or monitored, the problem must be assessed both inside and outside the organization. Public involvement in the process from the very beginning is key to identification of points of concern and uncertainty. With information in hand from a series of listening sessions, the cycle can more formally begin. Important components include designing a plan considered to be the preferred or best plan among several alternative plans, identifying reference areas to use as points of comparison, and implementing and monitoring the plan to learn from the management actions.

## AN EXAMPLE OF MONITORING AND USE OF ADAPTIVE MANAGEMENT

When we discussed the 1993 NFWP above, the economic and social impacts of regulating logging to conserve or foster late-successional habitat were not adequately addressed. The efforts of the NFWP's authors to end the stalemate between segments of the population who supported continued timber management on federal lands, and those who saw federal lands as refuges for late-successional and old-growth species, particularly the northern spotted owl, are a key component of the story.

Indeed, the objectives of the NFWP as a whole are threefold:

1. Protecting and enhancing habitat for mature and old-growth forests and related species

2. Restoring and maintaining the ecological integrity of watersheds and aquatic ecosystems
3. Producing a predictable level of timber sales, special forest products, livestock grazing, minerals, and recreational opportunities, as well as maintaining the stability of rural communities and economies

Using an adaptive management approach, a monitoring program was established to better understand the extent to which management attains these objectives and to more fully grasp the interplay among them. The monitoring program relies on both internal and external sources of data. For instance, internal data were collected directly by the regional monitoring team or by cooperators funded through the monitoring program. External data were collected by programs such as the U.S. Forest Service's Forest Inventory and Analysis Program. Data include information on populations and (occasionally) fitness of key species as well as information on the changes in area of old forests, socioeconomic conditions in the region, and watershed condition (Haynes et al. 2006). Recently, 10-year results were released and researchers can now make the first of these assessments (Haynes et al. 2006). This wealth of information is readily available to managers and the public, and it helps adapt past and inform new decisions made on both public and adjacent private lands in the Pacific Northwest (Spies et al. 2007).

## SUMMARY

Monitoring is done for a variety of reasons, but at its core, monitoring is done to provide information and make more informed decisions. In many instances, monitoring is done either as a legal requirement or in response to a crisis. As species become listed as threatened or endangered, as economically important species (e.g., deer) decline in number, or as pest species that jeopardize human health increase in number, immediate action and monitoring are often called for by managers and the public. If challenged in court then a judge can have considerable influence over the establishment and continuation of a monitoring program.

In other cases a manager, landowner, or stakeholder may simply realize that knowing how a resource is changing over time can mean that management may be more effective in the future. Foresters certainly take this approach by using continuous forest inventory, but wildlife managers also have recognized the importance of long-term monitoring data. Programs addressing trends in breeding birds, amphibians, and carbon dioxide, the EPA's program of monitoring and assessing water quality, and the NEON program all represent organized large-scale efforts to acquire data to more fully understand system responses to stressors and hence make more informed resource management decisions. With all monitoring programs, however, funding is important to consider. Funding can be tenuous, especially when monitoring is long term, and the individuals, agencies, or organizations responsible for the monitoring efforts often must spend considerable effort explaining the value of their monitoring programs to ensure that funding continues.

Whatever the impetus is for establishing a monitoring program, the objectives must be clear and specific, the questions treated as hypotheses, and the data collected in a rigorous and unbiased manner to ensure that they are able to inform future decisions. The true importance of these steps will become clear after a plan is implemented and difficult questions arise, including when or if to make changes in the monitoring protocol, when monitoring should end, and at what point the data initiate a change in management actions. All of these decisions are best made by managers and stakeholders working together.

## REFERENCES

Anthony, R.G., E.D. Forsman, A.B. Franklin, D.R. Anderson, K.P. Burnham, G.C. White, C.J. Schwarz, J.D. Nichols, J.E. Hines, G.S. Olson, S.H. Ackers, L.S. Andrews, B.L. Biswell, P.C. Carlson, L.V. Diller, K.M. Dugger, K.E. Fehring, T.L. Fleming, R.P. Gerhardt, S.A. Gremel, R.J. Gutierrez, P.J. Happe, D.R. Herter, J.M. Higley, R.B. Horn, L.L. Irwin, P.J. Loschl, J.A. Reid, and S.G. Sovern. 2004. Status and trends in demography of northern spotted owls, 1985–2003. Final Report to the Regional Interagency Executive Committee, Portland, OR.

Beeton, R.J.S., K.I. Buckley, G.J. Jones, D. Morgan, R.E. Reichelt, and T.E. Dennis. 2006. Independent report to the Australian Government Minister for the Environment and Heritage. Australian State of the Environment Committee, Canberra.

Cody, M.L., ed. 1985. *Habitat Selection in Birds*. Academic Press, Orlando, FL. 558 pp.

Czech, B. 2006. If Rome is burning, why are we fiddling? *Conservation Biology* 20:1563–1565.

Dwyer, W.L. 1994. Seattle Audubon Society, et al. v. James Lyons, Assistant Secretary of Agriculture, et al. Order on motions for Summary Judgment RE 1994 Forest Plan Seattle, WA: U.S. District Court, Western District of Washington.

Environment Canada. 1996. State of the Environment Report Yukon. Environment Canada, Whitehorse, Yukon.

GAO (Government Accountability Office and National Academy of Science). 2007. Highlights of a forum: Measuring our nation's natural resources and environmental sustainability. U.S. Government Printing Office Publication number GAO-08-127SP. Washington, D.C.

Guarderas, A.P., S.D. Hacker, and J. Lubchenco. 2008. Current status of marine protected areas in Latin America and the Caribbean. *Conservation Biology* 22:1630–1640.

Haynes, R.W., B.T. Bormann, D.C. Lee, and J.R. Martin, Jon R., tech. eds. 2006. Northwest Forest Plan—the first 10 years (1994–2003): Synthesis of monitoring and research results. U.S. Department of Agriculture, Forest Service, Pacific Northwest Research Station. Gen. Tech. Rep. PNW-GTR-651. Portland, OR. 292 pp.

Heinz Center. 2002. *The State of the Nation's Ecosystems: Measuring the Lands, Waters, and Living Resources of the United States*. Cambridge University Press, New York.

Lint, J. 2001. Northern spotted owl effectiveness monitoring plan under the Northwest Forest Plan: Annual summary report 2000. Northwest Forest Plan Interagency Monitoring Program, Regional Ecosystem Office, Portland, OR.

Lint, J. tech. coord. 2005. Northwest Forest Plan—the first 10 years (1994–2003): Status and trends of northern spotted owl populations and habitat. U.S. Department of Agriculture, Forest Service Gen. Tech. Rep. PNW-GTR-648. Portland, OR. 176 pp.

McComb, B.C. 2007. *Wildlife Habitat Management: Concepts and Applications in Forestry*. Taylor & Francis, CRC Press, Boca Raton, FL. 319 pp.

McComb, W.C., S.A. Bonney, R.M. Sheffield, and N.D. Cost. 1986. Snag resources in Florida—Are they sufficient for average populations of cavity nesters? *Wildlife Society Bulletin* 14:40–48.

Ottke, C., P. Kristensen, D. Maddox, and E. Rodenburg. 2000. *Monitoring for impacts: Lessons on natural resources monitoring from 13 NGOs*. Vol. I and II. World Resources Institute and Conservation International. Washington, D.C.

Sauer, J.R., J.E. Hines, and J. Fallon. 2007. The North American Breeding Bird Survey, Results and Analysis 1966–2006. Version 10.13.2007. U.S. Geological Survey Patuxent Wildlife Research Center, Laurel, MD.

Schroeder, R.L. 2006. A system to evaluate the scientific quality of biological and restoration objectives using National Wildlife Refuge Comprehensive Conservation Plans as a case study. *Journal for Nature Conservation* 14(3–4):200–206.

Sheffield, R.M., N.D. Cost, W.A. Bechtold, and J.P. McClure. 1985. Pine growth reductions in the Southeast. U.S. Department of Agriculture, Forest Service, Resource Bull. SE-83. 112 pp.

Songorwa, A.N. 1999. Community based wildlife management (CWM) in Tanzania: Are the communities interested? *World Development* 27(12):2061–2079.

Spies, T.A., K.N. Johnson, K.M. Burnett, J.L. Ohmann, B.C. McComb, G.H Reeves, P. Bettinger, J.D. Kline, and B. Garber-Yonts. 2007. Cumulative ecological and socio-economic effects of forest policies in coastal Oregon. *Ecological Applications* 17:5–17.

U.S Department of Agriculture, Forest Service; U.S. Department of the Interior, Bureau of Land Management [USDA USDI]. 1994. Final supplemental environmental impact statement on management of habitat for late-successional and old-growth forest related species within the range of the northern spotted owl. Volumes 1–2 and Record of Decision. U.S. Government Printing Office, Washington, D.C.

Walters, C. 1986. *Adaptive Management of Renewable Resources*. Macmillan, New York.

Walters, C.J., and C.S. Holling. 1990. Large-scale management experiments and learning by doing. *Ecology* 71:2060–2068.

# 2 Lessons Learned from Current Monitoring Programs

Ecological monitoring addresses a diversity of questions, interests, and organizational objectives, but all monitoring programs face similar challenges and obstacles. These can include, but are not limited to, biases in sampling design, logistical constraints, funding limitations, and the inevitable complexities associated with data analysis. There is much to learn from how past monitoring programs have successfully overcome these common challenges, and this chapter details the development and challenges of several large-scale monitoring programs. The following programs are not meant as an exhaustive review, but rather as an example of current monitoring strategies and initiatives focused on animal or plant populations.

Millions of dollars are spent every year on monitoring various species and communities at scales ranging from local projects to global initiatives. The 2003 United Nations Environmental Program list includes 65 major monitoring and research programs throughout the world involved in efforts related to climate change, pollutants, wetlands, air quality, and water quality (to name a few) (Spellerberg 2005). Likewise, across the United States there is a diversity of monitoring programs with varying goals, objectives, and institutional mandates. Some federal programs, such as the Biomonitoring of Environmental Status and Trends (BEST) program, are involved in monitoring the effects of environmental pollutants on wildlife populations occupying national wildlife refuges. Others, such as the USGS Breeding Bird Survey (BBS), have decades' worth of data that can be used to identify long-term trends in bird species on both regional and national scales, but do not relate these data to habitat elements per se, although numerous investigators have taken this step (e.g., Flather and Sauer 1996; Boulinier et al. 1998a; Cam et al. 2000; Boulinier et al. 2001; Donovan and Flather 2002). In addition to these federal efforts, nongovernment organizations such as the Wildlife Conservation Society have relied heavily on national and international monitoring efforts to provide a basis for understanding changes in the resources they are most interested in protecting. Finally, citizen-based monitoring programs, such as checklists and biological atlases, have been conducted throughout the world. Given the legal mandates associated with environmental compliance and accountability, monitoring efforts will continue to be the basis for making decisions about how and where to invest resources to achieve societal goals and agency mandates.

## FEDERAL MONITORING PROGRAMS

### THE BIOMONITORING OF ENVIRONMENTAL STATUS AND TRENDS (BEST)

The importance of monitoring animal populations to detect the effects of environmental contaminants has been a major environmental concern since the 1960s (Johnson et al. 1967). One such example occurred at the Kesterson National Wildlife Refuge during the 1980s, where large population declines and deformities in fish were linked to high selenium levels in agricultural drainwater used to irrigate wetlands on and off the refuge (Marshal 1986; Harris 1991). Selenium levels also were associated with alarming deformities in waterfowl hatchlings including twisted wings, swollen heads, and missing eyes. Environmental catastrophes like this increased the pressure on the U.S. Fish and Wildlife Service (FWS) to expand monitoring programs to assess existing and anticipate future effects of environmental contaminants on fish and wildlife populations and their habitats within national wildlife refuges (Figure 2.1). In response to this need, the Biomonitoring of Environmental Status and Trends (BEST) Program was initiated to develop a comprehensive approach for monitoring the nation's protected areas at multiple levels of biological complexity ranging from organisms to populations to communities. The overall goal of the program is to provide scientific information on the impacts of environmental contaminants for natural resource management and conservation planning. The consequences of environmental pollutants and contaminants are complex and may take years, if not decades, to manifest themselves in animal and plant populations. Therefore, clearly defined goals and objectives are a necessary first step for monitoring the ecological effects of environmental pollution.

**FIGURE 2.1** Algae accumulation is a common problem associated with environmental contaminants and agricultural runoff.

## What Is the Goal of the Monitoring Program and How Is It to Be Achieved?

BEST has three major goals: (1) measure and assess the effects of contaminants on selected species and habitats, (2) conduct research and activities directed at providing innovative biomonitoring methods and tools, and (3) deliver effective and efficient tools for assessing contaminant threats to species and habitats. The first goal of the BEST Project focuses on the occurrence, severity, and changes in environmental contaminants on wildlife populations. The primary audience for this information is resource managers attempting to identify regions of the country where contaminant threats to biological resources warrant further investigation. Unfortunately, the tools necessary for identifying biocontaminants and tracking their effects in wildlife populations are an inexact science, and so the second goal focuses on evaluating and testing monitoring methods within an adaptive framework. There is a general emphasis on monitoring methods that can be linked to demographic parameters (such as reproductive rates and survival). These types of methods are the most difficult and laborious of population parameters to estimate, and so BEST is continually investing in developing new methods of collecting the necessary data. The third goal focuses on exploiting information technologies, such as Internet-based data gathering methods, as distributing tools that facilitate the communication of monitoring principles, techniques, and results to others.

## Where to Monitor?

Given that the goals of BEST are so wide ranging, the early stages of the program encountered many obstacles common to incipient monitoring efforts. Challenges included selecting unbiased areas for monitoring, studying contaminant levels at different levels of biological organization, and choosing what exactly to measure. Ironically, one of the program's initial objectives of identifying existing or potential contaminant-related problems led to a biased selection of areas aimed at maximizing the potential for identifying contaminants and their ecological effects. That is, researchers were looking for sites with preexisting problems of contamination and so had no way of comparing changes in wildlife populations that could be effected by environmental contamination with areas that were not contaminated (i.e., control sites). Because of the inferential limitations of selecting only highly affected sites, a second network of lands was required to produce unbiased estimates. The first network of sites is used to describe the exposure and response of selected species to contaminants, and measure the changes in exposure and response over time. A second set of networks describes contaminants and their effects in important habitats used by species of concern. This second habitat-based network would not only describe the distribution of contaminants and their effects, but also describe indirect effects (e.g., reduction of prey items) and changes in habitat quality over time. Therefore, BEST adopted a monitoring approach that relies on multiple lines of evidence from different regions for identifying contaminant exposure at multiple ecosystem levels.

## What to Monitor?

After the identification of a site suffering environmental contamination, the larger and more difficult questions of what to measure and the techniques to use still remained. Researchers working on BEST decided to employ a two-tiered monitoring approach that includes a variety of methods for assessing environmental contamination including biomarkers, toxicity tests and bioassays, community health indices, and residue analyses. The first tier includes methods that are applicable to a wide variety of habitats and are readily available and inexpensive. The second tier includes more specialized (and also more expensive) methods than traditional Tier 1 methods. These methods provide greater insight for specific situations and would be more useful in determining cause-and-effect relationships for a selected species or habitat.

An example of this general approach, and one of BEST's most successful programs, is the Large River Monitoring Network (LRMN) which measures and assesses the effects of contaminants on resident fish in rivers throughout the United States. The LRMN serves as a searchable online database (http://www.cerc.usgs.gov/data/best/search/) where one can access data on fish health in multiple river basins by using a suite of organismal and suborganismal "endpoints." These endpoints are meant to monitor and assess the effects of environmental contaminants in fish populations and include variables such as age, length, weight, lesions, and a number of other biological markers. As a national monitoring program, BEST-LRMN is unique in that it utilizes these biomarkers to evaluate persistent chemicals in the environment and to detect changes before population effects may be evident. The online relational database allows anyone to access information organized by basin (e.g., Colorado Basin), species (e.g., brown trout), and gender. Since the initial application of the program in 1995, a considerable knowledge base has been developed regarding the characteristics and advantages for assessing the impacts of environmental contaminants on fish populations throughout the country.

## THE NORTH AMERICAN BREEDING BIRD SURVEY (BBS)

Similar to the BEST Program, environmental contaminants spurred the need to monitor bird populations throughout the United States. Rachel Carson's book *Silent Spring* brought national attention to the potential effects of dichloro-diphenyl-trichloroethane (DDT) on bird populations. Responding to the potential of pesticide effects on avian populations, Chandler Robbins and colleagues at the Patuxent Wildlife Research Center developed the North American Breeding Bird Survey (BBS) with the goal of monitoring bird populations over large geographic areas. Beginning in 1966, this pioneering work has resulted in the creation of one of the world's most successful long-term, large-scale, international avian monitoring programs (Sauer et al. 2005; Thogmartin et al. 2006; U.S. Geological Survey 2007).

### What Is the Goal of the Monitoring Program?

The mission of the BBS is to provide scientifically credible measures of the status and trends of North American bird populations at continental and regional scales to inform biologically sound conservation and management actions (U.S. Geological Survey 2007). Although this was an ambitious goal, clearly stating the objective early on has helped many to successfully use these data for many purposes. For example, BBS data have been instrumental in the development of methods to estimate population trends from survey data (Link and Sauer 1997a,b, 1998; Sauer et al. 2003; Alpizar-Jara et al. 2004; Sauer et al. 2004), quantifying the effects of habitat loss and fragmentation (Boulinier et al. 1998a, 2001), and studying community ecology at large geographic scales (Flather and Sauer 1996; Cam et al. 2000; La Sorte and McKinney 2007). A literature review in 2002 found that more than 270 scientific publications have relied heavily, if not entirely, on BBS data, making this one of most widely used and applied monitoring programs in the world.

### Where and How to Monitor?

The founders of this program had a monumental task in addressing some of the key questions in monitoring program design. How can a single program effectively describe and monitor over 420 bird species throughout the continental United States and Canada? Every square meter of a national landscape cannot be monitored, so which spots should be surveyed? How can it be done in a way that allows the data to be scaled up (or down)? The protocol they developed stipulates that each year during June, the height of the avian breeding season in the region, BBS participants skilled in avian identification collect bird population data along roadside survey routes. Each survey route is 41 km (24.5 mi) long with stops at 0.8-km (0.5-mi) intervals. At each stop, a 3-min point count is conducted. During the count, every bird heard or seen within a 0.4-km (0.25-mi) radius is recorded. Surveys start a half hour before local sunrise and take about 5 hours to complete. Over 4,100 survey routes are located across the continental United States and Canada. Predictably, this amount of work results in a vast and complicated database of information on bird populations.

### What to Monitor?

Although the decision of what exactly to monitor was largely determined by the stated objectives of the plan, researchers still faced a number of obstacles to collecting these data and providing one of the most important products of the BBS: annual estimates of population trends and relative bird abundances at various geographic scales for more than 420 bird species. For example, not all bird species are effectively sampled using roadside surveys. Birds vary in their detectability and some species avoid roads altogether; this had to be accounted for. Much thought and analysis, however, has been devoted to ensuring data quality and dealing with the associated biases of roadside sampling, and this is an ongoing area of research (Barker et al. 1993; Sauer et al. 1994; Kendall et al. 1996; Link and Sauer 1997a,b; Boulinier et al. 1998b; Link and Sauer 1998; Hines et al. 1999; Pollock et al. 2002; Alpizar-Jara et al. 2004; Sauer et al. 2004; Thogmartin et al. 2006; Link and Sauer 2007). In addition, the program attempts to randomly distribute routes in order to sample habitat types that are representative of the entire region. Other requirements such as consistent methodology and observer expertise, visiting the same stops each year, and conducting surveys under suitable weather conditions are necessary to produce comparable data over time and between geographic regions. A large sample size (number of routes) is needed to average local variations and reduce the effects of sampling error. Variation in counts can be associated with sampling techniques as well as the true (i.e., natural) variation in population trends. Indeed, the survey produces an index of relative abundance rather than a complete count of breeding

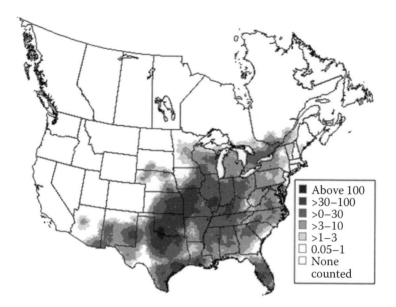

**FIGURE 2.2** (*A color version of this figure follows page 144.*) Abundance map for the eastern meadowlark based on Breeding Bird Survey data collected between the summers of 1994–2003. (From Sauer, J.R., J.E. Hines, and J. Fallon. 2007. The North American Breeding Bird Survey, Results and Analysis 1966–2006. Version 10.13.2007. U.S. Geological Survey Patuxent Wildlife Research Center, Laurel, MD.)

bird populations, and assumes that fluctuations in these indices of abundance are representative of the population as a whole. Another issue, quite separate from variation, is that the precision of abundance estimates will change with sample size. The density of BBS routes varies considerably across the continent, reflecting regional densities of skilled birders who tend to be associated with urbanization patterns. Consequently, abundance estimates in regions with fewer routes are less precise than estimates in regions with a large number of routes. The greatest densities of surveys are in the New England and Mid-Atlantic states, whereas densities are lower elsewhere in the United States.

Despite these complicated issues of sampling design and analysis (indeed, selecting *what* to monitor often entails much more than simply choosing a species!), the efforts of BBS researchers have resulted in a valuable source of information on bird population trends and an excellent source of ideas and lessons for the design of other broad-scale monitoring programs.

For instance, BBS data can be used to produce continental-scale maps of relative abundance. When viewed at continental or regional scales, these maps provide a reasonable indication of the relative abundances of species that are well sampled by the BBS (Figure 2.2). Analyzing population change on survey routes is probably the most effective use of BBS data, but these data do not provide an explanation for the causes of population trends. Population trend data have been used, however, to associate population declines with environmental effects such as habitat loss and fragmentation (Askins 1993; Boulinier et al. 1998b, 2001; Donovan and

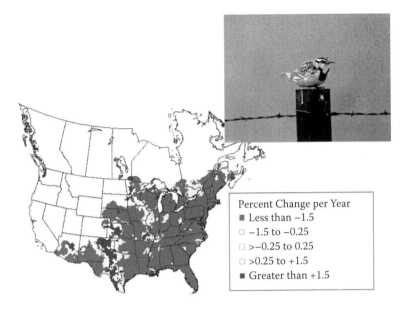

**FIGURE 2.3** (*A color version of this figure follows page 144.*) Trend map for the eastern meadowlark based on Breeding Bird Survey trend estimates collected between the summers of 1966–2003. (From Sauer, J.R., J.E. Hines, and J. Fallon. 2007. The North American Breeding Bird Survey, Results and Analysis 1966–2006. Version 10.13.2007. U.S. Geological Survey Patuxent Wildlife Research Center, Laurel, MD; photo inset by Laura Erickson is used with permission.)

Flather 2002). To evaluate population changes over time, BBS indices from individual routes are combined to obtain regional and continental estimates of trends. Few species, however, have consistent trends across their entire ranges, so spatial patterns in population trends are of particular interest to scientists and managers attempting to identify "hot spots" of regional declines. Route-specific trends can be smoothed to produce trend maps that allow for the identification of regions of increase and decline (Figure 2.3).

Although trends at the species level will always be a basic use of BBS data analyses, combining species into groups with similar life-history traits, known as *guilds*, provides additional insight into patterns of population trends (Askins 1993; Sauer et al. 1996; Hines et al. 1999). The concept of grouping species based on certain life-history characteristics (e.g., breeding habitat, migratory behavior, etc.) can be useful for identifying widespread environmental effects because individual species often differ widely in their response to environmental change. Consistent trends within an entire guild may be indicative of overall changes in an environmental resource (e.g., declining forest birds due to the loss of forests to development).

## ENVIRONMENTAL MONITORING AND ASSESSMENT PROGRAM (EMAP)

National monitoring programs all share the common challenge of developing a monitoring framework that can scientifically determine and track the condition of a natural resource distributed over thousands of kilometers. Sometimes this need to monitor a natural resource is a legal mandate for a federal or state agency. As an example, under the Clean Water Act the Environmental Protection Agency (EPA) has statutory responsibilities to monitor and assess inland surface waters and estuaries. To achieve this goal, the Environmental Monitoring and Assessment Program (EMAP) was created to develop the science needed for a statistical monitoring framework to determine the condition, and to detect trends in condition, both at the level of individual states as well as for all the nation's aquatic ecosystems (McDonald et al. 2002). Given its legal responsibility and need to produce legally defensible results, EMAP emphasizes a sampling design guided by both statistical and scientific objectives.

### What Is the Goal of the Monitoring Program?

The primary goal of EMAP is clear and was determined legislatively: to develop a sampling design that provides an unbiased, representative monitoring of an aquatic resource with a known level of confidence. The necessarily general nature of the objective informed a number of other steps in designing the monitoring program and outlined several key challenges. The need to be applicable across the landscape mandated that EMAP's sampling design rely on a multiscaled approach of collecting samples with state-based partners and aggregating those local data into broader state, regional, and national assessments. This approach of "scaling up" data from various locales requires EMAP's research goals to include (1) establishing the statistical variability of EMAP indicators when used in aquatic ecosystems in diverse ecological areas of the country, (2) establishing the sensitivity of these indicators to change and trend detection, and (3) developing indicators and designs that will allow the additional monitoring of high-priority aquatic resources (e.g., Great Rivers, wetlands). The key challenge in this monitoring strategy, and one that is shared with many other national assessments, is how to draw a representative sample from a small number of sites to provide an unbiased estimate of ecological condition over a larger geographic region.

### Where and How to Monitor?

Choosing sites and methods that adequately addressed the key challenge was not easy. EMAP researchers have spent considerable time and effort in developing a probability-based sampling design to estimate the condition of an aquatic resource over large geographic areas. Probability-based sampling designs have a number of requirements including a clearly defined population, a process by which every element in the population has the opportunity to be sampled with a known probability, and a method by which that sampling can be conducted in a random fashion (Cochran 1977). As is the case with any monitoring project covering a larger geographic region, including the BBS described earlier, samples should be distributed throughout the study area to be maximally representative. EMAP's design accomplishes this by taking samples at regular intervals from a random start (a systematic random design). To achieve this, EMAP uses hexagonal-shaped grids to add systematic sampling points across a study region (Figure 2.4; White et al. 1991). The grid is positioned randomly on the map of the target area, and sample locations from within each hexagonal grid cell are selected randomly. Why is this necessary? In short, the use of a sampling grid ensures an unbiased spatial separation of randomly selected sampling units (systematic random sample). Also, the grid allows for the potential

**FIGURE 2.4** The EMAP base grid overlain on the United States. There are about 12,600 points in the conterminous United States with approximately 27 km between points in each direction. A fixed position that represents a permanent location for the base grid is established, and the sampling points to be used by EMAP are generated by a slight random shift of the entire grid from this base location.

of dividing the entire target population into any number of subpopulations (or strata) of interest. Subsequent random sampling within these strata allows statistical inferences to be made about each subpopulation. As an example, stratified sampling is often used in a regional stream survey to enhance sampling effort in a watershed of special interest so that its condition can be compared with the larger regional area.

## What to Monitor?

Once the sampling design was established, the next question to address was a familiar one: what exactly is to be measured? Like BEST's efforts of monitoring environmental pollutants, an adequate response consisted of more than simply electing one indicator; there are many definitions of an "aquatic resource" that all have unique characteristics. At the coarsest level, EMAP addressed this by dividing aquatic resources into different water body or system types, such as lakes, streams, estuaries, and wetlands. Subsequently, they use a second level of strata, ecoregions, to capture regional differences in water bodies. The lowest level of strata in the EMAP design distinguishes among different "habitat types" within an aquatic resource in a specific geographic region. For example, portions of estuaries with mud-silt substrate will have much different ecological characteristics than portions of estuaries with sandy substrates. It is within this lowest stratum that sampling and monitoring is conducted.

EMAP's sampling design takes one of two very different routes depending on whether the aquatic ecosystem to be sampled is discrete or extensive (McDonald et al. 2002). A discrete aquatic system consists of distinct natural units, such as lakes. Population inferences for a discrete resource are based on numbers of sampling units that possess a measured property (e.g., 10% of the lakes are acidic). For discrete resources, EMAP often uses an intensification of the sampling grid. In some cases where the sampling units are a large enough area, the grid can be used directly by selecting those units in which one or more grid points fall (e.g., estuaries in a state). With this method, the probability that a unit gets into the sample (its inclusion probability) is proportional to the unit's area (e.g., larger lakes have a higher probability of being sampled). The inferences linking sampling data to the entire population are then in terms of area. Alternatively, a unit of a discrete resource can be treated as a point in space. For example, the center point of lakes could be used. This method of sampling is appropriate for inference in terms of numbers of units in a particular condition (e.g., 7% of Northeastern lakes are chronically acidic).

Extensive resources, on the other hand, extend over large regions in a more or less continuous and connected fashion (e.g., rivers), and do not have distinct natural units. In these cases, population inferences are based on the length or area of the resource. The nature of the ecosystem determines the particular sampling technique that is used. EMAP uses area sampling for extensive systems such as rivers or point sampling for discrete systems like estuaries. In area sampling, the extensive resource is broken up into disjoint pieces, much like a jigsaw puzzle, and sample selection is from a random selection of these pieces. The sampling values that are obtained are then used to characterize or represent the entire resource. To avoid any sampling bias, points are located at random within the extensive resource.

Only once sites and appropriate sampling techniques are selected, can an indicator of ecological condition of the aquatic system be chosen and sampled. Effective aquatic ecological indicators are central to determining the condition of aquatic resources, and EMAP has identified a number of ecological indicators (see McDonald et al. 2002 for a full list). In general, EMAP focuses on combining biological indicators that are able to be sampled through analysis of the fish, benthic macroinvertebrate, and plant communities. EMAP also makes extensive use of an index of biotic integrity (IBI, a multimetric biological indicator; Karr 1981) to evaluate the overall fish assemblage, which provides a measure of biotic condition.

## NONGOVERNMENTAL ORGANIZATIONS AND INITIATIVES

### Monitoring the Illegal Killing of Elephants (MIKE)

Between 1970 and 1989, half of Africa's elephants, over 700,000 individuals, were killed due to a surging international ivory trade (Douglas-Hamilton 1989; Blake et al. 2007). This decline prompted the Convention on the International Trade in Endangered Species of Wild Flora and Fauna (CITES) to list African elephants on Appendix I of the convention, banning the trade of tusks in international markets. Despite its protected status, the optimal approach to African elephant management and conservation remains a topic of great debate (Blake et al. 2007). In response to the need for better data regarding the status and trends of African elephants (Figure 2.5), the Monitoring the Illegal Killing of Elephants (MIKE) project was initiated in 1997. The overall goal of MIKE is to provide the information needed for

# Lessons Learned from Current Monitoring Programs

**FIGURE 2.5** African elephant populations in much of Africa have been decimated and now are only common in protected parks.

governments and agencies to make appropriate management and enforcement decisions, and to build institutional capacity for the long-term management of elephant populations. The MIKE program is funded by a diversity of agencies and NGOs, including the Wildlife Conservation Society, U.S. Fish and Wildlife Service, the European Union, and the World Wildlife Fund.

## What Is the Goal of the Monitoring Program?

MIKE has three specific program objectives including to (1) measure the levels and trends in the illegal hunting of elephants, (2) determine changes in these trends, and (3) determine the factors associated with these trends. Once again, the clarity of the objectives is central in developing other aspects of the monitoring program. In the case of MIKE, the breadth of the goals meant that a suite of factors needed to be investigated, including habitat type, elephant population levels, human conflicts, adjacent land uses, human access, water availability, tourism activities, civil strife, and development activities. The monitoring objectives of the program also emphasized a need for standardized methodologies, representative sampling, and collecting data on population trends and the spatial patterns of illegal killing (Figure 2.6). The nature of the objectives even clarified what the main benefit of this comprehensive monitoring scheme would be: an increased knowledge base of elephant numbers and movements, a better understanding of the threats to their survival, and an increased general knowledge of other species and their habitats.

## Where and How to Monitor?

Given the remote location of many of the habitats and populations of interest (Figure 2.7), site selection was of paramount importance for collecting representative

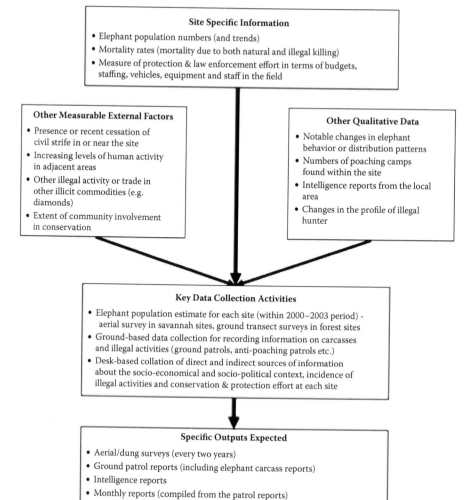

**FIGURE 2.6** Monitoring scheme for the MIKE initiative. (With permission from S. Blake.)

and unbiased data. A minimum of 45 sites in 27 states were initially selected in Africa and 15 sites in 11 states in Asia. The methods of site selection were based on a number of variables chosen to make the sites a representative sample, including habitat types, elephant population size, protection status of sites, poaching history, incidence of civil strife, and level of law enforcement. Statistical analysis and modeling have also been used to select sites based on geographical, environmental, and socioeconomic characteristics.

## What to Monitor?

In central Africa in 2003–04, Blake et al. (2007) used this approach to implement the first systematic, stratified, and unbiased survey of elephant populations within

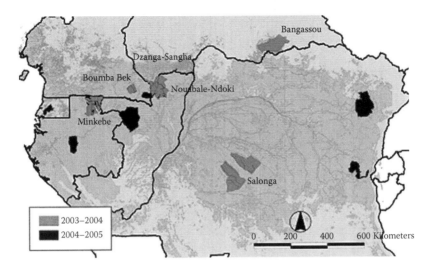

**FIGURE 2.7** The suite of MIKE sites in the equatorial forests of central Africa. (With permission from S. Blake.)

each MIKE site. The primary data source they elected to use were dung counts based on along line-transects. In addition to the standardized transects, they undertook reconnaissance surveys to provide supplementary information on the incidence of poaching and other human impacts. At each survey site, an attempt was made to sample elephant abundance across a gradient of human impact. Stratification of each site was based on the elephant sign encounter rate generated during MIKE pilot studies or on expected levels of human impact as a proxy for elephant abundance. Sample effort was designed to meet a target precision of 25% (coefficient of variation) of the elephant dung density estimate. Density estimates of forest elephants in MIKE survey sites were obtained via systematic line-transect distance sampling that used dung counts as an indicator of elephant occurrence. Advanced data analysis (using distance sampling) provided robust estimates of dung density, relative elephant density, and spatial distributions within each site (Figure 2.8).

The decision to record a diversity of variables allowed researchers to conduct analyses that addressed all of the MIKE objectives and provided other valuable insights. Blake et al. (2007) found that human activities were a major determinant of the distribution of elephants even within highly isolated national parks. In almost all cases the relative elephant abundance interpolated from transect data was the mirror image of human disturbance, and elephant abundance was consistently highest farthest from human settlement (Figure 2.8). They estimated that, despite international attention and conservation status, forest elephants in central Africa's national parks are losing range at an alarming rate. Twenty-two poached (confirmed) elephant carcasses were found from 4,478 km of reconnaissance surveys walked during the inventory period. The combined inventory, distribution, and reconnaissance data showed little doubt that forest elephants are under imminent threat from poaching and range restriction. This innovative monitoring scheme and analysis

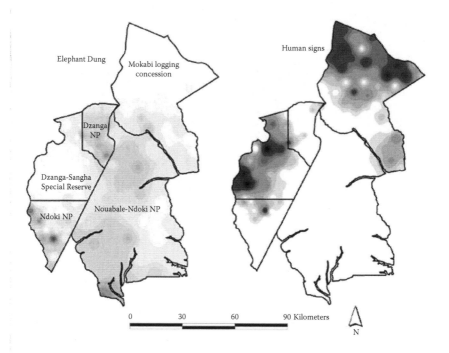

**FIGURE 2.8** Interpolated elephant dung count and human-sign frequency across the Ndoki–Dzanga MIKE Site. Increasing darkness of sites signifies increasing dung and human-sign frequency. (With permission from S. Blake.)

demonstrated that even with an international ban of the trade in ivory in place, forest elephant range and numbers are in serious decline.

## LEARNING FROM CITIZEN-BASED MONITORING

Volunteer-based monitoring efforts have a long history in the United States and throughout the world. For example, Audubon's Christmas Bird Count (CBC) is a volunteer-based annual survey of winter bird distributions throughout the United States that dates back to the early 1900s. Although biological surveys based on volunteer effort have a rich history, they are now being increasingly used for monitoring long-term and large-scale changes in animal and plant populations. The combination of current demand for broad-scale, long-term ecological data and an explosion of volunteer-based efforts has resulted in a fairly well-informed movement in which scientists have made great strides in increasing the scientific rigor of monitoring programs that involve citizens. This movement has become so popular in recent years that many of these programs and initiatives are falling under the global label of "citizen science."

### What Is the Goal of the Monitoring Program?

Although citizen science takes many forms and has many objectives (see Cornell Lab of Ornithology's citizen science programs http://www.birds.cornell.edu/), atlas surveys are a globally common example and are widely implemented for many species and taxa. Atlases consist of volunteers documenting the distribution (and often breeding status) of species within a survey grid covering an entire region of interest (Donald and Fuller 1998; Bibby et al. 2000). The goal of many atlas surveys focuses on documenting and monitoring temporal and spatial shifts in species distributions over long time periods (Donald and Fuller 1998). One such example of a regional atlas is the New York State Breeding Bird Atlas (Andrle and Carroll 1988; McGowan and Corwin 2008). This project is sponsored by the New York State Ornithological Association and the Department of Environmental Conservation in cooperation with the New York Cooperative Fish and Wildlife Research Unit at Cornell University, Cornell University Department of Natural Resources, and the Cornell Laboratory of Ornithology. The New York State Breeding Bird Atlas (BBA) is a comprehensive, statewide survey with the specific objective of documenting the distribution of all breeding birds in New York. As with all the preceding monitoring programs, stating the program objective informs and drives other aspects of the monitoring program. For instance, its breadth indicates that substantial planning is needed to collect occurrence data for multiple species over a wide range of habitats and develop a protocol that can be easily followed and adhered to by voluntary participants. Time is an essential component of any monitoring program, and atlases are often unique among other monitoring initiatives due to scope of their sampling. In the case of the New York State BBA, the surveys were conducted in two time periods: the first atlas project ran from 1980 to 1985 (Andrle and Carroll 1988) while New York's second atlas covered 2000 to 2005 (McGowan and Corwin 2008). Broad-scaled distributional surveys, such as atlases, are obviously an attempt to monitor long-term range changes that are beyond the scope of most monitoring programs.

**Where and How to Monitor?**

Given that it was necessary to account for the entire state to meet the objective of the BBA, researchers first had to determine how to make the scale manageable. For the New York State BBA, both surveys used a grid of 5,332 blocks 5 × 5 km that covered the entirety of New York State (125,384 km$^2$), representing one of the largest and finest resolution atlas data sets in the world (Gibbons et al. 2007). The state was stratified into 10 regions and one or two regional coordinators in each area were responsible for recruiting volunteers in each atlas effort and overseeing coverage of the blocks in their region. Once this system was prepared, efforts were needed to ensure that the volunteers reported quality data despite monitoring independently from one another in different locations. To do this, the researchers assigned atlas surveyors to one or more blocks and instructed them to spend at least 8 hours in the block, visiting each habitat represented, and recording data on at least 76 bird species. This "76 species" threshold was treated as a standard of "adequately surveyed" based on experience from previous atlases (Smith 1982). Measures such as these are integral to monitoring programs with multiple observers responsible for sampling different sites and species. Without some form of controlling and documenting variation in sampling effort, the data would be vulnerable to a number of biases.

**What to Monitor?**

The objective of the BBA required that volunteers would survey their atlas block(s) and record every bird species encountered and the observed breeding activity ranging from possible breeding (e.g., singing male in appropriate habitat type), probable breeding (e.g., pair observed in breeding habitat), and confirmed breeding (e.g., nest found). Although the BBA did not provide a definitive statement concerning the absence of a breeding record for a species not listed in a block, absence was interpreted by researchers to mean that species could not be found given adequate effort and observer ability, or that the species occurs in low enough densities to escape detection. In addition, atlasers were asked to submit data on effort including the total number of hours spent surveying and the number of observers. Mandating that volunteers record a variety of data, including data on sampling efforts, further reduced the possibility of biases (or at least allowed researchers to account for variation in effort during analyses).

One final important lesson that can be drawn from programs such as the BBA is that researchers must be transparent about the appropriate significance and uses of the data they generate. Whether or not atlases can be used as an effective tool for monitoring animal populations relates to the relationship between changes in regional occupancy (as measured by atlas surveys) and changes in local abundance (Gaston et al. 2000). In macroecology, this relationship is synonymous with the abundance-occupancy rule, which predicts that changes in regional occupancy will accurately reflect changes in local abundance. Relatively few studies have addressed the relationship between abundance and occupancy using atlas data, but those that have generally support the use of atlas data for monitoring large-scale population dynamics (Böhning-Gaese 1997; Van Turnhout et al. 2007; Zuckerberg et al. 2009). Once this information and relationship is made explicit, atlas data can be correctly used to make a number of observations and assessments. In the New York State BBA, with its two survey periods, researchers and managers can analyze the changes in regional distribution for over 250 bird species. Bird species demonstrated a wide variation in distributional changes from widespread increases (Figure 2.9) to startling range contractions (McGowan and Corwin 2008). Approximately half of all of the bird species in New York demonstrated significant

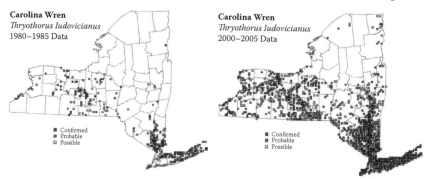

**FIGURE 2.9** (*A color version of this figure follows page 144.*) Changes in the distribution of the Carolina wren between two statewide atlases conducted in 1980–1985 and 2000–2005. This species has shown one of the most dramatic increases in occupancy of any species recorded during the atlas project.

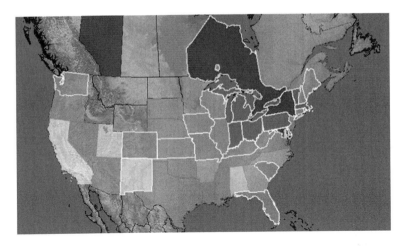

**FIGURE 2.10** The distribution and status of breeding bird atlases throughout the United States and Canada. The darker states have completed or are in the process of completing a second atlas. Map created using Breeding Bird Atlas Explorer (http://www.pwrc.usgs.gov/bba/index.cfm?fa=bba.Bbahome).

changes in their distribution between the two atlases. Of those with significant changes, 55% increased in their distribution. As a group, woodland birds demonstrated a significant increase in their average distribution between the two atlas periods while grassland birds showed the only significant decrease (Zuckerberg et al. 2009). Scrub-successional, wetland, and urban species showed no significant change in their distribution between the two atlas periods (McGowan and Corwin 2008; Zuckerberg et al. 2009). Within migratory groups, there were significant increases in the overall distribution of permanent residents and short-distance migrants while neotropical migrants showed no significant change. These trends suggest that certain regional factors of environmental change, such as reforestation or climate change, may be affecting entire groups of species.

Atlases offer an excellent opportunity for monitoring changes in large-scale and long-term population dynamics. Furthermore, the quality of their data will almost certainly increase as future advances in monitoring are applied to atlas implementation and analysis. For instance, improvements in occupancy estimation and modeling will undoubtedly be applied to projects such as the BBA to account for the varying detection probabilities of different species, and there are likely to be significant improvements in training models for participants to decrease observer bias even more (MacKenzie 2005, 2006). In addition, repeat atlases for several regions of the United States will be available in the near future (Figure 2.10). These databases will be critical for monitoring changes in species' distributions in response to relatively broad-scale environmental drivers such as regional climate change.

## SUMMARY

Despite their varied interests, funding sources, and target species/communities, all of these monitoring programs share many of the same components and obstacles.

Issues such as clearly defined monitoring goals and objectives, and where, how, and what to sample are common aspects of many monitoring programs and will be discussed at length throughout this book. Despite the challenges presented to them, these programs represent how careful planning, committed individuals, and thoughtful sampling design and analysis can help answer critical questions and thereby advance conservation goals for a variety of types of organizations. Whether it is predicting the effects of environmental contamination, estimating avian declines across an entire country, or monitoring elephant populations in remote African forests, these programs serve as encouraging reminders of the power and effectiveness of monitoring information for guiding conservation decision-making.

## REFERENCES

Alpizar-Jara, R., J.D. Nichols, J.E. Hines, J.R. Sauer, K.H. Pollock, and C.S. Rosenberry. 2004. The relationship between species detection probability and local extinction probability. *Oecologia* 141:652–660.

Andrle, R.F., and J.R. Carroll. 1988. *The Atlas of Breeding Birds in New York State*. Cornell University Press, Ithaca, New York.

Askins, R.A. 1993. Population trends in grassland, shrubland, and forest birds in eastern North America. *Current Ornithology* 11:1–34.

Barker, R.J., J.R. Sauer, and W.A. Link. 1993. Optimal allocation of point-count sampling effort. *Auk* 110:752–758.

Bibby, C. J., N. D. Burgess, D. A. Hill, and S. Mustoe. 2000. *Bird Census Techniques*. 2nd ed. Academic Press, San Diego, CA.

Blake, S., S. Strindberg, P. Boudjan, C. Makombo, I. Bila-Isia, O. Ilambu, F. Grossmann, L. Bene-Bene, B. de Semboli, V. Mbenzo, D. S'hwa, R. Bayogo, L. Williamson, M. Fay, J. Hart, and F. Maisels. 2007. Forest elephant crisis in the Congo Basin. *PLOS Biology* 5:945–953.

Böhning-Gaese, K. 1997. Determinants of avian species richness at different spatial scales. *Journal of Biogeography* 24:49–60.

Boulinier, T., J.D. Nichols, J.E. Hines, J.R. Sauer, C.H. Flather, and K.H. Pollock. 1998a. Higher temporal variability of forest breeding bird communities in fragmented landscapes. *Proceedings of the National Academy of Sciences of the United States of America* 95:7497–7501.

Boulinier, T., J.D. Nichols, J.E. Hines, J.R. Sauer, C.H. Flather, and K.H. Pollock. 2001. Forest fragmentation and bird community dynamics: Inference at regional scales. *Ecology* 82:1159–1169.

Boulinier, T., J.D. Nichols, J.R. Sauer, J.E. Hines, and K.H. Pollock. 1998b. Estimating species richness: The importance of heterogeneity in species detectability. *Ecology* 79:1018–1028.

Cam, E., J.D. Nichols, J.R. Sauer, J.E. Hines, and C.H. Flather. 2000. Relative species richness and community completeness: Birds and urbanization in the Mid-Atlantic states. *Ecological Applications* 10:1196–1210.

Cochran, W. G. 1977. *Sampling Techniques*. 3rd ed. John Wiley & Sons, New York.

Donald, P.F., and R.J. Fuller. 1998. Ornithological atlas: A review of uses and limitations. *Bird Study* 45:129–145.

Donovan, T.M., and C.H. Flather. 2002. Relationships among North American songbird trends, habitat fragmentation, and landscape occupancy. *Ecological Applications* 12:364–374.

Douglas-Hamilton, I. 1989. Overview of status and trends of the African elephant. In S. Cobb, editor. The ivory trade and the future of the African elephant: Prepared for the Seventh CITES Conference of the Parties, Lausanne, October 1989. Ivory Trade Review Group, Oxford, U.K.

Flather, C.H., and J.R. Sauer. 1996. Using landscape ecology to test hypotheses about large-scale abundance patterns in migratory birds. *Ecology* 77:28–35.

Gaston, K.J., T.M. Blackburn, J.J.D. Greenwood, R.D. Gregory, R.M. Quinn, and J.H. Lawton. 2000. Abundance-occupancy relationships. *Journal of Applied Ecology* 37:39–59.

Gibbons, D.W., P.F. Donald, H.-G. Bauer, L. Fornasari, and I.K. Dawson. 2007. Mapping avian distributions: The evolution of bird atlases. *Bird Study* 54:324–334.

Harris, T. 1991. *Death in the Marsh*. Island Press, Washington, D.C.

Hines, J.E., T. Boulinier, J.D. Nichols, J.R. Sauer, and K.H. Pollock. 1999. COMDYN: Software to study the dynamics of animal communities using a capture-recapture approach. *Bird Study* 46:209–217.

Johnson, R.E., T.C. Carver, and E.H. Dustman. 1967. Residues in fish, wildlife, and estuaries. *Pesticides Monitoring Journal* 1:7–13.

Karr, J.R. 1981. Assessment of biotic integrity using fish communities. *Fisheries* 6:21–27.

Kendall, W.L., B.G. Peterjohn, and J.R. Sauer. 1996. First-time observer effects in the North American Breeding Bird Survey. *Auk* 113:823–829.

La Sorte, F.A., and M.L. McKinney. 2007. Compositional changes over space and time along an occurrence-abundance continuum: Anthropogenic homogenization of the North American avifauna. *Journal of Biogeography* 34:2159–2167.

Link, W.A., and J.R. Sauer. 1997a. Estimation of population trajectories from count data. *Biometrics* 53:488–497.

Link, W.A., and J.R. Sauer. 1997b. New approaches to the analysis of population trends in land birds: Comment. *Ecology* 78:2632–2634.

Link, W.A., and J.R. Sauer. 1998. Estimating population change from count data: Application to the North American Breeding Bird Survey. *Ecological Applications* 8:258–268.

Link, W.A., and J.R. Sauer. 2007. Seasonal components of avian population change: Joint analysis of two large-scale monitoring programs. *Ecology* 88:49–55.

MacKenzie, D.I. 2005. What are the issues with presence-absence data for wildlife managers? *Journal of Wildlife Management* 69:849–860.

MacKenzie, D.I. 2006. *Occupancy Estimation and Modeling: Inferring Patterns and Dynamics of Species*. Elsevier, Burlington, MA.

Marshal, E. 1986. Selenium in western wildlife refuges. *Science* 231:111–112.

McDonald, M.E., S. Paulsen, R. Blair, J. Dlugosz, S. Hale, S. Hedtke, D. Heggem, L. Jackson, K.B. Jones, B. Levinson, A. Olsen, J. Stoddard, K. Summers, and G. Veith. 2002. Research strategy: Environmental monitoring and assessment. U.S. Environmental Protection Agency, Research Triangle Park, North Carolina.

McGowan, K.J., and K. Corwin. 2008. *The Second Atlas of Breeding Birds in New York State*. Cornell University Press, Ithaca, New York.

Pollock, K.H., J.D. Nichols, T.R. Simons, G.L. Farnsworth, L.L. Bailey, and J.R. Sauer. 2002. Large scale wildlife monitoring studies: Statistical methods for design and analysis. *Environmetrics* 13:105–119.

Sauer, J.R., J.E. Fallon, and R. Johnson. 2003. Use of North American Breeding Bird Survey data to estimate population change for bird conservation regions. *Journal of Wildlife Management* 67:372–389.

Sauer, J.R., W.A. Link, J.D. Nichols, and J.A. Royle. 2005. Using the North American Breeding Bird Survey as a tool for conservation: A critique of BART et al. (2004). *Journal of Wildlife Management* 69:1321–1326.

Sauer, J.R., W.A. Link, and J.A. Royle. 2004. Estimating population trends with a linear model: Technical comments. *Condor* 106:435–440.

Sauer, J.R., G.W. Pendleton, and B.G. Peterjohn. 1996. Evaluating causes of population change in North American insectivorous songbirds. *Conservation Biology* 10:465–478.

Sauer, J.R., B.G. Peterjohn, and W.A. Link. 1994. Observer differences in the North-American Breeding Bird Survey. *Auk* 111:50–62.

Smith, C.R. 1982. What constitutes adequate coverage? New York State Breeding Bird Atlas Newsletter 5:6.

Spellerberg, I.F. 2005. *Monitoring Ecological Change*. 2nd ed. Cambridge University Press, Cambridge, U.K.

Thogmartin, W.E., F.P. Howe, F.C. James, D.H. Johnson, E.T. Reed, J.R. Sauer, and F.R. Thompson. 2006. A review of the population estimation approach of the North American landbird conservation plan. *Auk* 123:892–904.

U.S. Geological Survey. 2007. Strategic Plan for the North American Breeding Bird Survey: 2006–2010: U.S. Geological Survey Circular 1307, 19 pp.

Van Turnhout, C.A.M., R.P.B. Foppen, R.S.E.W. Lueven, H. Siepel, and H. Esselink. 2007. Scale-dependent homogenization: Changes in breeding bird diversity in the Netherlands over a 25-year period. *Biological Conservation* 134:505–516.

White, D., A.J. Kimmerling, and W.S. Overton. 1991. Cartographic and geometric components of a global sampling design for environmental monitoring. *Cartography and Geographic Information Systems* 19:5–22.

Zuckerberg, B., W.F. Porter, and K. Corwin. 2009. The consistency and stability of abundance-occupancy relationships in large-scale population dynamics. *Journal of Animal Ecology* 78:172–181.

# 3 Community-Based Monitoring

From the time the Western European natural history organizations undertook formative field studies in the 18th century, to the sportsmen organizations of North America that helped spur the demise of market-hunting in the 19th and 20th centuries, to the indigenous peoples of the Amazon currently carrying out GIS mapping initiatives, citizens have often had a significant and meaningful role to play in conservation (Reiger 2001; Withers and Finnegan 2003; Tripathi and Bhattarya 2004; Fernández-Gimenez et al. 2008). Yet just as science in general comes in many shapes and sizes and under a variety of distinct monikers, the manifestations of scientific research that hinge upon citizen involvement are numerous and varied. Community-based monitoring is but one item on this long list that also includes community science, citizen science, participatory research, community co-management, and civic science (Fernández-Gimenez et al. 2008). The key differences among these endeavors is often found in the degree of influence that resource managers and scientists wield, the manner in which community or citizen is defined, and the specific questions or goals the stakeholders wish to address.

Community-based monitoring, broadly defined, is ecological monitoring that in some manner directly incorporates local community members and/or concerned citizens. The traditional approach is for scientists and resource managers to develop protocols that they consider most likely to generate rigorous data, and then transfer the necessary information to communities for them to carry out the protocols (Fernández-Gimenez 2008). A successful transfer of knowledge entails either the stratified sampling of communities and citizens to ensure that only those most apt to conduct science are invited to participate or that there is the provision of thorough training in a workshop format (Fernández-Gimenez 2008). The goals and objectives of such monitoring programs typically address the needs of resource agencies, scientists, and citizens who highly value the conventions of Western science (Fernández-Gimenez 2008). This approach, however, perhaps ideal in terms of the rigors of the scientific world, has waned in efficacy in recent years as community-based monitoring programs (CBMP) expand into more remote locales with communities and citizens who are less familiar and comfortable with the objectives and rigors of Western scientific inquiry (Sheil 2001; Spellerberg 2005). In order to implement programs effectively that are viable over the desired space and length of time in these new contexts, nontraditional, arguably less scientific designs have become more common (Fraser et al. 2006).

## A CONFLICT OVER BENEFITS

Ecological monitoring is complex and increasingly sophisticated with each new publication and technological development. To be able to generate convincing inferences grounded in strong data, monitoring program designs require a high level of scientific rigor, powerful statistical design and analysis, and the consideration of specific, science-based questions. This is particularly true as contemporary scientific research reveals the enormous extent of the uncertainties and complexities we confront when we endeavor to monitor or even understand ecosystems and leads us to question many past assumptions and mandate even more powerful, precise techniques (Kay and Regier 2000; Resilience Alliance Website 2008). It should surprise few that the interface between the newer, arguably less rigorous community-based monitoring program designs and the increasing demand for more rigorous science is an area ripe for tensions. Indeed, especially with tenure- and promotion-driven demands for rigor, many scientists are hesitant to view monitoring protocols as particularly useful or ecologically meaningful when they are designed to satisfy the objectives of citizens unfamiliar with Western science and its associated monitoring techniques. So why might it be worthwhile to continue working with and encouraging the design of community-based monitoring? Well, because a CBMP's contribution to science is but one of many important considerations; there are also a variety of economic, ethical, educational, and functional reasons to design and implement a CBMP. In some contexts, these reasons may be strong enough to compensate for deviation from the institutional ideal.

### Economic

At times, developing community-based monitoring programs in lieu of scientist-managed programs is either the best fiscal option or, given severe budget constraints, the *only* option. Natural resource agencies and universities have often been faced with financial constraints. The fiscal challenges have led to notable increases in community-based monitoring. In Canada, for instance, environmental agencies suffered budget declines of 30%–60% through the late 20th and early 21st century, an amount substantial enough to begin to compromise their capacity to remain viable institutions (Plummer and Fitzgibbon 2004). Confronted with the threat of becoming an institutional anachronism, considerable expense-cutting actions such as phasing out a number of its programs, including many monitoring initiatives, seemed all but inevitable (Whitelaw et al. 2003). Yet, given the agencies' and the public's mutual need for information about the local environment, rather than cutting programs altogether, more economically efficient alternatives were sought and found. Since the 1990s, natural resource management across Canada has been marked by the devolution of monitoring and resource management responsibilities to citizens and communities (Whitelaw et al. 2003). This strategy has effectively reduced costs, prevented data gaps in monitoring programs, and allowed resource agencies to retain a fairly comprehensive understanding of Canada's resource base despite their fiscal crisis (Whitelaw et al. 2003; Plummer and Fitzgibbon 2004).

# Community-Based Monitoring

**FIGURE 3.1** In Canada, as of 2003, a surprisingly large number of communities were committed to participating in the Canadian Community Monitoring Network. (Redrafted from Whitelaw, G., H. Vaughan, B. Craig, and D. Atkinson. 2003. *Environmental Monitoring and Assessment* 88(1–3):409–418.)

Although some of the motives of this devolution have been questioned (Plummer and Fitzgibbon 2004), Canadian community-based monitoring has emerged in a fascinating diversity of forms over the past few decades. A large number of communities are involved in the attempt to establish the Canadian Community Monitoring Network (Figure 3.1). From the successful monitoring of bowhead whales by the Inuit (Berkes et al. 2007), to Community Environment Watch's successful work with school groups (Sharpe et al. 2000), as well as a number of unviable efforts (Fraser et al. 2006; Sharpe and Conrad 2006), Canada's budget reductions have resulted in a scenario that is ripe for research and driven by an exciting need for scientists, resource managers, and communities to learn and work adaptively in the field of community-based monitoring. In the contemporary economy, monitoring in a sparsely populated country such as Canada would likely be more expensive and less extensive without these initiatives.

## ETHICAL

Ethical considerations can outweigh perceived scientific deficiencies and make community-based monitoring the most appropriate choice. Broadly speaking, the movement over the last several decades away from traditional, top-down techniques toward strategies that involve citizens has not been exclusive to community-based monitoring but has been occurring throughout the world of natural resource and protected area management and conservation (Phillips 2003). One of the fundamental causes of this shift has been the realization by many conservationists and resource professionals that traditional, command-and-control strategies are ineffective in many new conservation frontiers, such as in communities unfamiliar with Western concepts of science and monitoring (Phillips 2003). The need to better navigate the interface between the environment and humans has necessarily led to an array of interdisciplinary approaches to conservation science that incorporate anthropology, psychology, geography, and sociology, and encourage collaboration among researchers from these fields (Saunders 2003; Berkes 2004).

These new conservation partners often make powerful arguments based on democratic and educational theories, that resource managers and conservationists have ethical obligations to involve communities and citizens as comprehensively as possible in the decision-making processes related to our shared, finite resource base (Chase et al. 2004). Further, empirical data have shown that, relative to more exclusive approaches, supporting local governance and empowering communities in the context of resource monitoring and management can have a more desirable impact on social capital, particularly the long-term ability of community members to network and self-organize; can increase local satisfaction with monitoring and resource management in general; and can encourage more sustainable local land-use decisions (EMAN and CNF 2003). Given these positive impacts, in many cases it is the institutional obligation of resource professionals and conservationists to embrace new approaches that involve citizens and communities (Halvorsen 2001, 2003; Meretsky et al. 2006).

Such ethical obligations are often underscored in scenarios involving indigenous peoples. Many resource agencies have controversial pasts in which they evicted or excluded such communities from their traditional lands by forcefully designating the areas as public, erecting literal or figurative fences to forbid access, and assuming full control of management and monitoring (Spence 1999). In the contemporary landscape in which the presidents and prime ministers of developed nations have begun issuing formal apologies to indigenous peoples to atone for these historic injustices, continuing top-down monitoring programs would be inappropriate in many cases (Smith 2008). If resource managers and conservationists are to have any influence on monitoring initiatives on these traditional lands, it should be in the role as a facilitator between the Western science of ecological monitoring and the local ecological knowledge of indigenous communities and any such arrangements must be agreed upon by locals. This is increasingly recognized in conservation circles (Meffe et al. 2002; Phillips 2003).

## EDUCATION AND COMMUNITY-ENRICHMENT

The topic of human–environment bonds has received considerable attention in academia. For instance, there is an ongoing debate that deals with the causes and implications of the ebbing interaction between our country's youth and nature

(Louv 2006; Stanley 2007). Perhaps the most well-known contributions are those that explore the concept of "nature-deficit disorder" (Louv 2006, 2007). Although this concept remains largely inconclusive, actively nurturing human–environment bonds has indeed been linked to the attenuation of a variety of mental and physical health impairments such as obesity, attention-deficits, and depression to increases in creativity and community-interaction and to decreases in aggression (Louv 2007; Stanley 2007; Cornell Lab of Ornithology 2008). Further, many of these results are not exclusive to children, but have also been shown to extend to entire families and communities; the enhancement of these bonds should thus be pursued (Lowman 2006). Community-based monitoring constitutes one way to do this. Indeed, monitoring programs have been found to be an excellent vehicle for family and community-based nature education that fosters social learning and general increases in well-being, and inspires the construction of whole family and community conservation ethics (Lowman 2006; Fernández-Gimenez et al. 2008). As it has also been determined that the benefits of a healthy relationship with the environment accrue whether the bonds are between fishers and people or pigeons and people, this argument applies to a variety of settings, from urban to rural (Dobbs 1999; Cornell Lab of Ornithology 2008). It is not hard to imagine a situation in which the educational benefits or a high degree of community-enrichment could be embraced by scientists who are otherwise reluctant to establish a CBMP.

The Cornell Lab of Ornithology's Celebrate Urban Birds project is one example of a monitoring program with the goal of maximizing these benefits. The project trains citizens across the United States to identify 16 species of birds and then conduct 10-minute point counts for them and submit the data online (Cornell Lab of Ornithology 2008; K. Purcell pers. comm.). Although the monitoring protocol is designed in such a manner that it provides insight into the effects of urbanization on avian fauna, the argument could certainly be made that the principal objective of the Urban Birds project is to enrich communities via nature-based, experiential learning. Indeed, the group openly encourages participants to synthesize monitoring with urban-greening projects, artistic and musical events, and a variety of other activities designed to reinforce community-spirit and service; it also seeks to cross cultural boundaries by providing materials and resources in both Spanish and English languages. The Urban Birds project, while a monitoring program, values the education of urban communities in conservation-related topics and the improvement of their well-being at least as highly as it does data collection (Cornell Lab of Ornithology 2008).

Some practitioners utilize the educational potential of community-based monitoring to advance a particular conservation agenda (Dobbs 1999). Many of these initiatives are designed to prevent the development of sensitive natural areas and minimize the impact of sprawl (Dobbs 1999). Wildlife ecologist Susan Morse of Keeping Track® in Vermont, for instance, runs workshops in which she trains citizen groups organized by regional conservation agencies and land trusts to locate tracks, scat, and sign of a number of wide-ranging mammal species within their core habitat. Susan further provides the trainees with a primer on the importance of conservation planning (Figure 3.2; S. Morse, pers. comm., Keeping Track 2009). This background prepares the groups to conduct Keeping Track's science-based track and sign surveys

**FIGURE 3.2** Black bear sign found during a Keeping Track workshop. Participants are trained to seek out, photograph, and record data on sign such as this as a component of the community-based monitoring programs they undertake in their local ecosystems.

along established transects in their communities once per season on a long-term basis. The year-to-year detection of the selected mammal species' presence within unfragmented, core habitats augments understanding of the species' local habitat preferences and in some cases provides indices of relative abundance. Perhaps most importantly, for the involved stakeholders, such information attests to the ecological integrity and conservation worthiness of these habitats. Analyzing the data and considering this information in the context of current themes in conservation biology enables more informed decision-making about the appropriate placement of future development (Dobbs 1999; S. Morse, pers. comm.). Keeping Track is an originator of the idea that citizens can and should participate in the long-term collection of wildlife data with the specific purpose of informing conservation planning at community and eco-regional levels. The protocol is also designed and carried out to enhance the bonds between communities and their ecological surroundings by engaging them in a type of monitoring that maximizes their interaction with the local ecosystems and wildlife (Hass et al. 2000).

Establishing a network of concerned communities that use the same monitoring protocol also creates the potential to scale up the local data and thereby form a cogent argument for increased habitat connectivity for the target species on the regional and national scale. Indeed, Keeping Track has trained groups of citizens across the entire country (S. Morse, pers. comm.). Over the years, as data have accumulated, Morse has also developed more rigorous methods of monitoring via tracking, particularly through the detection of scent-marking sign such as felid retromingent scent posts and bear mark trees. These new methods have led Morse to believe that "we can powerfully use scent-marking in our track and sign surveys to predict where to find mammal sign and then deploy remote cameras to photo-capture individual resident animals over time" (S. Morse, pers. comm.). This could increase the power of the tracking-based monitoring, because if groups can identify individuals captured in the photos, conservation planners may be capable of differentiating between resident and nonresident wildlife. Such information would further inform development and conservation planning and facilitate the appropriate application of wildlife laws and regulation.

It is important to underscore that there is a fine line between incorporating environmental education and community-enrichment initiatives into monitoring programs and the incorporation of monitoring into environmental education and community-enrichment initiatives. It is the responsibility of scientists to be fully transparent about how a particular monitoring program should be developed and how data that results from the effort should be appropriately interpreted. In this same vein, it is integral that conservationists and resource managers clearly state the goals and objectives of education-related community-based monitoring programs before design and implementation to reduce the potential for conflict over time.

## Effectiveness

Community-based monitoring may simply be the most or only effective approach under some circumstances (Sheil 2001). This appears to be particularly true if the objective of an ecological monitoring program is related to guiding or influencing active management or conservation activities in rural, inhabited landscapes in which communities participate in resource-extraction or agriculture-based economies. Factors such as the intimacy of community relationships with the environment, geographic isolation of the ecosystem under consideration, or the contentious nature of interfering with or manipulating extractive behaviors from the top-down, may mean that some activities are more easily influenced using community-based rather than institutionally based monitoring programs (Sheil 2001). In fact, in some cases strict, top-down monitoring and management initiatives and associated regulations promote local resistance, resource depletion, the deterioration of sustainability, and the undermining of scientist–citizen relationships (Berkes 2007; Bjorkell 2008). This is often true when scientific information is used to manage landscapes in a manner that supersedes traditional, local programs or paints local-ecological knowledge as illegitimate (Huntington et al. 2006; Bjorkell 2008). Indeed, in certain contexts, community-based and collaborative approaches centered on local institutions and ideas are simply much more informative, more likely to result in effective management, governance, and conservation on a local scale, and more likely to generate monitoring programs that are viable over the desired space and time (Huntington et al. 2006; Bjorkell 2008). In Madagascar, for instance, arranging participatory wetland monitoring programs through local institutions allayed citizen concern that the government fishery agency was using its power to profit from local fisheries (Andrianandrasana et al. 2005). This, in turn, helped legitimize fishery laws and regulations that citizens had previously not respected due to the belief that government officials implemented them in their own self-interest.

The spatial scale of a monitoring program can also make a community-based protocol more effective than one operated by scientists. For projects that span entire regions, countries, or continents, the coordination of a sufficient number of ecologists, biologists, and resource managers to meet project objectives is usually impractical. However, organizing a network of citizens to undertake monitoring activities, although still a challenge, may be more practical. For example, the MEGA-Transect project along the 3,625-km-long Appalachian Trail, includes nearly 100 volunteers to handle equipment, gather data, and record observations to monitor environmental trends (Cohn 2008). This project, managed by researchers at the National Zoo's Conservation and Research Center in Front Royal, Virginia, also includes a 960-km citizen-run motion-sensor camera survey of the trail from Virginia to Pennsylvania (Cohn 2008). Without the aid of citizens and communities, such monitoring and data collection efforts would likely be unrealistic. The North American Breeding Bird Survey and the Breeding Bird Atlas programs discussed below as well as in Chapter 2 provide other examples.

## DESIGNING AND IMPLEMENTING A COMMUNITY-BASED MONITORING PROGRAM

Although this list of potential benefits is by no means exhaustive (see Fernandez-Gimenez et al. 2008, for instance), it is clear that community-based monitoring has the potential to yield rich, varied results, not all of which are grounded in science. This implicitly reveals that these programs often have a more diverse set of stakeholders than those run entirely by scientists. This can make designing and implementing an effective protocol for a CBMP a very difficult task. Indeed, communities in conjunction with the scientists, resource professionals, and practitioners working with them, are characterized by distinctive amalgamations of needs, desires, opportunities, and education levels that all interact in intricate ways over varying spatial and temporal scales. Just like the ecosystems in which they are embedded, such groups are not homogenous entities, but uniquely complex systems. In light of this, there is no single protocol for the most effective or desirable CBMP; rather, the components of each must be determined, based on the specific scientific, ecological, social, and cultural scenario in which it is to be implemented. The existence of different methodological approaches for designing and implementing CBMPs should therefore come as no surprise. It is possible to discuss two markedly different categories that vary in their degree of top-down input from scientists: prescriptive and collaborative. Prescriptive approaches to CBMP design are those in which science professionals craft a protocol to accurately capture ecological data and train citizens to carry it out (Fore et al. 2001; Engell and Voshell 2002). Data analysis is generally done by scientists, but can also be, and sometimes should be, undertaken by citizens (Fore et al. 2001; Engell and Voshell 2002; Lakshminarayanan 2007). In contrast, the collaborative approach is usually undertaken through the use of a framework that encourages scientists and communities to work jointly and interact as one larger community in the design

# Community-Based Monitoring

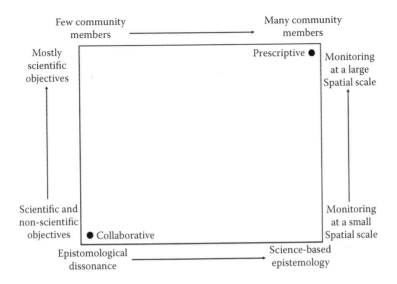

**FIGURE 3.3** This diagram places highly prescriptive and highly collaborative approaches to design where we consider them most appropriate according to the variables listed. Deviations from those two points will result in a combination of the two approaches. The variables listed are limited due to space constraints, and others should certainly be considered, including the presence or absence of a culture of volunteerism.

of a mutually acceptable and useful monitoring program tailored to their specific scenario. However, past and current efforts to design CBMPs rarely fit perfectly into either category; most are a fusion of both. Thus, the categories are actually two bookends of a continuum rather than discrete types. Locally autonomous monitoring programs are also legitimate and should be respected and institutionally supported, but they are not the main thrust of this chapter. The ideal mix of design techniques depends on a number of factors, including spatial scale and objectives of monitoring and the size, local expertise, and socioeconomic status of the community. Figure 3.3 may prove helpful as a starting point for practitioners and will serve as a useful framework for the remainder of this section.

## THE PRESCRIPTIVE APPROACH

The prescriptive approach to CBMP protocol design is largely focused on the rigor of the monitoring methods, the accuracy and precision of the collected data, and the power of data analysis. One example includes the many Water Watch Organizations within the United States (Fore et al. 2001). In the state of Washington, for instance, over 11,000 citizens have been trained to monitor stream ecosystems using the benthic-index of biological integrity: a measure of the diversity of a stream's invertebrate organisms often used as an indicator for other stream ecosystem characteristics (Fore et al. 2001). This indicator and its associated collection method was developed by scientists, and the participating citizens were trained by science professionals (Fore et al. 2001). In the particular case discussed by Fore et al. (2001), when scientists

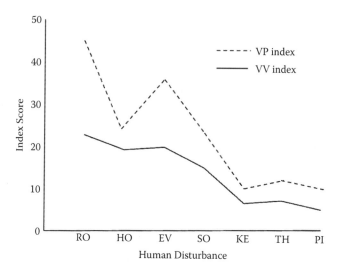

**FIGURE 3.4** Fore et al. (2001) assessed differences between the volunteer generated (VV) and science professional generated (VP) benthic indices of biological integrity as they relate to human disturbance near stream ecosystems. (Redrafted from Fore, L., K. Paulsen, and K. O'Laughlin. 2001. *Freshwater Biology* 46:109–123.)

later questioned the accuracy and precision of the citizen-derived data, they intervened in a largely top-down way by independently undertaking data collection and analysis, and then statistically analyzing the differences between their results and those of the citizens (Figure 3.4). Although they found no significant differences in any case in which the citizens had been properly trained, the process allowed scientists to augment the scientific value of the monitoring program by improving their ability to confidently interpret the citizen's data (Fore et al. 2001). Top-down, compliance monitoring such as this is generally supported by the scientific community and can be appropriate in the context of prescriptively designed CBMPs, thus it merits consideration (Fore et al. 2001). Another example is the New York State Breeding Bird Atlas (BBA), a project discussed in Chapter 2. Once again, the BBA is a statewide survey in which citizens sample habitat in New York to document the distribution of all breeding birds in New York that was conducted in two time periods: the first from 1980 to 1985 (Andrle and Carroll 1988) and the second from 2000 to 2005 (McGowan and Corwin 2008). Volunteers in this project are given a handbook and other information created by scientists to assist them with atlasing. This is part of a concentrated effort on the part of the researchers to prescribe a particular set of protocols that achieve consistent coverage within each atlas block so that changes in species distributions can be considered true ecological patterns as opposed to some deviation in sampling methodology due to observer bias or differences in training between the two time periods.

While many practitioners refer to programs of this type strictly as "citizen science" rather than community-based monitoring, we have included such initiatives under the umbrella term of community-based monitoring for two reasons. The first is to provide clarity, in the sense that citizen science initiatives are by no means limited to ecological monitoring. The second is that the design and implementation processes of prescriptive plans are largely top-down, and can therefore result in excessively top-down, hierarchical designs in which participants are "used" by scientists rather than collaborated with (Lakshminarayanan 2007). This has historically been the case with some geographically broad programs in which community volunteers learned from and appreciated local data collection, but were entirely excluded from the scientists' meta-analyses (Lakshminarayanan 2007). In the past, this treatment has been justifiably interpreted as skepticism about a public's intellect; has insulted participants who invested considerable emotion, time, and effort into assisting scientists; and has led to the cessation of monitoring (Lakshminarayanan 2007). As these failed programs were classified as citizen-science initiatives, it seems useful to describe prescriptive monitoring programs as community-based monitoring here as a reminder that, in terms of both the long-term viability as well as the ethical basis of the program, it is necessary to interact with the public as social groups and to acknowledge their intellect, efforts, and emotions whenever they are involved. Ways for science professionals to more actively accomplish this include assuming the role of facilitator rather than expert during training; undertaking data calibration, collection, and analysis in a way that embraces the concepts of open access and freedom; and establishing a reliable system for citizens to provide feedback to scientists (Meffe et al. 2002; Lakshminarayanan 2007). That being said, the term *citizen science* should not be categorically rejected or criticized, and there are numerous inspiring, culturally sensitive, and scientifically impressive citizen-science initiatives. Furthermore, it is important to keep in mind that the terminology used to classify science involving a public will vary depending on the source (see Bacon et al. 2005; Cooper et al. 2007, 2008; Fernández-Gimenez 2008).

### In What Context Does It Work?

As mentioned above, the programs designed in this manner focus nearly exclusively on the methods, accuracy, and precision of the science. Consequently, they are often only appropriate and able to engage community volunteers over the long-term in communities that already ascribe a high value to Western scientific inquiry (Cooper et al. 2008). Indeed, although the long-term capacity and willingness of citizens to participate in data collection must be considered in program design, it is often only necessary to do so in terms of the program's temporal and economic logistics and the scientific utility of data because the epistemological harmony between citizens and scientists makes a science-focused design mutually valuable. It has been argued that communities that are congruous with prescriptive designs are those embedded within societies that offer many opportunities for citizens to enter into a scientific profession, but only after years of training and/or the attainment of academic degrees (Cooper et al. 2007, 2008). In this context, rigorous, scientifically focused CBMPs provide a desired opportunity for citizens socialized to appreciate scientific fields but trained in others and therefore unable to access science positions, to legitimately contribute to science without the extensive education process. In this sense, local expertise is not necessarily integral to the program because citizens will be open to learning the protocols and adhering to instructions. Communities with a socioeconomic status that fosters a culture of volunteerism and provides a considerable amount of leisure time will be particularly likely to fit this description (Danielsen et al. 2009). In scenarios lacking these social and/or economic characteristics, citizens may need to be incentivized economically or otherwise for undertaking ecological monitoring designed with the prescriptive approach (Andrianandrasana et al. 2005).

It may also be the case that prescriptive designs are needed when the monitoring program's spatial scale, number of participants, and quantity of data are large. The BBA, for instance, enlisted over 1,200 volunteers throughout New York State during both time periods, which resulted in 361,594 records for 246 species in 1980 BBA and 383,051 records for 251 species in 2000. In such scenarios, the design and implementation of a monitoring program, as well as the data collection and analysis, would likely be chaotic without the strict oversight of scientists, and the ability of researchers to confidently scale up data from the local level would be limited. On a related note, when the goals of a monitoring program include using the data for publication in scientific journals or satisfying a government or institutional mandate, a prescriptive approach may be the only means of attaining them.

## THE COLLABORATIVE APPROACH

The second prevalent approach to CBMP protocol design and implementation is fully collaborative. As mentioned above, collaborative protocols are often approached with the use of a framework that aims to make initially separated communities and researchers into collaborators. Many such frameworks have been proposed and are readily accessible in academic journals, practitioner's manuals, and anthologies. The Southern Alliance for Indigenous Resources (SAFIRE) is one example that led to the creation of a design that is adapted to the specific context and appears to have been drafted with rural communities of Africa in mind (Figure 3.5; Fröde and Masara 2007). This framework creates a forum in which science professionals and citizens are given the opportunity to have significant input. Indeed, it has been suggested

Step 1: Preliminary ecological research action forms the base of any action in a project area. Research will mainly consist of assessing ecological aspects in project feasibility studies and resource assessments.

Step 2: Management plans for the resource use are developed in a participatory way on the base of the ecological assessments, and their implementation is started.

Step 3: The plan for community-based ecological monitoring can be developed by communities and field staff.

Step 4: The practical aspects of community-based ecological monitoring are set up, based on this plan.

Step 5: Ecological monitoring is implemented.

Step 6: The ecological monitoring process forms the base for the adaptation of management procedures—the ultimate goal of the monitoring process.

**FIGURE 3.5** SAFIRE's six steps to creating a community-based monitoring program that is at once acceptable to scientists and resource managers as well as local communities. (Redrafted from Fröde, A., and C. Masara. 2007. Community-based ecological monitoring. Manual for Practitioners. SAFIRE—Southern Alliance for Indigenous Resources. Harare, Zimbabwe. p. 64.)

that the most meaningful and durable plans that lead to active conservation activities and enjoy multilayered support are designed in collaborative ways that work with all involved viewpoints. These successful programs must also be sensitive to the differential expertise of involved parties in terms of their natural science aptitude and organizational capacity (Sheil 2001; Cooper et al. 2007, 2008; Lakshminarayanan 2007). Sometimes, the most important task for science professionals in the collaborative approach is to maintain a sense of equality during the planning and implementation stages. The varying levels of expertise in both monitoring and social empowerment must emerge from collaborations between science and citizens and be incorporated into the CBMP.

### In What Context Does It Work?

This strategy seems more appropriate than a prescriptive design when there is a measurable degree of epistemological dissonance between the visions of monitoring held by the researcher and the community; this is likely to be the case when a community desires a significant amount of autonomy relative to its natural resources, when there are stark cultural differences between locals and researchers, and when the socioeconomic status of a community makes its members unwilling and/or unable to volunteer simply to meet a scientist's objectives.

One manifestation of such dissonance is the possession of distinct ideas about acceptable survey methods or indicators. For instance, Jensen et al. (1997) mentioned how some First Nation communities in Canada use the taste of game to monitor the status and health of nearby wildlife populations. As an indicator on which to base game management or policy, this would be unlikely to convince most scientists and politicians. At the same time, a protocol that strictly addresses concerns such as statistical power of methods and biodiversity indicators may not be valued enough by a community in a developing country for them to carry out monitoring, especially if their local livelihoods depend on occupations with high time and energy requirements. Conflicts regarding the desired objectives or other functional aspects of monitoring programs can also arise from epistemological dissonance. In discussing the bowhead whale monitoring program described above, Berkes et al. (2007) mentioned that "the scientific objectives were about conserving populations and species, the Inuit objectives were about Inuit-bowhead relationships and access to the resource." In other cases, a community may wish to focus primarily on a program's educational or culturally enriching components or perhaps even the economic benefits of monitoring game populations, while a scientist or resource manager may wish to optimize statistical power for a publication or gather data to conserve a species. In nearly all cases with such dissonance, if having a monitoring program that is locally sustainable and institutionally recognized is the ultimate goal, then involved parties must be willing to accept outside influences and compromises and a collaborative approach is necessary. If the dissonance cannot be reconciled, fundamental differences in worldview and values must be respected rather than forcefully altered or denigrated (Sheil 2001; Berkes et al. 2007). In such cases, a locally autonomous rather than collaborative approach may arise and, as mentioned above, should enjoy an adequate degree of institutional support.

The collaborative approach also seems more appropriate when the scale of monitoring and the size of the community are at a very local level and scientists have the ability to run workshops to facilitate monitoring-related decision-making processes with all participants. It merits noting, however, that efforts to scale-up data collected across many small communities with protocols designed collaboratively have successfully influenced national resource management (Danielsen et al. 2005).

Despite our specific suggestions, newly proposed CBMP design frameworks vary in how and which ecological and social characteristics are considered and are nearly ubiquitous. There is an extensive body of literature describing and/or advocating alternative approaches to CBMP design and implementation; just as with the monitoring programs themselves, even the approach to design and implementation must be adapted to each specific context (Fleming and Henkel 2001; Fraser et al. 2006; Fröde and Masara 2007; Conrad and Daoust 2008). Once again, important variables to consider when determining an approach to CBMP design include the spatial scale and the objectives of monitoring and the size of the community involved (Figure 3.3).

## SUGGESTIONS FOR SCIENTISTS

As briefly indicated above, there is much debate about community-based monitoring within the scientific community, usually in terms of its scientific utility. This is particularly the case in scenarios that mandate collaborative design approaches, as they commonly incorporate both biodiversity conservation and livelihood objectives (Sheil 2001; Fraser et al. 2006). Resistance to this is widespread due to arguments such as that social objectives dilute the all-important conservation objectives and that mixing social-benefits with science ineluctably dilutes the objectivity and therefore rigor of the scientific data collected (Berkes 2007; D. Kramer, pers. comm.). Nonetheless, many past monitoring programs have integrated citizens yet failed to integrate local values and livelihood indicators so were not viable over the long-term (Sheil 2001). A number of problems arise in such circumstances, including volunteer "burnout," a lack of observer objectivity, or simply a dearth of interest and therefore irregular participation that leads to data fragmentation (Sharpe and Conrad 2006). If it is important to scientists that monitoring is conducted in a region where they cannot monitor themselves, where resources are intimately linked to local livelihoods, or where Western science is simply not the local priority, scientists will likely have to be flexible and incorporate local epistemologies if any biodiversity objectives are to be attained. Furthermore, working in this flexible way has proven worth the effort. Along with the potential benefits described above, scientists generate scientific data and build healthy relationships with citizens. Further, communities gain the capacity and institutional support to monitor locally valued resources and the opportunity to legitimize their worldviews and opinions among science professionals, whose activities may have previously been viewed as threats to local livelihoods (Huntington et al. 2006). Open, constructive bonds between scientists and society have an important role to play in the contemporary conservation landscape. Some suggestions and strategies for successfully reaching the necessary compromises include resolving the underlying conflicts between scientists and nonscientific monitoring, using participatory action research, and approaching plan design and implementation in a way that embraces systems-thinking (Greider and Garkovich 1994; Castellanet and Jordan 2002; Bacon et al. 2005; Walker et al. 2006b).

## Resolving the Underlying Conflict

Resolution of the ideological discord between our ideas of science, nature, and monitoring and those of a local community can inspire scientists to think and work more flexibly. There are several ways to resolve the conflicts. One is by understanding that ecological monitoring is socially constructed. Applying the core of social constructionism to monitoring is simple enough: our particular needs, values, and interests are what have conceived and reified our processes for monitoring (Boghossian 2001). In the absence of our particular culture, the processes of monitoring have been conceived and developed in distinct forms by other societies with different needs, values, and interests. To these other societies, their monitoring processes are viewed as valid and highly valued, in the same way that ours are to us. The difference and one of the primary sources of conflict, therefore, has a cultural base. Realizing this can assist scientists in achieving a philosophy that facilitates the acceptance of local ecological knowledge and social indicators into ecological monitoring programs. In the context of social constructionism, denying the inclusion of disparate needs, values, and interests in monitoring with the argument that it invalidates the process is to erroneously and dogmatically perceive our socially particular monitoring processes and our culture as sacrosanct and inherently superior to those of a local community.

A second way to resolve the conflict is to confront the so-called publish or perish culture of academia. The tradition within the institution is to mandate that professionals publish with regularity in "high-impact" journals in order to attain tenured positions or improve or maintain their standing (Cohen 2006). Given strict publication requirements and the potential for traditional opinions among peer reviewers, there may be hesitancy on the part of some professionals to make the compromises needed to work with communities in a flexible way. Indeed, doing so may hinder publication and put one's job security at risk. Publicly addressing these potentially negative impacts may lead to a reconsideration of the traditional metrics for evaluating work published in well-regulated interdisciplinary and transdisciplinary publications. Eventually such developments could result in a system that allows for the frequent communication via publication that is integral to our field while encouraging rather than discouraging university scientists to work more collaboratively with communities. A number of alternatives to the traditional system are currently being explored, and although there is the potential for huge benefits, the shift is a cautious one (Casati et al. 2007).

## Participatory Action Research

Participatory action research (PAR) is a style of research that "promotes broad participation in the research process and supports action leading to a more just or satisfying situation" for all stakeholders (Bacon et al. 2005). In the context of ecological

| Research Objectives | Action Research Objectives |
|---|---|
| 1. Characterize different types of shade on the coffee farms | 1. The process will provide information used by small-scale farmers to improve management and conserve natural resources |
| 2. Assess the role of shade tree products and other benefits to farm households | 2. The process will support small farm households and expand social and market networks |
| 3. Assess effects of different types of farmer cooperatives on shade management | 3. Researchers provide training and support to farmers in their conservation and production activities |
| 4. Assess environmental benefits of small coffee farms emphasizing native tree species | |

**FIGURE 3.6** Objectives generated during a PAR project in El Salvador that clearly takes the needs, desires, interests, and values of diverse stakeholders into account. (Redrafted from Bacon, C., E. Mendez, and M. Brown. 2005. Participatory action research and support for community development and conservation: Examples from shade coffee landscapes in Nicaragua and El Salvador. Center for Agroecology & Sustainable Food Systems. Research Briefs. No. 6.)

research, this goal is normally attained by designing protocols that ensure that both researchers and other stakeholders are able to improve their respective situations (Bacon et al. 2005). This often involves workshops and extensive, fully transparent dialog between researchers and the community with the purpose of generating goals and objectives that meet the expectations of the maximum number of participants (Figure 3.6) and mandates that an atmosphere of equality is fostered in which the researchers and citizens become components of a larger community linked by the research process itself (Castellanet and Jordan 2002). Indeed, to attain a comprehensive PAR experience, the endeavor must be undertaken "with and by local people" instead of on or for them (Cornwall and Jewkes 1995). PAR techniques are particularly useful in collaborative approaches to CBMP design because they are explicitly implemented to foster a fully collaborative project. Indeed, the techniques are designed to remove scientists from the linear mold of conventional research-thinking in which they assume the controlling role of the expert and encourage them to seek outside input (Castellanet and Jordan 2002). It is clear that they have the power to facilitate the acknowledgement and acceptance of the value of integrating nonscientific components into the ecological monitoring program and can encourage scientists to approach subsequent research more holistically.

## Systems Thinking

At times, scientists do not accept the input of communities or individual citizens into ecological monitoring plans due to a tendency to think at the national and global scales and to neglect important variables at the local scale. This is predictable in that we, as Western scientists, are often trained to value and prioritize the variables that are most highly regarded by university and government scientists, such as scientific rigor and statistical power, not those valued by locals for their relevance to livelihoods or social well-being. If communities and citizens are to be fully integrated into monitoring programs, then this is a flawed approach, because CBMPs are complex, adaptive, social-ecological systems. A social-ecological system is an "ecological system intricately linked with and affected by one or more social systems" (Anderies et al. 2004). It is sometimes stated that true social-ecological systems are those comprised of multiple social systems that affect one another through independent interactions with the biophysical or ecological system (Anderies et al. 2004). To be complex implies that the system is comprised of multiple subsystems at multiple scales and that those at smaller scales are embedded within those at larger scales. This interrelated structure means that an action undertaken in one subsystem causes feedbacks or reactions in the others. These feedbacks and reactions, in turn, result in the readjustment of the system as a whole (Folke 2006). Finally, adaptability indicates that a system has the capacity to adjust itself in order to increase or maintain survival in the face of environmental perturbation. In other words, an adaptive system will adjust to novelties in the environment in order to retain an appropriate, functional structure despite those novelties. In the context of social systems, it is sometimes argued that adaptability is further defined by the capacity of the actors within the system to influence how such adjustments play out (Walker et al. 2006a).

Most community-based ecological monitoring programs fit this definition. First, they involve ecological systems that are intricately linked, through monitoring activities, to multiple social systems (i.e., the local community, the scientific community, the government, and systems comprised of interacting combinations of individuals from these larger systems). To reveal how they also meet the remaining criteria, let's look at two examples:

1. Changes in a local community's attitudes toward monitoring cause an alteration in which aspects of the biophysical world are sampled. This change at the local scale affects the quantity and quality of the data that reach scientists at the national scale, leading them to alter their methods of analysis and interpretation so that they retain the capacity to confidently report the results to the government.
2. Decreases in resource agency funding caused by a global recession lead scientists to deprioritize the biophysical system surrounding a particular community and to decrease fiscal support for monitoring efforts there. These national and international occurrences impact the capacity of a community to monitor and lead them to reduce the quantity of indicators monitored so that they can continue to afford to monitor their most valued resources.

In both cases, the social systems exhibit an ability to affect one another through otherwise independent interactions with the biophysical system. The social systems also clearly act at different scales, yet not in an isolated manner; some are encompassed within others and all are interdependent to the extent that seemingly independent actions at one scale result in a chain of events that reverberates through the other scales and ultimately leads to the readjustment of the monitoring plan as a whole. Finally, the human actors affected by changes at other scales, such as scientists affected by local changes or locals impacted by national and international changes, undertake actions to ensure that the system retains certain necessary functions despite the readjustment. The system is therefore complex, adaptive, and a model of social-ecological synthesis.

Viewing CBMPs in this manner, as linked arrangements of mutually important cogs, inevitably leads to the conclusion that all components of CBMPs have the potential to impact one another and thereby the CBMP as a whole, so *all* are important, thus *all* have to be considered. In contexts conducive to collaborative approaches, such broad thinking will probably underscore the importance of incorporating locally valued components not normally valued by Western scientists, such as those more economically or socially based, into the plan (Walker et al. 2002; Bosch et al. 2007). At the same time, it will also likely prevent an excessively local focus, which *can* result when scientists and resource agencies overcompensate for past, top-down designs (Giller et al. 2008). There are a variety of systems-based modes of thought, such as resilience-thinking and ecosystem management, and extensive bodies of related literature that can help not simply scientists, but all involved parties to attain a more holistic view of monitoring (Meffe et al. 2002; Walker et al. 2006b).

## SUMMARY

Citizen and community involvement in natural resource and conservation science are not novel phenomena. Rather, they have long histories from which many lessons can be drawn and applied to CBMPs. In addition to this, many of the lessons learned from previous noncommunity-based ecological monitoring programs and the methods and suggestions contained within this book are essential even when monitoring is community based. One important lesson provided by both of these sources is that the goals and objectives of the plan should be clearly generated and stated before plan design and implementation begins. This may be particularly significant in the context of community-based monitoring as the goals and objectives often diverge markedly from the norm.

Perhaps the most important lesson is that the ideal protocol for monitoring an ecosystem with a community will be adapted to the particular scenario under consideration. Strategies for designing a CBMP protocol can be broadly classified as prescriptive and collaborative, yet choosing an appropriate strategy is a key determinant of the design's long-term viability, thus previously outlined frameworks should be viewed as two options on a continuum and used as suggestions rather than methodologies to be adhered to; indeed, design strategy itself must also be unique to the protocol's context.

In many cases, particularly where more collaborative approaches are needed, it is likely that we, as Western scientists, will have to accept social values, livelihood indicators, and epistemologies distinct from those heralded within our field into traditional monitoring protocols if we are to attain both local and institutional acceptance and viability and fulfill the objectives of all participants. For community-based monitoring initiatives with which we are involved to reach their full potential, therefore, we, as ecologists, biologists, and resource managers, must endeavor to think and work in more expansive, interdisciplinary ways.

## REFERENCES

Anderies, J.M., M.A. Janssen, and E. Ostrum. 2004. A framework to analyze the robustness of social-ecological systems from an institutional perspective. *Ecology and Society* 9(1):Art 18.

Andrianandrasana, H.T., J. Randriamahefasoa, J. Durbin, R.E. Lewis, and J.H. Ratsimbazafy. 2005. Participatory ecological monitoring of the Alaotra wetlands in Madagascar. *Biodiversity and Conservation* 14:2757–2774.

Andrle, R.F., and J.R. Carroll. 1988. *The Atlas of Breeding Birds in New York State.* Cornell University Press, Ithaca, New York.

Bacon, C., E. Mendez, and M. Brown. 2005. Participatory action research and support for community development and conservation: Examples from shade coffee landscapes in Nicaragua and El Salvador. Center for Agroecology & Sustainable Food Systems. Research Briefs. No. 6.

Berkes, F. 2004. Rethinking community-based conservation. *Conservation Biology* 18(3):621–630.

Berkes, F. 2007. Community-based conservation in a globalized world. *Proceedings of the National Academy of Sciences of the United States of America* 104(39):15188–15193.

Berkes, F., M.K. Berkes, and H. Fast. 2007. Collaborative integrated management in Canada's north: the role of local and traditional knowledge and community-based monitoring. *Coastal Management* 35(1):143–162.

Bjorkell, S. 2008. Resistance to top-down conservation policy and the search for new participatory models: The case of Bergo-Malax' Outer Archipelago in Finland. Chapter 8 in *Legitimacy in European Nature Conservation Policy: Case Studies in Multi-Level Governance.* Keulartz, J., and G. Leistra (eds.). The International Library of Environmental, Agricultural and Food Ethics, Netherlands.

Boghossian, P. 2001. What is social construction? Times Literary Supplement, January 2001.

Bosch, O.J.H., C.A. King, J.L. Herbohn, I.W. Russel, and C.S. Smith. 2007. Getting the big picture in natural resource management—Systems thinking as "method" for scientists, policy makers and other stakeholders. *Systems Research and Behavioral Science* 24(2):217–232.

Casati, F., F. Giunchiglia, and M. Marchese. 2007. Publish and perish: Why the current publication and review model is killing research and wasting your money. *Ubiquity* 8(3). http://www.acm.org/ubiquity/views/v8i03_fabio.html

Castellanet, C., and C.F. Jordan. 2002. *Participatory Action Research in Natural Resource Management: A Critique of the Method based on Five Years' Experience in the Tranzamazónica Region of Brazil.* Taylor & Francis, New York. p. 231.

Chase, L.C., D.J. Decker, and T.B. Lauber. 2004. Public participation in wildlife management: What do stakeholders want? *Society and Natural Resources* 17(7):629–639.

Cohen, J. 2006. Publish or perish—Is open access the only way forward? *International Journal of Infectious Diseases* 10:417–418.

Cohn, J. 2008. Citizen science: Can volunteers do real research? *BioScience* 58(3):192–197.

Cooper, C.B., J. Dickinson, T. Phillips, and R. Bonney. 2007. Citizen science as a tool for conservation in residential ecosystems. *Ecology and Society* 12(2):11.

Cooper, C.B., J.L. Dickinson, T. Phillips, and R. Bonney. 2008. Science explicitly for nonscientists. *Ecology and Society* 13(2):RESP. 1.

Conrad, C.T., and T. Daoust. 2008. Community-based monitoring frameworks: Increasing the effectiveness of environmental stewardship. *Environmental Management* 41(3):358–366.

Cornell Lab of Ornithology 2008. Celebrate Urban Birds Web site: http://www.birds.cornell.edu/celebration/

Cornwall, A., and R. Jewkes. 1995. What is participatory research? *Social Science and Medicine* 41(12):1667–1676.

Danielsen, F., N.D. Burgess, A. Balmford, P.F. Donald, M. Funder, J.P.G. Jones, P. Alviola, D.S. Balete, T. Blomley, J. Brashares, B. Child, M. Enghoff, J. Fjeldsa, S. Holt, H. Hubertz, A.E. Jensen, P.M. Jensen, J. Massao, M.M. Mendoza, Y. Ngaga, M.K. Poulsen, R. Rueda, M. Sam, T. Skielboe, G. Stuart-Hill, E. Topp-Jogensen, and D. Yonten. 2009. Local participation in natural resource monitoring: A characterization of approaches. *Conservation Biology* 23(1):31–42.

Danielsen, F., A.E. Jensen, P.A. Alviola, D.S. Balete, M. Mendoza, A. Tagtag, C. Custodio, and M. Enghoff. 2005. Does monitoring matter? A quantitative assessment of management decisions from locally-based monitoring of protected areas. *Biodiversity and Conservation* 14(11):2633–2652.

Dobbs, D. 1999. On the track of something good. *Magazine of the National Audubon Society* 101(3):36–39.

Ecological Monitoring and Assessment Network Coordinating Office and the Canadian Nature Federation (EMAN and CNF). 2003. Improving local decision making through community based monitoring: Toward a Canadian community monitoring network. Available Online: http://www.ccmn.ca/english/library/ccmn.pdf

Engell, S.R., and J.R. Voshell Jr. 2002. Volunteer biological monitoring: Can it accurately assess the ecological condition of streams? *American Entomologists*. Fall issue:164–177.

Fernández-Gimenez, M.E., H.L. Ballard, and V.E. Sturtevant. 2008. Adaptive management and social learning in collaborative and community-based monitoring: A study of five community-based forestry organizations in the western USA. *Ecology and Society* 13(2):Art. 4.

Fleming, B., and D. Henkel. 2001. Community-based ecological monitoring: A rapid appraisal approach. *APA Journal* 67(4):456–466.

Folke, C. 2006. Resilience: The emergence of a perspective for social–ecological systems analyses. *Global Environmental Change* 16(3):253–267.

Fore, L., K. Paulsen, and K. O'Laughlin. 2001. Assessing the performance of volunteers in monitoring streams. *Freshwater Biology* 46:109–123.

Fraser, E.D.G., A.J. Dougall, W.E. Mabee, M. Reed, and P. McAlpine. 2006. Bottom up and top down: Analysis of participatory processes for sustainability indicator identification as a pathway to community empowerment and sustainable environmental management. *Journal of Environmental Management* 78(2):114–127.

Fröde, A., and C. Masara. 2007. Community-based ecological monitoring. Manual for Practitioners. SAFIRE—Southern Alliance for Indigenous Resources. Harare, Zimbabwe. p. 64.

Giller, K.E., C. Leeuwis, J.A. Andersson, W. Andriesse, A. Brouwer, P. Frost, P. Hebinck, I. Heitkonig, M.K. van Ittersum, N. Koning, R. Ruben, M. Slingerland, H. Udo, T. Veldkamp, C. van de Vijver, M.T. van Wijk, and P. Windmeijer. 2008. Competing claims on natural resources: What role for science? *Ecology and Society* 13(2):Art. 34.

Greider, T., and L. Garkovich. 1994. Landscapes: The social construction of nature and the environment. *Rural Sociology* 59(1):1–24.

Halvorsen, K.E. 2001. Assessing public participation techniques for comfort, convenience, satisfaction, and deliberation. *Environmental Management* 28(2):179–186.

Halvorsen, K.E. 2003. Assessing the effects of public participation. *Public Administration Review*. 63(5):535–543.

Hass, C.C., S.C. Morse, and H.G. Shaw. 2000. Keeping Track® Project and Data Management Protocol. Keeping Track®, Inc. Huntington, VT.

Huntington, H.P., S.F. Trainor, D.C. Natcher, O.H. Huntington, L. DeWilde, and F.S. Chapin III. 2006. The significance of context in community-based research: Understanding discussions about wildfire in Huslia, Alaska. *Ecology and Society*. 11(1):40.

Jensen, J., K. Adare, and R. Shearer. 1997. Canadian Arctic contaminants assessment report. Indian and Northern Affairs Canada, Ottawa.

Kay, J.J., and H.A. Regier. 2000. Uncertainty, complexity, and ecological integrity: Insights from an ecosystem approach. Chapter 8 in *Implementing Ecological Integrity Restoring Regional and Global Environmental and Human Health: Restoring Regional and Global Environmental and Human Health.* Crabbe, P. (ed.). NATO Scientific Publications, Kluwer Academic Publishers, Dordrecht, The Netherlands. p. 492.

Keeping Track®. 2009. Web site: http://www.keepingtrack.org/

Lakshminarayanan, S. 2007. Using citizens to do science versus citizens as scientists. *Ecology and Society* 12(2):RESP. 2.

Latour, B. 2004. Why has critique run out of steam? From matters of fact to matters of concern. *Critical Inquiry* 30(2):225–248.

Louv, R. 2006. *Last Child in the Woods: Saving Our Children from Nature-Deficit Disorder.* Algonquin Books, Chapel Hill, NC. p. 334.

Louv, R. 2007. Leave no child inside: The growing movement to reconnect children and nature, and to battle "nature deficit disorder." *Orion.* March/April 2007.

Lowman, M. 2006. No child left indoors. *Frontiers in Ecology and the Environment* 4(9):451.

McGowan, K.J., and K. Corwin 2008. *The Second Atlas of Breeding Birds in New York State.* Cornell University Press, Ithaca, NY.

Meffe, G.K., L.A. Neilson, R.L. Knight, and D.A. Schenborn. 2002. *Ecosystem Management: Adaptive, Community-Based Conservation.* Island Press, Washington, D.C.

Meretsky, V.J., R.L. Fischman, J.R. Karr, D.M. Ashe, J.M. Scott, R.F. Noss, and R.L. Schroeder. 2006. New directions in conservation for the National Wildlife Refuge System. *Bioscience* 56(2):135–143.

Phillips, A. 2003. Turning ideas on their head: The new paradigm for protected areas. In *Innovative Governance: Indigenous Peoples, Local Communities and protected Areas.* H. Jaireth and D. Smyth (eds.). Ane Books, New Delhi. pp. 1–28.

Plummer, R., and J. Fitzgibbon. 2004. Co-management of natural resources: A proposed framework. *Environmental Management* 33(6):876–885.

Reiger, J.F. 2001. *American Sportsmen and the Origins of Conservation.* Third ed. Oregon State University Press, Corvallis. 338 p.

Resilience Alliance Website. 2008. http://www.resalliance.org/

Saunders, C.D. 2003. The emerging field of conservation psychology. *Human Ecology Review* 10(2):137–149.

Sharpe, A., and C. Conrad. 2006. Community based ecological monitoring in Nova Scotia: Challenges and opportunities. *Environmental Monitoring and Assessment.* 13(1–3):305–409.

Sharpe, T., B. Savan, and N. Amott. 2000. Testing the waters. *Alternatives* 26:30–33.

Sheil, D. 2001. Conservation and biodiversity monitoring in the tropics: Realities, priorities, and distractions. *Conservation Biology* 15(4):1179–1182.

Smith, T. 2008. The letter, the spirit, and the future: Rudd's apology to Australia's Indigenous people. Australian Review of Public Affairs. *Digest.* March issue.

Spellerberg, I.F. 2005. *Monitoring Ecological Change.* Cambridge University Press, Cambridge, U.K. 391 pp.

Spence, M.D. 1999. *Dispossessing the Wilderness: Indian Removal and the Making of National Parks.* Oxford University Press, USA. 200 pp.

Stanley, M. 2007. Redeeming the nature of childhood. *Children, Youth, and Environments* 17(2):208–212.

Tripathi, N., and S. Bhattarya. 2004. Integrating indigenous knowledge and GIS for participatory natural resource management: State-of-the-Practice. *The Electronic Journal on Information Systems in Developing Countries* 17(3):1–13.

Walker, B., S. Carpenter, J. Anderies, N. Abel, G.S. Cumming, M. Janssen, L. Lebel, J. Norberg, G.D. Petersen, and R. Pritchard. 2002. Resilience management in social-ecological systems: A working hypothesis for a participatory approach. *Conservation Ecology* 6(1):14.

Walker, B.H., L. Gunderson, A. Kinzig, C. Folke, S. Carpenter, and L. Schultz. 2006a. A handful of heuristics and some propositions for understanding resilience in social-ecological systems. *Ecology and Society* 11(1):Art. 13.

Walker, B.H., D.A. Salt, and W.V. Reid. 2006b. *Resilience Thinking: Sustaining Ecosystems and People in a Changing World.* Island Press, Washington, D.C. p. 174.

Whitelaw, G., H. Vaughan, B. Craig, and D. Atkinson. 2003. Establishing the Canadian community monitoring network. *Environmental Monitoring and Assessment* 88(1–3):409–418.

Withers, C.W.J., and D.A. Finnegan. 2003. Natural history societies, fieldwork, and local knowledge in nineteenth-century Scotland: Towards a historical geography of civic science. *Cultural Geographies* 10(3):334–353.

# 4 Goals and Objectives Now and Into the Future

Societal values are often the primary factors influencing the goals and objectives of a monitoring or management plan (Elzinga et al. 2001; Yoccoz et al. 2001). Consequently, it is important to understand what goals society has for the resources involved. There are many guidelines available that document how best to identify, engage, and understand stakeholders in an issue; empower them in decision-making as the plan is developed; and integrate them as key partners in the adaptive management process. Yet these are far from simple tasks, and even if societal values are fully understood and integrated, these values change, sometimes abruptly. Societies, cultures, and the expectations of their members evolve as surely as do species and ecological communities. This presents a daunting challenge for those charged with developing a monitoring plan, because the selection of the species and habitat elements, and the scales over which they are measured, must be selected now in the absence of knowing if these will be the correct parameters to have measured 5, 10, 20, or 100 years from now (Figure 4.1).

In light of this, program managers should work with stakeholders to identify easily understood indicators of state variables (e.g., populations) and their state systems (e.g., habitat) to increase the odds that they will inform the decisions of future managers in a meaningful way. This can be difficult as there are often many options to choose from. Whitman and Hagan (2003) developed a matrix of 137 indicator groups by 36 evaluation criteria as a means of indexing biodiversity responses to forest management actions. A good rule of thumb for ensuring that yours are easily understood is to keep in mind that as indicators begin to span multiple species, multiple times, and multiple areas, clearly articulating goals, objectives, and uses for a program becomes increasingly complex. Although the numerous obstacles and complicated decisions are daunting, making the effort to understand and incorporate societal values is often the only way to develop indicators of change that will be meaningful to society and meet specific monitoring goals and objectives.

## TARGETED VERSUS SURVEILLANCE MONITORING

Before the process of setting goals and objectives begins, one should have a clear idea of what monitoring is and is not. In their review of monitoring for conservation, Nichols and Williams (2006) argue that monitoring should be equivalent to any scientific endeavor, complete with clearly defined hypotheses that should be produced

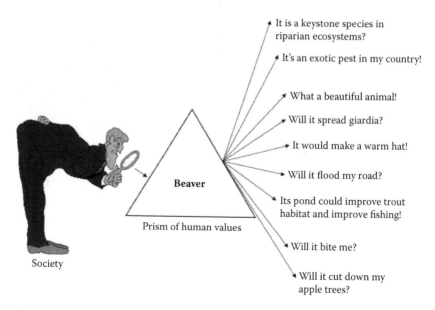

**FIGURE 4.1** Society places many values on natural resources, and those values change over time. How would you develop a socially relevant monitoring plan for beaver populations? R. M. Muth (unpubl. data) suggested that society views resources through a prism of values. One segment of society may be most concerned with the incidence of Giardia in a beaver population, another with the degree of damage to property, and another with their impact on wetland development. A monitoring plan that is carried out over time must consider to the degree possible those values that society may find important in future years as human populations grow, economic indicators change, and technology advances.

through deductive logic and be postulated well before any data are collected. They go on to discuss what monitoring is by contrasting two distinct approaches: targeted monitoring versus surveillance/"omnibus" monitoring (Nichols and Williams 2006). Targeted monitoring requires that the monitoring design and implementation be based on a priori hypotheses and conceptual models of the system of interest. In contrast, they suggest that surveillance monitoring lacks hypotheses, models, or sound objectives (Nichols and Williams 2006).

Surveillance monitoring, however, is the more common of the two and often involves data collection with little guidance from management-based hypotheses. In many cases, these types of programs focus on a large number species and locations under the assumption that *any* knowledge gained about a system is useful knowledge. Surveillance monitoring has been criticized as "intellectual displacement behavior" because it lacks management-oriented hypotheses and clearly defined objectives (Nichols 2000). The primary, often unstated, goal of most surveillance programs is the continuation of past monitoring efforts and the identification of general population trends. Once a trend is detected—usually a decline—management options such as immediate conservation action or undertaking research to identify the cause of these declines are generally implemented (Nichols and Williams 2006). The main limitations of this approach are a dependence on statistical hypothesis testing for initiating management actions (i.e., an insignificant trend would lead to no management), time lags between an environmental change and a population response, costs and resource availability, and a lack of information on the causes of decline (Nichols and Williams 2006).

Despite its limitations, however, surveillance monitoring should not be viewed as a wasted effort. Many of the large-scale monitoring programs covering large geographic regions and estimating changes in numerous species or communities, such as those discussed in Chapter 2, could be considered forms of surveillance monitoring. Proponents of these types of programs emphasize the potential to identify unanticipated problems. For example, the omnibus surveillance of multiple species throughout a region may identify significant population changes for a particular species that are unexpected and perhaps counterintuitive. These changes would then be a starting point for more intensive monitoring and future hypotheses aimed at identifying the magnitude and causes of these changes. Before reaching this stage, however, surveillance monitoring is arguably necessary. It would be difficult to develop adequate hypotheses for a program that monitors the patterns and changes of the hundreds of bird populations throughout the United States, as the USGS Breeding Bird Survey does (Sauer et al. 2006).

In general, targeted monitoring puts less emphasis on finding and estimating population trends and a greater emphasis on monitoring priority species based on taxonomic status, endemism, sensitivity to threats, immediacy of threats, public interest, and other factors (Elzinga et al. 2001; Yoccoz et al. 2001; Nichols and Williams 2006). Targeted monitoring avoids the largest potential pitfall of surveillance monitoring—that significant parameters are missed because they were not identified early in the planning process. In most scenarios that warrant the implementation of a monitoring program, more specific parameters are integral to attaining the goals and objectives. Therefore, we typically advocate the use of targeted monitoring and the development of clear models, hypotheses, and the objectives it entails.

## INCORPORATING STAKEHOLDER OBJECTIVES

Once the concept of monitoring is clearly defined, you can begin to explore which monitoring priorities and objectives are best for your program. Both are often influenced by multiple stakeholders who bring to the table their own goals, needs, assumptions, and predictions that can conflict, coincide, or be mostly unrelated to yours. One primary goal of any monitoring program, therefore, is to incorporate the concerns and predictions of each party. This can be more art than science and is sometimes a challenge to carry out effectively, but it is certainly an important exercise to perform early in the stages of monitoring. Excluding interested parties from the process may require a recrafting of monitoring objectives after data have been collected, which could undermine an entire program.

The effort to include multiple stakeholders in the early stages of monitoring could be thought of as a preemptive attempt at conflict resolution. Any monitoring program must make several decisions with respect to the objectives, the scale of data collection, and what type of data are to be collected. Your ideas of what final decisions might result from the program are entirely subjective. To set things in stone without considering other points of view can result in serious conflicts with many legal and social implications. Conflicts among stakeholders are common in natural resource management. These conflicts are often the result of differing perceptions, varying interpretations of the law, and self-interests that hold the potential to be reconciled with one another—and with scientifically rigorous monitoring (Anderson et al. 2001). Thus, eliciting input from stakeholders in an a priori fashion and attempting to resolve conflicts before they become problems is always recommended. Anderson et al. (1999) suggested the following general protocol for conflict resolution in natural resource management that can be easily adapted for incorporating stakeholders in wildlife monitoring.

## Participants

A good first step is designating a group of people or committee for identifying stakeholders. Having an unbiased group of people, possibly representing different stakeholder groups, to oversee whom to invite to the table leads to greater credibility and transparency. Indeed, the careful consideration of participants may be one of the most important steps in any monitoring plan.

## Data

Once a group of stakeholders comes to the table there is likely to be a wide-ranging discussion on what types of data are relevant for monitoring. This is an important step in identifying information needs, assessing the potential costs and feasibility of collecting different data types, and agreeing on important state variables. In addition, stakeholders may already have data in their possession that they would be willing to submit for analysis. If stakeholders are already bringing data to the table, it is advisable that all parties sign a "certification" stating that the data have been checked for errors and come complete with metadata.

## Analysis

A lofty goal for any initial discussion on monitoring might include an agreement on what data analyses will or will not be used. Although directions may shift as the analytical process proceeds, an early discussion on potential approaches and important assumptions (e.g., independence, parametric assumptions, and representative sampling) can be extremely useful.

## Results

It is important for the stakeholders to agree on the interpretation and reporting of results. In many cases two groups of stakeholders could read the same scientific result and reach two different conclusions with different management implications. A clear understanding of the possible results and their interpretation will avoid confusion in interpretation down the road. Falsely assuming that all stakeholders understand analytical results may lead to the creation of a power hierarchy where those more comfortable with quantitative analysis have greater sway or are dismissed because they do not appreciate the more practical aspects of the monitoring plan.

## No Surprise Management

Communication is a key component to any successful collaboration. Changes in project goals, objectives, data acquisition, data analysis, and sampling strategies should be updated to a group of stakeholders on a regular basis. Meetings should occur frequently enough that people can discuss ongoing or unexpected trends, and deliberate, but not excessively often so that there's nothing new to discuss. Web site updates and Webinars can be a useful way of engaging a large number of stakeholders regularly with less impact on their time.

## IDENTIFYING INFORMATION NEEDS

Information collected should be designed to answer specific questions at spatial and temporal scales associated with the life history of the species and the scope of the management activities that could affect the species (Vesely et al. 2006). Identifying which factors to measure is usually best understood within a conceptual framework that articulates the inter-relationships among state variables (e.g., number of seedlings), processes influencing those variables (e.g., drought), and the scale of the system of interest (e.g., grassland ecosystem). Initially, such a conceptual model represents a shifting competition of hypotheses regarding the current state of our knowledge of a particular system and target species or communities (Figure 4.2).

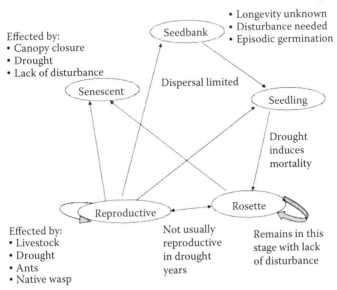

**FIGURE 4.2** Example of a conceptual diagram of population change. (Redrafted from Elzinga, C.L., D.W. Salzer, J.W. Willoughby, and J.P. Gibbs. 2001. *Monitoring Plant and Animal Populations.* Blackwell Science, Malden, MA.)

The development of a conceptual model is intellectually challenging and may take months, but this initial step is critical for developers of the monitoring protocol.

When developing a conceptual model, consider the following:

- It should represent your current understanding of the system that you intend to monitor.
- It should help you understand how the system works. What are the entities that define the structure of the system? What are the key processes? This often yields a narrative model—a concise statement of how you think the system works (i.e., a hypothesis).
- It should describe the state variables. What mechanisms and constraints will be included? Which will be excluded? What assumptions will be made about the system? At what spatial and temporal scales does the system operate? This often results in the construction of a schematic model, perhaps a Forrester diagram (a "box and arrow" model).

The conceptual model should allow the key states or processes that are most likely to be affected by management actions to be identified for monitoring. This will provide a framework for generating hypotheses about how the system works and inform the next step in designing the monitoring program: to develop a set of monitoring objectives that is based on these hypotheses, plus the results of your stakeholder outreach efforts.

## THE ANATOMY OF AN EFFECTIVE MONITORING OBJECTIVE

Developing a conceptual model and understanding stakeholder values leads to the identification of important state variables and processes from which you can derive a set of effective and well-designed management objectives. The objectives serve as the foundation of the monitoring program. A hastily constructed set of management objectives will ultimately limit the scope and ability of a monitoring program to achieve its goals. A well-constructed set will provide the details for how, when, and who will measure the variables that are necessary for successful monitoring. As part of the larger framework, objectives force critical thinking, identify desired conditions, determine management and alternative management scenarios, provide direction for what and how to monitor, and provide a measure of management success or failure (Elzinga et al. 2001). There are three types of objectives that are pertinent to monitoring (Elzinga et al. 2001; Yoccoz et al. 2001; Pollock et al. 2002), and are reviewed in the following text.

### SCIENTIFIC OBJECTIVES

Scientific objectives are developed to gain a better understanding of system behavior and dynamics (Yoccoz et al. 2001). In this case, a set of a priori hypotheses is developed to predict changes in state variables in response to environmental change. For example, a set of hypotheses regarding the population dynamics of shrubland songbirds in Connecticut may identify several state variables (e.g., bird abundance, presence/absence, reproductive success) and how those variables may change due to changing environmental conditions (e.g., drought, disturbance, land use change). In this case, several hypotheses are generated that readily translate to monitoring objectives. The key to using scientific objectives is to develop competing hypotheses and predictions that can be compared to patterns resulting from data analyses.

## Management Objectives

Management objectives incorporate the predicted effects of management actions on system responses. These objectives describe a desired condition, identify appropriate management steps if a condition is or is not met, and provide a measure of success (Elzinga et al. 2001). Not unlike scientific objectives, management objectives should be developed using a priori hypotheses of how a species or population will respond to a given management action. The data collected are then compared to these predictions.

## Sampling Objectives

Sampling objectives describe the statistical power that one is attempting to achieve through their management objectives. Many management objectives will seek to estimate the condition and/or a change in a target population (e.g., a 10% increase in juvenile survivorship), but the degree to which that estimate approximates the true condition will, in part, be a function of its statistical power. Consideration of statistical power is critical within a monitoring framework because of the implications of missing a significant effect (Type II error) and not initiating management when it is necessary to do so. In a monitoring program, the perceived condition of a system relates to a target or a threshold in a current state to a desired state. These targets or thresholds are reflected in management objectives. For example, a threshold objective would be limiting the coverage of a wetland site by an invasive species such as common reed to less than 20%. Once that condition (i.e., <20%) is reached or exceeded, a management action would be initiated. Management objectives may also relate to an active change in an existing state. That is, a change objective would be decreasing the cover of the common reed to less than 10%. In both of these cases, a sampling objective describes the statistical precision and variation associated with estimating that condition or change, oftentimes using a confidence level (e.g., be 90% confident the coverage of common reed is estimated to within ±5%). Sampling objectives relate to the statistical power of your sampling scheme and better inform you about your power to detect significant change (Gibbs et al. 1998).

Whether it is a scientific, management, or sampling objective, all monitoring objectives should consist of several key ingredients (Figure 4.3):

## What?

The most important component of a monitoring objective identifies what will be monitored. Will you be monitoring a species or a group of species? Is the focus a specific population? Is the focal species an indicator species serving as a surrogate

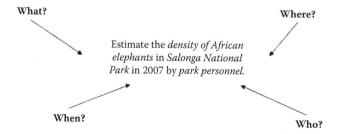

**FIGURE 4.3** The components of a monitoring objective.

for another species or habitat type? Also, be clear as to the parameter associated with this target that you will be measuring. Are you collecting information on abundance, occurrence, reproductive success, demographics, or density?

## WHERE?

The site or geographic area should be clearly delineated. Oftentimes, managers cannot apply a monitoring program over the entire area of interest. In these cases, sample areas must be specified, and the results of these sampling areas used to draw inference to the rest of the study area (Pollock et al. 2002). It is important that these sampling areas be defined objectively, be representative of the larger study area, be free of biases, and represent an appropriate spatial scale relative to the species or processes of interest.

## WHEN?

Providing a time frame is critical for achieving a monitoring objective. Time frames should incorporate species' life history characteristics (e.g., breeding season, flowering, longevity), logistical constraints, and political schedules. Short time frames are always preferred due to changes in budget, management adaptation, and the uncovering of unexpected information (Elzinga et al. 2001), but may not be useful if dealing with long-lived organisms. When monitoring programs must be necessarily conducted over long time periods, considerable attention should be given to the likelihood of continued funding to support the program.

## WHO?

Often overlooked is the fact that the earlier you can identify who is conducting the monitoring, the more likely that the program will be implemented correctly. This avoids the inevitable ambiguity of "passing of the buck." In addition to the stakeholders' responsibility for conducting the monitoring efforts is their involvement in the design and interpretation of the data. When continued funding is needed to maintain a monitoring program over long periods of time, key stakeholders include those involved in budgeting. At all stages of monitoring program managers should continually be thinking of who in society could be affected by these results, and ensure that they are kept informed.

# ARTICULATING THE SCALES OF POPULATION MONITORING

Another important aspect of setting goals and objectives is to define the scales of space and time over which monitoring should occur. Most populations occupy a landscape representing different habitat patches of varying quality. As a result, individuals in a population are interdependent on a number of areas that are influenced by constantly changing environmental conditions. Population trends are dependent not only on the quality of individual patches and areas, but also by the spatial and temporal distribution of suitable and unsuitable habitat patches. Management and development can have an immediate impact on the accessibility of habitat for a species, but the impact can also be delayed if the habitat changed is utilized seasonally or only in a particular ecological context, or if the manager or developer incorporates a species' needs into her endeavors. Consequently, a hierarchical approach that takes a broad perspective that is not restricted by politically and management-imposed boundaries and allows an examination of population and habitat change across multiple scales should be considered when developing a monitoring plan.

## PROJECT OR SITE SCALE

Managers and regulators often want to know what effect a specific management action will have on a population. Indeed, many of the factors that influence populations occur at the site level. For forests this might be a stand, for a town it could be an individual property, or for agriculture, a field. Factors such as resource availability, predation, parasitism, and competition that change as a result of management at these local scales can affect a number of demographic processes including fertility, survivorship, mortality, and dispersal within focal populations. Although these processes are important, the ability to repeatedly monitor them at the site level can be logistically and economically difficult. Even when it is possible, however, the most critical question that pertains to studying of species in small areas must still be addressed: Are the changes that are documented at the site level reflective of those occurring in the same species over a wider geographic area? For species with a restricted geographic range such as some plants and fish stocks, this scale may be an accurate representation of population change. In general, if the species is influenced by the same factors (e.g., weather, resources, predation, topography) throughout its geographic range, then fluctuations on a small scale will often accurately represent fluctuations on a wider scale. However, if factors are site specific, as they often are, then fluctuations in occurrence and abundance will vary from place to place (Holmes and Sherry 1988). In this case, monitoring population change at a larger scale will be more meaningful. It is important to note, however, that as the scale gets larger, identifying the causes of any observed changes can be increasingly difficult.

## LANDSCAPE SCALE

Since many animal populations are dynamic, occupy a heterogeneous landscape, and use multiple resources across that landscape, it is important that population monitoring incorporates the context of any particular site. Indeed, monitoring efforts must incorporate areas surrounding the forest stand, field, or other site, to include water bodies, ownership, other habitat types, barriers, and corridors. This is especially important for species that have large home ranges and/or use multiple sites during their life history, or for species that depend on dynamic habitats that are influenced by weather, season, or succession.

The arrangement of resources over a large geographic area can be critical for sustaining a population. Monitoring over large complex landscapes may increase your ability to detect long-term changes in abundances, predation rates, extinction rates, patch dynamics, metapopulation dynamics, disturbance effects, and rates of human influences on focal species (Noss 1990). However, monitoring at this scale also presents difficulties in design and logistics, because landscapes, not patches, become the sampling units (McGarigal and McComb 1995; McGarigal and Cushman 2002; Meffe et al. 2002; Pollock et al. 2002).

At the landscape scale, population trends are dependent not only on the quality or extent of individual habitat patches, but also on the spatial distribution, pattern, and connectivity of suitable and unsuitable patches (Meffe and Carroll 1997). Landscape pattern, in turn, is influenced by the type of patches, size of patches, length of surrounding edge, barriers between patches, and nature of corridors (McComb 2001; McComb 2007). Fragmentation, for instance, can lead to the isolation of some important habitat patch types, which can lead to disruptions in species dispersal, and eventually to extinction despite management to conserve a species at a site scale. Whether an area is connected or fragmented ultimately depends on the habitat requirements and dispersal abilities of the organism. Nonetheless, at times even sampling over large landscapes cannot answer key questions because no information is provided concerning the effect of the context of the landscape on a species' population dynamics. Consequently, sampling at even larger spatial scales may be needed for some species.

## RANGEWIDE SCALE

A larger perspective is especially important for sensitive and management-indicator species because many have developed morphological and behavioral adaptations that are unique to certain geographic locations (Cody 1985). Rangewide data and trends provide a perspective into localized population changes. For example, populations of blue-winged warblers have been declining in Connecticut at a rate of 3.4% per year over the last 40 years (Figure 4.4). However, the concern over this rate of decline is greatly amplified because an estimated 13% of the global population of this species is found in Connecticut (Figure 4.5; Rosenberg and Wells 1995).

There are few examples of successful long-term monitoring programs that document changes in populations throughout entire geographic ranges, but the North American Breeding Bird Survey, Waterfowl Harvest Surveys, Salmon Escapement Monitoring, USDA Forest Service's Forest Health Monitoring Program, and Agricultural Production Monitoring have produced useful results. Ideally, all monitoring protocols would be designed to detect population changes in a species throughout its range; however, this is often logistically and financially prohibitive. This is particularly true for the many species that may have broad geographic ranges. Yet even if active monitoring cannot be undertaken on a rangewide scale, a monitoring program

Goals and Objectives Now and Into the Future

**FIGURE 4.4** Estimated declines of blue-winged warblers in Connecticut. (Adapted from Sauer, J.R., J.E. Hines, and J. Fallon. 2006. The North American Breeding Bird Survey, Results and Analysis 1966–2006. Version 6.2.2006. USGS Patuxent Wildlife Research Center, Laurel, MD; photo inset by Laura Erickson is used with permission.)

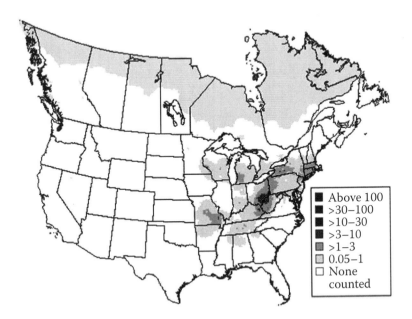

**FIGURE 4.5** (*A color version of this figure follows page 144.*) Distribution of blue-winged warblers in the United States and Canada. (From Adapted from Sauer, J.R., J.E. Hines, and J. Fallon. 2006. The North American Breeding Bird Survey, Results and Analysis 1966–2006. Version 6.2.2006. USGS Patuxent Wildlife Research Center, Laurel, MD.)

must make an attempt to understand the context of any population changes it documents in the most informed way possible.

## ORGANISM-CENTERED PERSPECTIVE

The most important spatial scale to a species is defined by its life history and habitat requirements. The different habitat requirements between groups of birds, insects, and mammals are clear, but even within these larger groups, species that share similar preferences for food, water, and cover may have vastly different requirements for space. Species with similar requirements for vegetation and other resources are often affected differently by management actions due to differences in area sensitivity and home range sizes. An organism-centered view suggests that there is no general definition or perspective of habitat pattern. Therefore, the effects of management actions that alter the spatial arrangement of habitat patches across a landscape will ultimately vary from one species to the next. For example, a 10-ha clearcut can have a profound influence on a small forest passerine with a restricted territory size, such as a black-and-white warbler, but have relatively no impact on a Cooper's hawk whose territory includes hundreds of hectares.

Area requirements and sensitivity to patch size are not the only factors influencing a species' response to management. The dispersal capabilities of species are a critical component that determines whether or not a population is directly effected by loss of habitat resulting from management or inordinately affected by fragmentation of its habitat. Species migrate between habitats that are separated by ecological and anthropogenic barriers. However, each species differs in its perception of these "gaps" in habitat and therefore in its ability to successfully cross them (With 1999) (Figure 4.6). A landscape is fragmented if individuals cannot move from patch to patch and are isolated within a single area. Simulations have suggested that species with limited dispersal capabilities are much less likely to successfully cross habitat "gaps" to other habitat clusters relative to species with a higher ability to disperse (With 1999). Whether or not management causes fragmentation, and how monitoring should address this effect, must be addressed from an organismal perspective.

## DATA COLLECTED TO MEET THE OBJECTIVES

After creating your conceptual model of population persistence for your species and putting this into the proper spatial context, specific questions regarding the potential impacts of management on a species should emerge. The scope of the monitoring program should lead the investigators to identify a set of questions that can be addressed by different data types. For instance, consider the following five questions and the decisions that could be made to meet the information needs associated with each (from Vesely et al. 2006):

1. Given our lack of knowledge of the distribution of a clonal plant species, we are concerned that timber management plans could have a direct impact on remaining populations that have not yet been identified on our management district. How will we know if a timber sale will impact this species?

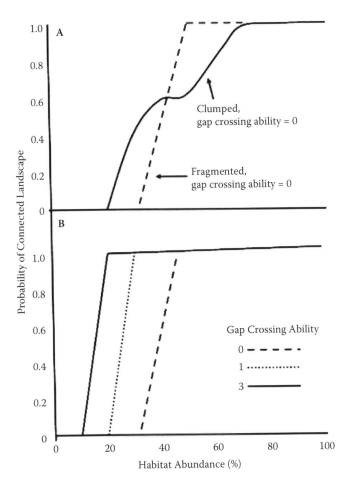

**FIGURE 4.6** Theoretical connectivity for two types of fragmented landscapes (A) and three species with differing gap-crossing abilities (B). (Redrafted from With, K.A. 1999. In *Forest Fragmentation: Wildlife and Management Implications.* J.A. Rochelle, A. Lehmann, and J. Wisniewski (eds.). Koninklijke Brill, Leiden, The Netherlands. With permission.)

In this example the plant species may have a geographic range extending well beyond the timber sale boundaries and over multiple national forests, but populations of this species are patchily distributed, and their abundance is poorly known. Based on our conceptual model of persistence, therefore, we are concerned that population expansion and persistence may be highly dependent on movement of propagules among subpopulations and that additional loss of existing patches may exacerbate the loss of the population over a significant portion of its range. Consequently, the primary goal of a monitoring effort should be to identify the probability of occurrence of the species in a timber sale. A *survey* of all (or a random sample of) impending timber sales will provide the land manager with additional information with regard to the distribution of the species. Although information may be collected that is related to fitness of the clone (size, number of propagules, etc.), the primary information need is to estimate the probability of occurrence of the organism prior to and following management actions. Indeed, this survey and manage approach also lends itself well to development of a secondary monitoring approach that utilizes a manipulative experiment. Explained in more detail later, identification of sites where the species occurs can provide the opportunity for random assignment of manipulations and control areas to understand the effect of management on the persistence of the species. This approach may be particularly important when dealing with species, such as cryptic or infrequently apparent species, where the probability of not detecting individuals is relatively high despite the species being present on the site.

> 2. Given the uncertainty in the distribution of a species of a small mammal species over a management area, we are concerned that a planned timber harvest could have an undue impact on a large proportion of individuals of this species within the management area. How do we obtain an unbiased estimate of the abundance of the species over the entire planning area to understand if the proposed management activities indeed have the potential to impact a significant portion of the population?

In this example, the species' geographic range extends well beyond the boundaries of the management area, but the manager is concerned that the sites under consideration for management may be particularly important for the species' persistence within her management area. The manager therefore needs to understand the dynamics of the population within the sites, but also the population dynamics within the entire management area, as well as the interplay between these levels to understand the full potential for adverse effects on the species. Based on survey information it is clear that the species occurs in areas that are planned for harvest. But do they occur elsewhere in the management area? With an *unbiased estimate of abundance* that extends over the area (or forest, or watershed, etc.) one can estimate (with known levels of confidence) if the proposed management activities might affect 1% of the habitat or population for this species or 80% of the habitat or population. Consider the differences in management direction given these two outcomes. Collecting inventory information following standardized protocols over management units provides the manager with a context for proposed management actions and is integral to a successful hierarchical approach.

> 3. Given the history of land management on a refuge, how will the future management actions described by the current Comprehensive Conservation Plan (CCP) influence the abundance and distribution of a subpopulation of a salamander species that we know occurs on our refuge?

In this example, the species, again, has a geographic range that extends well beyond the boundaries of the refuge, but there is concern that the relatively immobile nature of subpopulations of this species may make the animals on the refuge highly important in contributing to its rangewide persistence. If the refuge's subpopulation is adversely impacted by management and has shown historical declines in abundance as a result of the past management activities, the CCP may need to be amended. The goal, therefore, is to establish the current status of the species on the refuge and allow managers to detect trends in abundance over time. Changes in abundance or even occurrence may be difficult to detect at the project scale (e.g., road building), because individuals are patchily distributed, but if data are collected cumulatively over space and time, impacts could become apparent. Consequently, this *status and trends* monitoring approach should extend over that portion of the refuge where the species is known or likely to occur and provide an estimate of abundance of the species at that scale at several different time periods. It is important to distinguish between an estimate of abundance over a large area (inventory) and a total count of all individuals in an area (census). Inventories, when conducted following sampling guidelines, and accounting for detection probabilities, can produce estimates with known levels of confidence. Censuses often are not cost effective unless the species occurs in very low numbers and the risk of regional or rangewide extinction is high.

In short, the focus of this monitoring effort would be to document changes in abundance over time over a spatial scale that encompasses the subpopulation of concern. One final consideration is that, if at all possible, the abundance estimates should be specific to age and sex cohorts to allow managers to identify potential impacts on population demographics. For instance, reduction in the oldest or youngest age classes, or of females, may provide information on recruitment rates that is significant enough to cause changes in management actions before a significant change in total abundance occurs.

> 4. Assume that concern has been expressed for a species of neotropical migrant bird whose geographic range extends across an ecoregion. The monitoring plan needs to assess if the history of land management throughout the ecoregion and the multiple plans for future management applicable to the region are contributing to changes in populations over time. In other words, are multiple types of management having an effect on the population?

In this example we are dealing with a species that is probably widely distributed, reasonably long lived, and spends only a portion of its life in the area affected by proposed management. One could develop a status and trends monitoring framework for this species, but the data resulting from that effort would only indicate an association (or not) with time. It would not allow the manager to understand the *cause and effect* relationship between populations and management actions.

In this case there are several strata that must be identified relative to the management actions. Can the ecoregion be stratified into portions that will not receive management and others that will receive management? If so, then are the areas in each stratum large enough to monitor abundance of those portions of the populations over time? Monitoring populations in both strata prior to and following management actions imposed within one of the strata would allow the managers to understand if changes occur in the most important response variables from the conceptual model due to management. For instance, if populations in both managed and unmanaged areas declined over time, then the managers might conclude that population change is independent of any management effects and some larger pervasive factor is leading to decline (e.g., climate change, changes in habitat on wintering grounds). On the other hand, should populations in the unmanaged stratum change at a rate different from that on the managed stratum, then the difference could be caused by management actions, and lead managers to change their plan.

5. Finally, say that the conceptual model suggests that the most likely factor affecting the change in population of a wide-ranging raptor is nest site availability. At the ecoregion scale, population density is low and the probability of detecting a change in abundance or fitness at that scale is likewise very low. Rather, managers may wish to monitor habitat elements that are associated with demographic characteristics of the species. How might a monitoring protocol be developed that would allow managers to use habitat elements as an indicator of the capability of an ecoregion to contribute to population persistence?

An unbiased estimate of the availability of habitat elements assumed to be associated with a demographic characteristic of the species and an estimate of the demographic characteristic assumed to be associated with the habitat elements are needed to develop *wildlife habitat relationships*. Ideally, monitoring of the habitat elements as well as the demographic processes can be conducted to assess cause and effect relationships (see above), but with rare or wide-ranging species, this may not be possible. In these cases, testing a range of relationships through use of information theoretic approaches can help you identify the "best" relationship, given the limitations of the data (Burnham and Anderson 2002). Regardless of the resulting monitoring design, it is important that the monitoring framework for the vegetation component of the habitat relationship is implemented at spatial and temporal scales consistent with those used by the species of interest.

## WHICH SPECIES SHOULD BE MONITORED?

If you have several options for species or groups of species that, if monitored, will yield data that meet your objectives, how should you decide which to monitor? Where should you focus time and money? Although the species selected will oftentimes be driven by the values of the stakeholders associated with land use and land management in the area of interest, sometimes characteristics of the species themselves help focus the list. The following categories of species are some of those most commonly viewed as worthy of special consideration and therefore particularly useful for practitioners when selecting the species to be monitored:

1. Level of risk—The perceived or real level of risk of loss of the species from the area now and into the future. Risk can be based on previously collected data, expert opinion, and stakeholder perceptions.
2. Regulatory status—Species listed under state and/or federal threatened or endangered species legislation.
3. Government Rare Species or Communities classification—Those species or plant communities designated by federal or state agencies as in need of special consideration.
4. Restricted to specific seral stages—Species sensitive to loss of a vegetative condition such as a stage of forest, wetland, or grassland succession. Species associated with seral stages or plant communities that are underrepresented relative to a reference condition or the historic range of variability often rise to the top when identifying focal species.

5. Sensitivity to environmental change/gradients—Species sensitive to environmental gradients such as distance from water, altitude, soil conditions, or characteristics. Under current climate change scenarios, for instance, species associated with high altitudes or high latitudes are of particular concern.
6. Ecological function—Species that are particularly important in modifying the processes and functions of an ecosystem. For instance, gophers expose soil in grasslands and voles move mycorrhizal fungal spores in forests.
7. Keystone species—Species whose effects on one or more critical ecological processes or on biological diversity are much greater than would be predicted from their abundance or biomass (e.g., beaver, large herbivores, predators).
8. Umbrella species—Species whose habitat requirements encompass those of many other species. Examples include species with large area requirements or those that need multiple vegetative conditions, such as raptors, bears, elephants, or caribou.
9. Link species—Species that play critical roles in the transfer of matter and energy across trophic levels or provide a critical link for energy transfer in complex food webs (e.g., insectivorous birds) or which through their actions influence trophic cascades effects (Ripple and Beschta 2008).
10. Game species—Species that are valued by segments of society for recreational harvest.
11. Those for which we have limited data or knowledge—Monitoring may provide an information base necessary to understand if continued monitoring is needed.
12. Public/regulatory interest—Some species are simply of high interest to the general public because of public involvement (e.g., bluebirds, wood ducks, rattlesnakes). These can include species that are desirable as well as those that interfere with people's lives.

## INTENDED USERS OF MONITORING PLANS

Monitoring plans can be useful for a variety of users including agency managers and planners, the general public, politicians (to ensure adherence to local, state, and federal legislation), nongovernmental organizations with similar missions, and also industries with Habitat Conservation Plans on adjacent or nearby lands. Different components of a monitoring plan often are useful to different stakeholders. For instance, a survey prior to a management action may allow a manager to alter the management action to accommodate a species found on the site during the survey. A declining trend in a focal species population in the managed portion of a site compared to an unmanaged portion may allow a land planner to make changes in an adaptive management framework over the site. While at a larger scale, regional declines in a species on public land holdings may lead to legislation or agreements that span ownership boundaries across the species' geographic range to encourage recovery. Expected products will be dependent on the questions that are asked. Occurrence, abundance, fitness, range expansion/contraction—each may be appropriate to address certain questions. Whatever the measure, whatever the question, and whatever the expected product, the results must be effectively communicated so that the manager, planner, or politician can make an informed decision regarding the likely effects of a management action or legislation on the long-term persistence of the species.

## SUMMARY

Some biologists distinguish between targeted and surveillance monitoring. Targeted monitoring requires that the monitoring design and implementation be based on a priori hypotheses and conceptual models of the system of interest. Surveillance monitoring oftentimes lacks hypotheses, models, or specific objectives. The structure of each approach is driven largely by societal values. Stakeholder involvement in identification of indicators and thresholds for changes in management efforts is a key step in developing a monitoring plan. Suggested steps in stakeholder involvement include:

- Identify the participants.
- Agree on the types of data needed.
- Agree on the types of analysis to be used.
- Agree on how the results will be interpreted.
- Agree on "no surprises" as to which stakeholders contribute to management decisions.

Developing a conceptual model and understanding stakeholder values can help identify important state variables and processes from which management objectives will emerge. Objectives should consider the following questions: what, where, when, and who? Objectives also have a scale component. Deciding if the results will address questions at the project, landscape, or geographic range scales influences the utility of the information that is gathered. Alternatively monitoring organisms may be most appropriate. Finally, where there is a choice as to which species to monitor, the values of the stakeholders may guide species selections. Rare species invariably rise to the top of a list, but economically important species, keystone species, or species that are indicative of ecosystem stresses may also be selected, depending on the stakeholder interests.

## REFERENCES

Anderson, D.R., K.P. Burnham, A.B. Franklin, R.J. Gutiérrez, E.D. Forsman, R.G. Anthony, and T.M. Shenk. 1999. A protocol for conflict resolution in analyzing empirical data related to natural resource controversies. *Wildlife Society Bulletin* 27:1050–1058.

Anderson, D.R., K.P. Burnham, and G.C. White. 2001. Kullback-Leibler information in resolving natural resource conflicts when definitive data exist. *Wildlife Society Bulletin* 29:1260–1270.

Burnham, K.P., and D.R. Anderson. 2002. *Model Selection and Inference: A Practical Information-Theoretic Approach*. Springer-Verlag, New York.

Cody, M.L. 1985. *Habitat Selection in Birds*. Academic Press, Orlando, FL.

Elzinga, C.L., D.W. Salzer, and J.W. Willoughby. 1998. Measuring and monitoring plant populations. Technical Reference 1730-1. Bureau of Land Management, National Business Center, Denver, CO.

Elzinga, C.L., D.W. Salzer, J.W. Willoughby, and J.P. Gibbs. 2001. *Monitoring Plant and Animal Populations*. Blackwell Science, Malden, MA.

Gibbs, J.P., S. Droege, and P. Eagle. 1998. Monitoring populations of plants and animals. *Bioscience* 48:935–940.

Holmes, R.T., and T.W. Sherry. 1988. Assessing population trends of New Hampshire forest birds: Local and regional patterns. *Auk* 105:756–768.

McComb, B.C. 2007. *Wildlife Habitat Management: Concepts and Applications in Forestry*. CRC Press/Taylor & Francis, Boca Raton, FL.

McComb, W.C. 2001. Management of within-stand features in forested habitats. Pages 140–153 in *Wildlife habitat Relationships in Oregon and Washington*. D.H. Johnson and T.A. O'Neil (eds.). Oregon State University Press, Corvallis, OR.

McGarigal, K., and S.A. Cushman. 2002. Comparative evaluation of experimental approaches to the study of habitat fragmentation effects. *Ecological Applications* 12:335–345.

McGarigal, K., and W.C. McComb. 1995. Relationships between landscape structure and breeding birds in the Oregon Coast Range. *Ecological Monographs* 65:235–260.

Meffe, G.K., and C.R. Carroll. 1997. *Principles of Conservation Biology*. 2nd ed. Sinauer, Sunderland, MA.

Meffe, G.K., L.A. Nielsen, R.L. Knight, and D.A. Schenborn. 2002. *Ecosystem Management: Adaptive, Community-Based Conservation*. Island Press, Washington, D.C.

Nichols, J.D. 2000. Monitoring is not enough: On the need for a model-based approach to migratory bird management. Pages 121–123 in R. Bonney, D. N. Pashley, R.J. Cooper, and L. Nichols eds. *Proceedings of Strategies for Bird Conservation: The Partners in Flight Planning Process*. Proceedings RMRS-P-16. Online at: http://birds.cornell.edu/pifcapemay.

Nichols, J.D., and B.K. Williams. 2006. Monitoring for conservation. *Trends in Ecology and Evolution* 21:668–673.

Noss, R.F. 1990. Indicators for monitoring biodiversity: A hierarchical approach. *Conservation Biology* 4:55–364.

Pollock, K.H., J.D. Nichols, T.R. Simons, G.L. Farnsworth, L.L. Bailey, and J.R. Sauer. 2002. Large scale wildlife monitoring studies: Statistical methods for design and analysis. *Environmetrics* 13:105–119.

Ripple, W.J., and Beschta, R.L. 2008. Trophic cascades involving cougar, mule deer, and black oaks in Yosemite National Park. *Biological Conservation* 141:1249–1256.

Rosenberg, K.V., and J.V. Wells. 1995. Important geographic areas to Neotropical migrant birds in the Northeast. Report to United States Fish and Wildlife Service, Hadley, MA.

Sauer, J.R., J.E. Hines, and J. Fallon. 2006. The North American Breeding Bird Survey, Results and Analysis 1966–2006. Version 6.2.2006. USGS Patuxent Wildlife Research Center, Laurel, MD.

Vesely, D., B.C. McComb, C.D. Vojta, L.H. Suring, J. Halaj, R.S. Holthausen, B. Zuckerberg, and P.M. Manley. 2006. Development of Protocols to Inventory or Monitor Wildlife, Fish, or Rare Plants. U.S. Department of Agriculture, Forest Service. General Technical Report WO-72, Washington, DC.

Whitman, A.A., and J.M. Hagan. 2003. Biodiversity Indicators for Sustainable Forestry. National Center for Science and the Environment, Washington, D.C.

With, K.A. 1999. Is landscape connectivity necessary and sufficient for wildlife management? Pages 97–115 in *Forest Fragmentation: Wildlife and Management Implications*. J.A. Rochelle, A. Lehmann, and J. Wisniewski (eds.). Koninklijke Brill, Leiden, The Netherlands.

Yoccoz, N.G., J.D. Nichols, and T. Boulinier. 2001. Monitoring of biological diversity in space and time. *Trends in Ecology and Evolution* 16:446–453.

# 5 Designing a Monitoring Plan

Design of a monitoring plan is a process (Figure 5.1) that will ideally lead you through problem identification to development of key questions, a rigorous sampling design, and analyses that can assign probabilities to observed trends. Finalizing a plan designed as an outcome of this process is a precursor to initiation of data collection. This is probably the single most important step in the monitoring plan. Once you have decided on the design for the monitoring plan, and begun collecting data, there is strong resistance to changing the plan because many changes will render the data collected thus far of less value. So design it correctly from the outset to minimize the need for changes later.

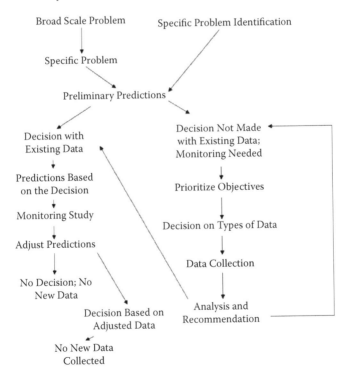

**FIGURE 5.1** The inventorying and monitoring process. (Redrafted from Jones 1986.)

## ARTICULATING QUESTIONS TO BE ANSWERED

It is important to view monitoring as comparable in many ways to conducting a scientific investigation. The first step in the process is to develop a conceptual framework for our current understanding of the system, complete with literature citations to support assumptions. Clearly no monitoring program will have all of the information needed to completely develop a conceptual model for the system under consideration. Available information will have to be extracted from the literature, from other systems, and from expert opinion. Nonetheless, the conceptual model needs to be developed in order to identify the key gaps in our knowledge and allow a clear articulation of the most pertinent questions (Figure 5.2).

As you develop the monitoring plan you should pay particular attention to some terms that are commonly used to define the problem and the approach. Within the context of land management and biodiversity conservation, these terms might guide you to the kind of monitoring design that you will choose to use.

These terms relate to the experimental design:

- Cause and effect—Will you be able to infer the cause for observed changes?
- Association or relationship—Will you be able to detect associations between pairs of variables such as populations and changes in area of a habitat type?
- Trend or pattern—Will patterns over space and/or time be apparent?
- Observation or detection—What constitutes having "observed' an individual?

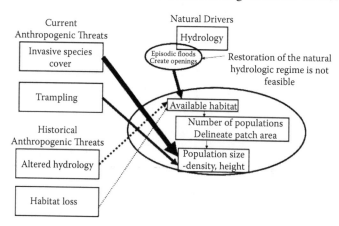

**FIGURE 5.2** Conceptual model developed as the basis for monitoring of San Diego ambrosia, one of many species identified as important within San Diego's Multiple Species Conservation Program. (Redrafted from Hierl, L.A., J. Franklin, D.H. Deutschman, and H.M. Regan. 2007. Developing Conceptual Models to Improve the Biological Monitoring Plan for San Diego's Multiple Species Conservation Program. Department of Biology, San Diego State University, and California Department of Fish and Game, Sacramento, CA.) The goal for managers is to maintain 90% of the base population. Drivers presumably influencing population persistence are highlighted.

# Designing a Monitoring Plan

These terms relate to the response variable that you will measure to assess one of the above:

- Occurrence—Was the species present, absent, or simply not detected?
- Relative abundance—Did you observe more individuals in one place or time than another?
- Abundance—How many individuals per hectare (or square kilometer) are estimated to be present?
- Fitness—Is the species surviving or reproducing better in one place or time than another?

These terms relate to the scope of inference for the effort:

- Stand, harvest unit, field, pasture, project, farm, district, watershed, forest, region—Defines the grain and extent of the spatial scale of the potential management effects.
- Home range, subpopulation, geographic range, stock, clone—Defines the grain and spatial extent associated with the focal species.
- Frequency of management or exogenous disturbances affecting the system—Helps define the sampling interval.
- Return interval between disturbances or other events likely to effect populations of the focal species—Helps define the duration of the monitoring framework.
- Disturbance intensity or the degree of change in biomass or other aspects of the system as a function of management or exogenous disturbances—Helps understand how effect sizes should be defined and hence the sampling intensity sufficient to detect trends or differences.

Once you have articulated questions based on the conceptual model for the system, then you should use terms from each of the groups above to further define the monitoring plan. Detail and focus are important aspects of a well-designed monitoring system. Use of vague or unclear terms, broad questions, or unclear spatial and temporal extents will increase the risk that the data collected will not adequately address the key questions at scales that are meaningful. Further, clearly articulated questions not only ensure that data collected are adequate to address specific key knowledge gaps or assumptions, but they also provide the basis for identifying thresholds or trigger points that initiate a new set of management actions.

If the above terms are considered when the monitoring plan is being designed, and trigger points for management action are described clearly prior to monitoring, then it should be apparent that the universe of questions that could be addressed by monitoring is very broad. Of course, your challenge is to identify the key questions that address the key processes and states in an efficient and coordinated manner over space and time. Given a conceptual model developed for a system, there is a range of questions that could be addressed through monitoring. Prioritization of these questions allows the manager to focus time and money on the key questions.

## INVENTORY, MONITORING, AND RESEARCH

The questions of concern may be addressed using inventory, monitoring, or research approaches (Elzinga et al. 1998). Inventory is an extensive point-in-time survey to determine the presence/absence, location, or condition of a biotic or abiotic resource. Monitoring is a collection and analysis of repeated observations or measurements to evaluate changes in condition and progress toward meeting a management objective. Detecting a trend may trigger a management action. Research has the objective of understanding ecological processes or in some cases determining the cause of changes observed by monitoring. Research is generally defined as the systematic collection of data that produces new knowledge or relationships, and usually involves an experimental approach, in which a hypothesis concerning the probable cause of an observation is tested in situations with and without the specified cause. Some biologists make a strong case that the difference between monitoring and research is subtle and that monitoring should also be based on testable hypotheses. Nonetheless, these three approaches to gaining information are highly complementary and not really very discrete. And all three approaches are needed to effectively manage an area without unnecessary negative effects.

## ARE DATA ALREADY AVAILABLE?

You may already have some data that have been collected previously or from a different area. Can you use these data? Should you? What constitutes adequate data already in hand, or how do we know when data are adequate to address a question? Well, that depends on the question! For example, if we want to be 90% sure that a species does not occur in a patch or other area to be managed in some manner in the next year, how many samples are required to reach that level of confidence? Developing a relationship between the amount of effort expended and the probability of detecting species "x" in a patch can provide insight into the level of effort needed to detect a species 90% of the time when it indeed does occur in the patch. This requires multiple patches and multiple samples per patch over time to place confidence intervals on probabilities (Figure 5.3). Where multiple species are the focus of monitoring, a species-area chart can be quite helpful. For example in Figure 5.4, sampling an area less than 7 hectares in size is not likely to result in a representative list of species for the site.

These sorts of questions require quite different data than would be required to answer the question: What are the effects of management "x" on species "y"? Note that the term *effect* is used in this example, so the experimental design is ideally in the form of a manipulative experiment (Romesburg 1981). In this case, we would want to have both pre- and post-treatment data collected on a sample of patches that do and do not receive treatment. In the following example, two of the treatments clearly had an effect on the abundance of white-crowned sparrows in managed stands in Oregon (Figure 5.5). Results such as these are based on specific questions.

# Designing a Monitoring Plan

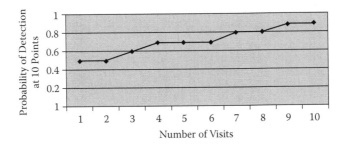

**FIGURE 5.3** A hypothetical cumulative probability of detection. Note that with increasing sampling effort, the probability of detecting a species increases to a plateau at about 90% with nine visits. Hence, future efforts at detecting this species should include at least nine visits. Clearly more visits are needed to detect rare species than common species.

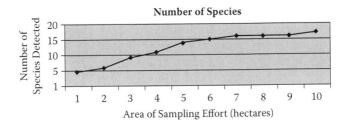

**FIGURE 5.4** A hypothetical species-area curve for one patch type. Note that when an asymptote is reached then sampling an area of that size is most likely to capture the most species, until a new patch type is reached, then an abrupt increase in species maybe noted.

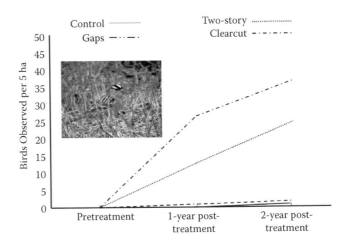

**FIGURE 5.5** Change in white-crowned sparrow detections following silvicultural treatments illustrating cause and effect monitoring results. (Redrafted from Chambers, C.L., W.C. McComb, and J.C. Tappeiner. 1999. *Ecological Applications* 9:171–185; inset photo by Laura Erickson, used with permission.)

It is the development of the question that is important, and the question should evolve from the conceptual model of the system. Clearly, the development of a conceptual model to describe the system states, processes, and stressors should be based to the degree possible on data. So although currently available data are valuable, they must address the question of interest in a manner that is consistent with the conceptual model. It is important to recognize that not all data are equal. Consider the following questions when evaluating the adequacy of a data set to address a question or to develop a conceptual framework:

1. Are samples independent? That is, are observations in the data set representing management units to which a treatment has been applied? Using a forest example, taking 10 samples of densities of an invasive species from one patch is not the same as taking one sample from 10 patches (Hurlbert 1984). In the former example, the samples are subsamples of one treatment area; in the latter, there is one sample in each of 10 replicate units. If the species under consideration has a home range that is less than the patch size, then the patches are reasonably independent samples. If the species under consideration has a home range that spans numerous patches, then the selection of patches to sample should be based on ensuring, to the degree possible, that one animal is unlikely to use more than one managed patch.
2. How were the data collected? What sources of variability in the data may be caused by the sampling methodology (e.g., observer bias, inconsistencies in methods, etc.)? If sample variability is too high because of sampling error, then the ability to detect differences or trends will decrease. Further if the samples taken are biased, then the resulting conclusions will be biased, and decisions made based on those conclusions may be inappropriate.
3. Were sites selected randomly? If not, then there may be (likely is) bias introduced into the data that should raise doubts in the minds of the scientists, managers, and stakeholders with regard to the accuracy of the resulting relationships or differences.
4. What effect size is reasonable? Even a well-designed study may simply not have the sample size adequate to detect a difference or relationship that is real simply because the study was constrained by resources, rare responses, or other factors that increase the sample variance and decrease the effect size that can be detected. Again, how this is dealt with depends on the question being asked. Which is more important, to detect a relationship that is real or to say that there is no relationship when there really isn't? In many instances, where monitoring is designed to detect an effect of a management action, the former is more important. In that case, the alpha level used to detect differences or trends may be increased (from, say, 0.05 to 0.10), but you will be more likely to say a relationship is real when it is really not. Alternatively, you may want to use Bayesian analysis or meta-analysis to examine the data and see if these techniques shed light on your question. See Chapter 11 for a more in-depth discussion of these analytical techniques.

5. What is the scope of inference? From what area were samples selected? Over what time period? Are the results of the work likely to be applicable to your area? As the differences in the conditions under which the data were collected increase compared to the conditions in your area of interest, the less confidence you should have in applying the results in your context.

If, after considering the above factors, you feel that the data can be used to reliably identify known from unknown states and processes in the conceptual model, then you should have a better idea where the model relies heavily on assumptions, weak data, or expert opinion. These portions of the conceptual model should rise to the top during identification of the question that monitoring should be designed to address.

Provided that the cautions indicated above are explored, it is reasonable and correct to use data that are already available to inform and focus the questions to be asked by a monitoring plan. Existing data are commonly used to address questions. For instance, Sauer et al. (2001) provided a credibility index that flags imprecise, small sample size, or otherwise questionable results. Yellow-billed cuckoos have shown a significant decline in southern New England over the past 34 years (Figure 5.6), but the analysis raises a flag with regards to credibility because of a deficiency in the data associated with low abundance (<1.0 bird/route). Further, an examination of the data would indicate that the one estimate in 1966 may be an outlier and may have an overriding effect on the results. In this example, it would be useful to delete the 1966 data and rerun the analyses to determine if the relationship still holds.

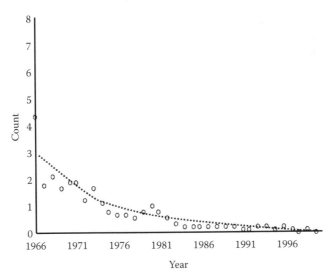

**FIGURE 5.6** Trends in abundance of yellow-billed cuckoos over its geographic range, 1966–1996. (Redrafted from Sauer, J.R., J.E. Hines, and J. Fallon. 2001. The North American Breeding Bird Survey, Results and Analysis 1966–2000. Version 2001.2, USGS Patuxent Wildlife Research Center, Laurel, MD.) Note that data summaries are also available for smaller areas such as ecoregions and states. These data may address a portion of a conceptual model and be valuable in designing a monitoring plan.

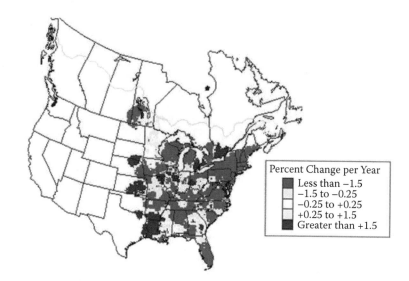

**FIGURE 5.7** (*A color version of this figure follows page 144.*) Changes in abundance of yellow-billed cuckoos over time varies from one portion of its geographic range to another. (From Sauer, J.R., J.E. Hines, and J. Fallon. 2001. The North American Breeding Bird Survey, Results and Analysis 1966–2000. Version 2001.2, USGS Patuxent Wildlife Research Center, Laurel, MD.) Care should be given to ensure that the data relate to the scope of inference of the monitoring plan that is being developed. Local trends may be informative, while regional trends may not; if your area of interest were in Louisiana, then national trends would be misleading.

Regardless of the outcome of this subsequent analysis, the data will likely prove useful when developing a conceptual model of population change for the species. When the proper precautions are taken, such data may allow the manager to focus on more specific questions with regard to a monitoring plan. For instance, within the geographic range of the species, what factors may be causing the predicted declines? The BBS data reveal that the species is not predicted to have declined uniformly over its range (Figure 5.7). This information may provide an opportunity to develop hypotheses regarding the factors causing the declines.

Recorded trends in abundance for eastern towhees (Figure 5.8) provide another clear example. In this case, declines are apparent throughout the entire northeastern United States. Based on our knowledge of changes in land use in the northeast and the association of this species with early successional scrub vegetation, we would hypothesize that the declines are a direct result of the regrowth of the eastern hardwood forests and subsequent loss of shrub-dominated vegetation. Indeed, monitoring the reproductive success of the species prior to and following vegetation management designed to restore shrub vegetation might allow detection of a cause-and-effect relationship that would lead to a change in monitoring for the species over the northeastern portion of its range. If changes in abundance were related to changes in land cover (and not to parasitism by brown-headed cowbirds or other effects), then an alternative monitoring framework could be developed. Infrequent monitoring of populations with frequent monitoring of the availability of the habitat elements important to the species (shrub cover) may be adequate to understand

# Designing a Monitoring Plan

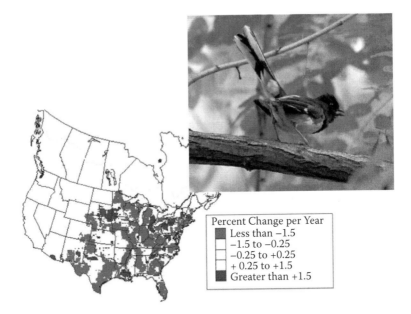

**FIGURE 5.8** (*A color version of this figure follows page 144.*) Predicted changes in abundance of eastern towhees over its geographic range, 1966–1996. (From Sauer, J.R., J.E. Hines, and J. Fallon. 2001. The North American Breeding Bird Survey, Results and Analysis 1966–2000. Version 2001.2, USGS Patuxent Wildlife Research Center, Laurel, MD; photo inset by Laura Erickson is used with permission.)

the opportunities for population recovery (or continued decline). This reveals the potential benefits of applying previously generated data. Generally, costs associated with monitoring habitat availability are less than costs associated with monitoring populations or population fitness, so if a cause-and-effect relationship were detected between vegetation and bird populations, then managers could see considerable savings in their monitoring program.

In summary, although there are general guidelines with regards to what constitutes reliable data, adequate data for one question may be inadequate for another. Use of existing data and an understanding of data quality can allow development of a conceptual model where states, processes, and stressors can be identified with varying levels of confidence. Those factors that are based on assumptions or weak data and which seem quite likely to be influencing the ability to understand management effects should become the focus when developing the questions to be answered by the monitoring plan.

## TYPES OF MONITORING DESIGNS

Before we provide examples of the types of monitoring designs, recall that there are several main points consistent among all designs:

1. Are the samples statistically and biologically independent?
2. How were the data collected? What sources of variability in the data may be caused by the sampling methodology (e.g., observer bias, inconsistencies in

methods, etc.)? If sample variability is too high because of sampling error, then the ability to detect differences or trends will decrease.
3. Were sites selected randomly? If not, then there may be (likely *is*) bias introduced into the data, which should raise doubts in the minds of the plan developers with regard to the accuracy of the resulting relationships or differences.
4. What effect size is reasonable to detect?
5. What is the scope of inference?

Once these questions are addressed, then there are additional considerations, depending on the type of monitoring that will be conducted.

## INCIDENTAL OBSERVATIONS

Opportunistic observations of individuals, nest sites, or habitat elements can be of some value to managers, but often are of not much use in a monitoring framework except to provide preliminary or additional information to a more structured program. For instance, global positioning system (GPS) locations of a species observed incidentally over a 3-year period could be plotted on a map, and some information can be derived from the map (known locations). The problem with using these data points in a formal monitoring plan is that they are not collected within an experimental design. There undoubtedly are biases associated with where people are or are not likely to spend time; with detectability among varying vegetative, hydrologic, or topographic conditions; and with detectability of different age or sex cohorts of the species. Consequently, this information should be maintained, but rarely would it be used as the basis for a formal monitoring design.

## INVENTORY DESIGNS

Assume that you are concerned that proposed management actions will impact a species that could be present in your management area. How sure do you want to be that the species occurs on the proposed management area? Do you want to know with 100% confidence? Or can you be 95% sure? How about 90%? The answer to that question will dictate both the sampling design and the level of intensity with which you inventory the site to estimate presence and (presumed) absence.

First and foremost, you will need to decide what will constitute an independent sample for this design. Are you most concerned that a species does/does not occur on a proposed set of managed patches? If so, sampling of all or a random sample of proposed managed patches will be necessary. But what if the managed patches are clustered and the species of interest has a home range that overlaps some of the managed patches? Are you likely to detect the same individual on multiple units? If so, is it necessary to detect it on all units within the species home range? Not likely. For species with home ranges larger than a managed patch, you will probably want to sample an area of sufficient size and intensity to have a high probability of detecting the species within that home-range-sized area. Randomly selected areas should be independent with respect to minimizing the potential for detecting the same individual on multiple sites. Random sampling of sites then would be constrained by eliminating those sites where double counting of individuals is likely.

In an inventory design, generally you wish to be "x" confident that you have detected the species if it is really there. Consequently, a pilot study that develops the relationship between the probability of detecting an individual and sampling intensity will be critical. Consider the use of remotely activated recorders for assessing presence/absence of singing male birds in an area. How many mornings would you need at each site to detect a species that does occur at the site? A pilot study may entail 20 or 30 sites (more if the species is rare) to allow you to graph the cumulative detections of the species over the number of mornings sampled (Figure 5.2). Eventually, an asymptote should be reached, and one could then assume that additional sampling would not likely lead to additional detections. Confidence intervals can be placed on these estimates, and the resulting estimated level of confidence in your ability to detect the species can (and should be) reported. Results of pilot work (or published data) such as this often allow more informative estimates of the probability of a species occurring in an area to be generated in the subsequent sampling during the monitoring program. This is especially the case when estimated levels of confidence can be applied to the monitoring data.

There are several issues to consider when developing a monitoring plan that is designed to estimate the occurrence or absence of a species in an area. First, the more rare or cryptic the species, the more samples that will be needed to assess presence, and the sampling intensity can become logistically prohibitive. In that case other indicators of occurrence may need to be considered.

Consider the following possibilities when identifying indicators of occurrence in an area:

1. Direct observation of a reproducing individual (female with young).
2. Direct observation of an individual, reproductive status unknown. (Are these first two indicators of occurrence or something different, such as indicators that management will have an impact?)
3. Direct observation of an active nest site.
4. Observation of an active resting site or other cover.
5. Observation of evidence of occurrence such as tracks, seeds, pollen.
6. Identification of habitat characteristics that are associated with the species.

Any of the above indicators could provide evidence of occurrence and hence potential vulnerability to management, but the confidence placed in the results will decrease from number 1 to number 6 for most species based on the likelihood that the fitness of individuals could be affected by the management action.

Finally, we strongly suggest that statistical techniques that include data that influence the probability of detection be used. MacKenzie et al. (2003) provide a computer program, PRESENCE, that uses a likelihood-based method for estimating site occupancy rates when detection probabilities are <1 and incorporates covariate information. They also provide information on survey designs that allow use of these approaches (MacKenzie and Royle 2005).

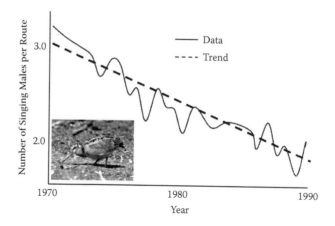

**FIGURE 5.9** Long-term trends of numbers of American woodcock observed on singing grounds in the eastern United States, 1968–1995. (Redrafted from Bruggink, J.G., and W.L. Kendall. 1995. American woodcock harvest and breeding population status, 1995. U.S. Fish and Wildlife Service, Laurel, MD. 15 pp; photo inset from U.S. Fish and Wildlife Service, Moosehorn National Wildlife Refuge.)

## STATUS AND TREND MONITORING DESIGNS

Long-term monitoring of populations to establish trends over time are coordinated over large areas under several efforts. For instance the U.S. Geological Survey (USGS) Biological Resources Division (BRD) has been involved in monitoring birds and amphibians, as well as contaminants and diseases that may affect populations of these species. Results of these efforts as well as supporting research information provide the basis for understanding why we might be seeing trends in biological resources over large areas (e.g., USGS Status and Trends Reports). Such monitoring systems provide information on changes in populations, but they do not necessarily indicate why populations are changing. For instance, consider the changes in American woodcock populations over a 27-year period (Figure 5.9).

Clearly, the number of singing males has declined markedly over this time period. This information is very important in that it indicates that additional research may be needed to understand why the changes have occurred. Are singing males simply less detectable than they were in 1968? Are populations actually declining? If so, are the declines due to changes in habitat on the nesting grounds? Wintering grounds? Migratory flyways? Is the population being overhunted? Are there disease, parasitism, or predator effects that are causing these declines? Are these declines uniform over the range of the species or are there regional patterns of decline? Analysis of regional patterns indicates that the declines may not be uniform (Figure 5.10). Indeed, declines in woodcock numbers are apparent in the northeastern United States, but not throughout the Lake states. So it would seem that causes for declines are probably driven by effects that are regional. Woodcock provide a good example of the need to consider the scope of inference in design of a status and trends monitoring plan. The Breeding Bird Survey data indicate that there are areas where declines have been significant, and the work by Bruggink and Kendall (1995) indicates that

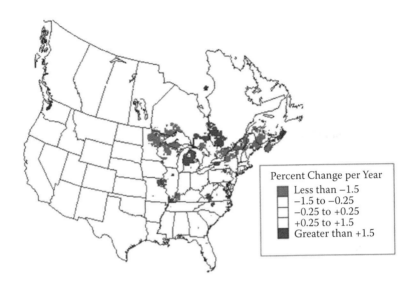

**FIGURE 5.10** (*A color version of this figure follows page 144.*) Predicted changes in American woodcock abundance over the species range. (From Sauer, J.R., J.E. Hines, and J. Fallon. 2001. The North American Breeding Bird Survey, Results and Analysis 1966–2000. Version 2001.2, USGS Patuxent Wildlife Research Center, Laurel, MD.)

the magnitude of the declines in some areas is perhaps even greater than might be indicated by the regional analyses. Such hierarchical approaches provide opportunities for understanding the potential causes for change in abundance or relative abundance of a species at a more local scale. This insight may allow managers to be more effective in identifying the causes of declines at local scales in a manipulative manner where local, intensive monitoring and research can result in the discovery of a cause and effect relationship.

Consequently the design of a status and trends monitoring plan should carefully consider the scope of inference, generally a large portion of the geographic range, and may necessitate the coordination of monitoring over large areas and multiple agencies. Site-specific status and trends analyses will probably be of limited value in many instances because the fact that species "x" is declining at site "y" is probably not as important as knowing why the species is declining at site "y." In addition, for those species with a metapopulation structure (ones with subpopulations maintained by dispersal and recolonization), observed changes at local sites may simply reflect source-sink dynamics in the metapopulations and not the trend of the population as a whole. Using status and trends monitoring to detect increases or declines in abundance is perhaps best applied to identify high priority species for more detailed monitoring, or to allow development of associations with regional patterns of vegetation or urban development. These associations allow the opportunity for a more informed development of hypotheses that can then be tested in manipulative experiments to identify causes for changes.

One aspect of status and trends monitoring that must be considered carefully is the sampling intensity needed to detect a change in slope over time. Consider the

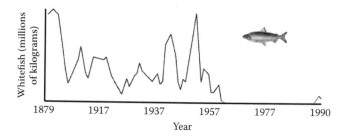

**FIGURE 5.11** Catch of whitefish from Lake Erie, 1879–1992. (Redrafted from Baldwin, N.A. et al. 2002. Commercial fish production in the Great Lakes 1867–2000. http://www.glfc.org/databases/commercial/commerc.asp; photo inset from the National Oceanic and Atmospheric Administration.)

data in Figure 5.9. Annual data were quite variable over the 27-year period, but a trend is still detectable because the slope was so steep. Annual variability caused by population fluctuations and sampling variance can prevent detection of a statistically significant change in slope. This problem is exacerbated when the slope is not so dramatic. Consider the trend for whitefish harvest in Lake Erie (Figure 5.11). Annual variability in whitefish harvest was so variable from year to year that detection of a statistically significant trend line is very difficult. Commercial harvest as an indicator of populations is clearly biased due to variability in harvest intensity, resource value, and techniques. Use of traditional time-series regression analyses to detect trends may not be the best approach. Use of Bayesian analyses may be more informative (Wade 2000). Nonetheless, when commercial harvests reach zero (or nearly so), a threshold is reached where dramatic action must be taken to recover the stocks. Hence, there are several factors that must be considered carefully in the design of status and trends plans:

1. What is the spatial scale over which you wish to understand if populations are declining or increasing? If the extent of the monitoring effort is not the geographic range of a species or sub-species, then what portion of the geographic range will be monitored and how will the information be used?
2. Is the indicator (e.g., occurrence, abundance, fitness) that is selected as unbiased as possible and not likely to vary among time periods except those caused by fluctuating populations?
3. Given the inherent variability in the indicator that is being used, how many samples will be needed each time period to allow detection of a slope of at least "x" percent per year over time (Gibbs and Ramirez de Arellano 2007)?
4. Given the inherent variability in the indicator that is being used, at what point in the trend is action taken to recover the species or reverse the trend (what is the trigger point)? Note that this must be well before the population reaches an undesirably low level, because the manager will first have to understand why the population is declining before action can be taken.
5. Will the data be used to forecast results into the future? If so, recognize that the confidence intervals placed on trend lines diverge dramatically from the

# Designing a Monitoring Plan

line beyond the bounds of the data. Forecasting even a brief period into the future is usually done with little confidence unless the underlying cause and effects are understood.

Finally, it is important to recognize that quite often when data are analyzed using time-series regression, autocorrelation among data points (the degree of dependence of one data point on preceding data) is not only likely, it should be expected. Although Chapter 11 provides extensive guidance in analyzing these data, briefly we describe why these factors must be considered during plan design.

Parametric analyses are designed to reject the null hypothesis with an estimated level of certainty (e.g., we reject the null hypothesis that the slope is 0 and conclude that there is a decline). With rare species, variable indicators, or sparse data, we may lack the statistical power necessary to reject that hypothesis, even when a trend line seems apparent. In the Figure 5.12, we may choose to use alternative analyses (Wade 2000) to assess trends in sample B, or increase our alpha level, thereby increasing our likelihood of rejecting the null hypothesis. Changing alpha will increase the risk of stating that there is a trend when there really is not; however, this may be an acceptable risk. If we risk losing a species entirely or losing it over a significant part of its geographic range, we will probably be willing to make an error that results in stating that a population is declining when it really is not. This is an important point, because two individuals approaching trend data with a different perspective (alpha level) may reach very different conclusions.

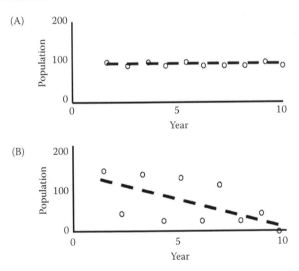

**FIGURE 5.12** Examples of two trends each resulting in different conclusions using traditional parametric analyses. (Redrafted from Wade, P.R. 2000. *Conservation Biology* 14:1308–1316.) The trend of line A is statistically significant ($P < 0.05$) because the variability of points around the line is so low, but the slope is barely negative ($b = -0.03$). The trend line in B is not significant ($P > 0.05$), but the slope is clearly negative ($b = -0.10$). Analysis of these data using traditional time-series analysis may miss biologically meaningful trends.

## CAUSE AND EFFECT MONITORING DESIGNS

Typical research approaches will either assess associations between response variables and predictor variables, or will use a structured experimental design that will assess cause and effect. Monitoring plans can also use this approach. Consider a trend line for white-tailed deer in Alabama that showed a significant decline. Consider also that the manager had just stopped using prescribed burning to manage the area. She might conclude that the cessation of burning caused the decline. But if data from a nearby unburned area were also available and that trend line was also negative, then she might reach a different conclusion and begin looking for diseases or other factors causing a more regional decline.

If the plan being developed is designed to understand the short- or long-term effects of some management action on a population, then the most compelling monitoring design would take advantage of one of two approaches to assess responses to those actions: retrospective comparative mensurative designs or Before–After Control-Impact (BACI) designs.

### Retrospective Analyses and ANOVA Designs

Monitoring conducted over large landscapes or multiple sites may use a comparative mensurative approach to assess patterns and infer effects (e.g., McGarigal and McComb 1995; Martin and McComb 2003). This approach allows comparisons between areas that have received management actions and those that have not, and often is designed using an analysis of variance (ANOVA) approach. Retrospective designs that compare treated sites to untreated sites raise questions about how representative the untreated sites are prior to treatment. In this design, the investigator is substituting space (treated versus untreated sites) for time (pre- versus posttreatment populations). The untreated sites are assumed to be representative of the treated sites before they were treated. With adequate replication of randomly selected sites, this assumption could be justified, but often large-scale monitoring efforts are costly, and logistics may preclude both sufficient replication and random selection of sites. Hence, doubt may persist regarding the actual ability to detect a cause-and-effect relationship, or the power associated with such a test may be quite low. Further, lack of random selection may limit the scope of inference from the work only to the sites sampled.

McGarigal and McComb (1995) compared bird detections among six levels of mature forest abundance and two levels of forest pattern using a retrospective ANOVA design (Figure 5.13). Logistics restricted them to only three replicates of each condition, and sites were selected randomly from among those available, though in most cases few options were available. Further, there was the potential for areas that had been managed to have different topography, site conditions, or other factors that made those sites more likely to receive management than those sites that did not. The patterns of association between bird species and landscape conditions were significant in many instances, and consistent with what we know about the life history of the species. But despite the fact that most of the species in Table 5.1 have been associated with older or closed-canopy forests in previous studies (lending support for the fact that their relative abundance should increase with increasing area of old forest), we cannot make any definitive conclusion. Rather, we can only hypothesize that these

# Designing a Monitoring Plan

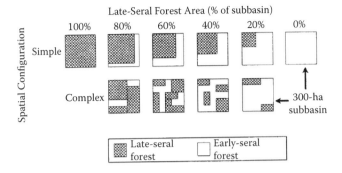

**FIGURE 5.13** The comparative mensurative experimental design used by McGarigal and McComb (1995) and Martin and McComb (2003) to assess relationships between landscape structure and composition and the relative abundance of breeding birds, small mammals, and amphibians in the Oregon Coast Range. (With permission from K. McGarigal.)

**TABLE 5.1**
**Selected Results of Analyses of Bird Species Showing a Positive Association Between the Area of Late Successional Forest and the Relative Abundance of Each Species in the Oregon Coast Range**

| | ANOVA | | Regression | |
|---|---|---|---|---|
| Species | F | P | $R^2$ (%) | P |
| Gray jay | 14.2 | <0.001 | 63 | <0.001 |
| Brown creeper | 13.4 | <0.001 | 59 | <0.001 |
| Winter wren | 8.6 | <0.001 | 53 | <0.001 |
| Varied thrush | 7.8 | <0.001 | 24 | 0.007 |
| Pileated woodpecker | 2.51 | 0.068 | 18 | 0.019 |
| Hermit warbler | 1.4 | 0.271 | 12 | 0.063 |

*Note:* These relationships suggest that increasing late successional forest in a watershed may lead to increases in abundance of these species, but this is not a cause and effect result. Note that the strength of the association varies among the species.

*Source:* Data from McGarigal, K., and W.C. McComb. 1995. *Ecological Monographs* 65:235–260.

species should respond positively to increasing areas of old forest. Monitoring data on these species collected over time can be used to test that hypothesis.

## Before–After Control Versus Impact (BACI) Designs

Although the BACI design is usually considered superior to the ANOVA design, BACI designs often are logistically challenging. The BACI design allows monitoring to occur on treated and untreated sites both before and after management has occurred (e.g., Chambers et al. 1999). BACI designs allow detection of cause-and-effect relationships, but within monitoring frameworks, they can often suffer from non-random assignment of treatments to sites. Often the location and timing of management

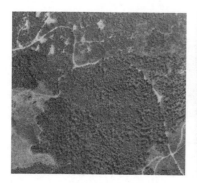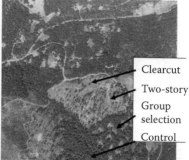

**FIGURE 5.14** One of three replicate areas prior to and following the management actions that included clearcut with reserves, two-story, and group selection harvests. A central control area can also be seen in the post-treatment photo.

actions do not lend themselves to strict experimental plans. In the following example, Chambers et al. (1999) used a BACI design experiment that included silvicultural treatments in three replicate areas on a working forest (Figure 5.14). The results from this effort again produced predictable responses, but the responses could clearly be linked to the treatments. In Figure 5.5, white-crowned sparrows were not present on any of the pretreatment sites, but were clearly abundant on the clearcut and two-story stands following treatment. The treatments caused a response in the abundance of this species. Conversely, hermit warblers declined in abundance on these two treatments but remained fairly constant on the control and group selection treatments (Figure 5.15). Consequently, we would predict that future management actions such as these would produce comparable results on this forest. However, it is important to remember that the scope of inference is this one forest; extrapolation to other similar sites should be done with caution. Further, to fully understand why these changes

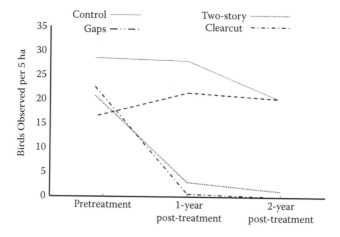

**FIGURE 5.15** Change in hermit warbler detections following silvicultural treatments illustrating cause and effect monitoring results. (Redrafted from Chambers, C.L., W.C. McComb, and J.C. Tappeiner. 1999. *Ecological Applications* 9:171–185.)

might have been observed, ancillary data on habitat elements important to these species also could be collected and if the treatments caused changes in important habitat elements, then the reasons for the effects become clearer. These habitat relationships analyses can be particularly informative when developing predictive models of changes in abundance or fitness of an organism based on management actions or natural disturbance.

### EDAM: Experimental Design for Adaptive Management

Choosing a design for monitoring can be facilitated using software specifically developed to assess statistical rigor of the plan (e.g., MONITOR program), which can guide design of adaptive management efforts. Anderson (1998) developed EDAM software to guide an adaptive management experiment that may never be repeated, but instead may lead directly to economically significant management decisions. Traditional design recommendations alone may not be helpful in this scenario because manipulating design variables, such as sample size or length of the experiment, often involves very high costs. Also, various stakeholders may have different points of view concerning the impact of possible incorrect inferences (Anderson 1998). There are two versions of the EDAM model: a Before–After Control-Impact Paired Series (BACIPS) experiment on a landscape scale, and an analysis of variance (ANOVA) experiment on a patch scale:

> "Each model represents a management experiment on a forest ecosystem, which can be examined from different stakeholders' points of view. Given a set of design choices, the model demonstrates all possible outcomes of the experiment and their likelihood of occurrence, with a special emphasis on their future economic and ecological impact. The experiments in the EDAM models are thus not primarily statistical exercises; instead, the adaptive management experiment is a three-stage interaction between people and nature that may stretch far into the future:
>
> Stage 1 Design and implementation. The experimenters plan and carry out a management plan that probes the forest ecosystem experimentally.
> Stage 2 Analysis. The experimenters learn, i.e., they make an inference (correct or incorrect) about the ecosystem based on data from the experiment.
> Stage 3 Management response. The experimenters respond to the inference with new management actions, which, as they are projected into the future, will effect both the forest ecosystem and various stakeholders"
> (Anderson 1998).

Once you have considered the appropriate monitoring design to answer your key questions(s), then you need to make several additional decisions regarding what you will measure, how you will select sample sites, what level of precision you require, and what your scope of inference will be. These then become steps in the development of a monitoring plan.

## BEGINNING THE MONITORING PLAN

Within a monitoring plan, the conceptual framework should provide the basis for a written description of the system, how it works, and what we do and do not know about it. The culmination of this discussion should be the identification of the key questions that drive the monitoring plan. These key questions should address the processes and stressors about which there are information gaps that keep us from understanding or predicting responses of organisms to management actions. The first step in the process is to identify, clearly and concisely, whatever questions are to be answered by the monitoring plan. Questions must be clearly focused. Once they have been articulated, and if existing data are insufficient to address the key question(s), then the following steps should be explained and justified, based on the conceptual model.

## SAMPLE DESIGN

In a research setting we may decide to state our key question as a hypothesis and alternative hypotheses. We can take the same approach in monitoring as well. Indeed, if numerous alternative hypotheses are to be compared using an information theoretic approach, these alternative hypotheses should be generated at this stage so that models reflecting them can be compared.

- Are you predicting occurrence?
- Are you assessing trends over time?
- Are you assessing patterns over space?
- Are associations with habitat elements important?
- Do you hope to understand a cause-and-effect relationship?

The answers to these questions should allow you to develop hypotheses and identify the most useful study design for testing them.

## SELECTION OF SPECIFIC INDICATORS

Once you have defined the question of interest, then you will need to decide what will be measured as a reliable indicator of change to address the key question. Noon et al. (1999) described an attribute (indicator) as "… simply some aspect of the environment which is measurable. When an attribute is measured it takes on a (usually) numeric value. Since the exact value of an attribute is seldom known with certainty, and may change through time, it is properly considered a *variable*. If the value of this attribute is indicative of environmental conditions that extend beyond its own measurement, it can be considered an *indicator*. Not all indicators are equally informative—one of the key challenges to a monitoring program is to select for measurement those attributes whose values (or trends) best reflect the status and dynamics of the larger system."

Questions that will aid you in undertaking this task include:

- What should I measure?
- Why was one indicator better than another?
- What are the benefits and costs associated with each of the potential indicators?

## SELECTION OF SAMPLE SITES

This step may seem trivial, but there are a set of questions that should be addressed when designing a monitoring plan. These include:

- How will you ensure that sites are independent from one another with respect to the response variable? Are sites sufficiently separated over space to ensure that the same individual is unlikely to be recorded on multiple sites in the same year?
- How did you select the number of sites? Given the pilot study data, was the sample size adequate to detect a desired effect? The process of deciding sample size should be well documented. See Gibbs and Ramirez de Arellano (2007) for assistance with these questions.
- How will you mark the sites on the ground? Sampling locations should be permanently marked and GPS locations recorded, and the degree of accuracy associated with the coordinates should be provided. Will you use flagging? How long will it last? Will flagging or more permanent marks bias your sample? For instance, if you are monitoring nest success or fish abundance along streams, could predators be attracted to the marks?
- How will you ensure that the same sites will be found in the future and that the data are collected in the same manner at the same sites?

### Detecting the Desired Effect Size

An excellent tool to assist you in assessing the power associated with detecting a trend that is real is the MONITOR program (Gibbs and Ramirez de Arellano 2007). Estimates of variability in the indicators of populations of plants and animals will be necessary to estimate sample sizes and power, so these estimates will have to come from previous studies in similar areas or from pilot studies (Gibbs et al. 1998). Because of the economic impacts of conducting monitoring programs without adequate power to detect trends, it is always wise to consult a statistician for assistance with this task. This step is extremely important because the results will provide insight into the sample sizes and precision of data needed to detect trends. If it is logistically or financially unfeasible to achieve the desired level of power, then the entire approach should be revisited.

### The Proposed Statistical Analyses

Will the analytical approaches available allow you to detect occurrence, trends, patterns, or effects given the data that you will collect? Is the analysis appropriate if data are not independent over space or time? Data collected over time to detect trends in Before–After Control-Impact (BACI) often have data that are not independent from one time period to the next. Repeated measures designs are often necessary to ensure that estimates of variance between or among treatments reflect this lack of independence (Michener 1997). Concerns regarding independence of data or alternative data analysis approaches (e.g., Bayesian) must be addressed during the planning stages. Too often data are collected and then a statistician is consulted, when the data could be so much more useful if the statistician were consulted at the planning stage for the program.

# Designing a Monitoring Plan

Indicator selection is challenging. The U.S. Environmental Protection Agency (EPA), U.S. Forest Service, and other agencies have tested various indicators for monitoring ecosystems, but there is little consensus on which indicators are best or how best to quantify them. Rather than these specifics, such agencies tend to generate lists of useful indicator characteristics to aid in site-specific indicator selection. The National Park Service (2007), for example, suggests that indicators are most useful when they:

- Have dynamics that parallel those of the ecosystem or component of interest
- Are sensitive enough to provide an early warning of change
- Have low natural variability
- Provide continuous assessment over a wide range of stress
- Have dynamics that are easily attributed to either natural cycles or anthropogenic stressors
- Are distributed over a wide geographical area and/or are very numerous
- Are harvested, endemic, alien, species of special interest, or have protected status
- Can be accurately and precisely estimated
- Have costs of measurement that are not prohibitive
- Have monitoring results that can be interpreted and explained
- Are low impact to measure
- Have measurable results that are repeatable with different personnel

The context of a monitoring program should be carefully considered in the selection of indicators. The largest monitoring program to date has been the EPA's Environmental Monitoring and Assessment Program (EMAP). In their review of EPA's EMAP program, the National Research Council (NRC 1995) discussed the relative merits of retrospective monitoring (EMAP's basic monitoring approach) versus predictive or stressor-oriented monitoring. Retrospective, or effects-oriented monitoring, seeks to find effects by detecting changes in the status or condition of some organism, population, or community. This would include trends in plant or animal populations, and it takes advantage of the fact that biological indicators integrate conditions over time. In contrast, predictive or stressor-oriented monitoring seeks to detect the cause of an undesirable effect (a stressor) before the effect occurs or becomes serious (NRC 1995). Stressor-oriented monitoring will increase the probability of detecting meaningful ecological changes, but it is necessary to know the cause–effect relationship so that if the cause can be detected early, the effect can be predicted before it occurs. The NRC (1995) concluded that in cases where the cost of failing to detect an effect early is high, use of predictive monitoring and modeling is preferred over retrospective monitoring. They concluded that traditional retrospective monitoring was inappropriate for environmental threats such as exotic species effects and biological extinctions because of the large time lag required for mitigation, and recommended that EPA investigate new indicators for monitoring these threats. Planners should keep in mind these approaches and analyses when identifying both the indicators and the likely outcomes of the monitoring effort.

## The Scope of Inference

The sample sites selected for the study will have to be defined to address an appropriately broad or narrow range of conditions. There is a tradeoff that should be discussed in the plan. That tradeoff is to monitor over a large spatial extent, so that results are broadly applicable, versus sampling over a narrow spatial extent to minimize among-site variance. The variability in the indicator likely will increase as the spatial extent of the study increases. As the variance of the indicator increases, the probability of detecting a difference between treatments or of detecting a trend over time will decrease.

If it is feasible to do so, collecting preliminary data from a broad range of sites may help you decide what an appropriately broad scope of inference would be before variance increases dramatically and would therefore influence the power of your test. For instance, if data on animal density were collected from 20 sites extending out from some central location, and the variance over number of samples was calculated, then the variance should stabilize at some number of samples. Even after stabilization, adding more samples from areas that are sufficiently dissimilar from the samples represented at the asymptote may lead to a jump in variance caused by an abrupt change in some environmental characteristic. This change in the variance of the indicator over space may represent an inherent domain of spatial scale for that indicator. Sampling beyond that domain, and certainly extrapolating beyond that domain, may not be warranted. Indeed, broadcasting from the monitoring data (extrapolating to other units of space outside of the scope of inference) and forecasting (predicting trends into the future from existing trends) must be done with great care because the confidence limits on the projections increase exponentially beyond the bounds of the data.

Consideration of the scope of inference is a key aspect of a monitoring program, but it is often not given adequate attention during the design phase. Instead, managers may wish to extrapolate beyond the bounds of the data after data have been collected, when the confidence in their predictions would be much greater if the scope had been considered during program design. Concepts of statistical power, sample size estimation, effect size estimation, and scope of inference are discussed in more detail in later sections.

## SUMMARY

Development of a monitoring plan must begin with articulating the key questions that are the result of careful construction and development of a conceptual framework. Once questions have been clearly articulated, then you can identify the type of monitoring design that you would like to use to address the questions. Carefully considering the appropriate response variables or indicators will allow you to ensure that your results will indeed answer your key questions. In addition, decisions about your scope of inference and the types of statistical analyses that you might like to conduct should be made during the design process; this is the topic of the next chapter.

## REFERENCES

Anderson, J.L. 1998. Experimental Design for Adaptive Management: A Decision Model Incorporating the Costs of Errors of Inference Before-After-Control-Impact, Landscape-Scale Version. School of Resource and Environmental Management, Simon Fraser University, Burnaby, BC.

Baldwin, N.A., R.W. Saalfeld, M.R. Dochoda, H.J. Buettner, and R.L. Eshenroder. 2002. Commercial fish production in the Great Lakes 1867–2000. http://www.glfc.org/databases/commercial/commerc.asp.

Bruggink, J.G., and W.L. Kendall. 1995. American woodcock harvest and breeding population status, 1995. U. S. Fish and Wildlife Service, Laurel, MD. 15 pp.

Chambers, C.L., W.C. McComb, and J.C. Tappeiner. 1999. Breeding bird responses to 3 silvicultural treatments in the Oregon Coast Range. *Ecological Applications* 9:171–185.

Elzinga, C.L., D.W. Salzer, and J.W. Willoughby. 1998. Measuring and Monitoring Plant Populations. Bureau of Land Management Technical Reference 1730-1, BLM/RS/ST-98/005+1730. Denver, CO.

Gibbs, J.P., S. Droege, and P. Eagle. 1998. Monitoring populations of plants and animals. *BioScience* 48:935–940.

Gibbs, J.P., and P. Ramirez de Arellano. 2007. Program MONITOR: Estimating the Statistical Power of Ecological Monitoring Programs. Version 10.0. URL: www.esf.edu/efb/gibbs/monitor/monitor.htm

Hierl, L.A., J. Franklin, D.H. Deutschman, and H.M. Regan. 2007. Developing Conceptual Models to Improve the Biological Monitoring Plan for San Diego's Multiple Species Conservation Program. Department of Biology, San Diego State University, and California Department of Fish and Game, Sacramento, CA.

Hurlbert, S.J. 1984. Pseudoreplication and the design of ecological field experiments. *Ecological Monographs* 54:187–211.

Jones, K.B. 1986. Data types. Pages 11–28 in *Inventory and Monitoring of Wildlife Habitat*. A.Y. Cooperrider, R.J. Boyd, and H.R. Stuart (eds.). USDI BLM, Service Center, Denver, CO. 858 pp.

MacKenzie, D.I., J.D. Nichols, J.E. Hines, M.G. Knutson, and A.D. Franklin. 2003. Estimating site occupancy, colonization and local extinction when a species is detected imperfectly. *Ecology* 84:2200–2207.

Mackenzie, D.I., and J.A. Royle. 2005. Designing occupancy studies: General advice and allocating survey effort. *Journal of Applied Ecology* 42:1105–1114.

Martin, K.J., and B.C. McComb. 2003. Amphibian habitat associations at patch and landscape scales in western Oregon. *Journal of Wildlife Management* 67:672–683.

McGarigal, K., and W.C. McComb. 1995. Relationships between landscape structure and breeding birds in the Oregon Coast Range. *Ecological Monographs* 65:235–260.

Michener, W.K. 1997. Quantitatively evaluating restoration experiments: Research design, statistical analysis, and data management considerations. *Restoration Ecology* 5:324–337.

National Park Service. 2007. Example criteria and methodologies for prioritizing indicators. USDI National Park Service. http://science.nature.nps.gov/im/monitor/docs/CriteriaExamples.doc. Accessed February 4, 2007.

NRC (National Research Council). 1995. *Review of EPA's Environmental Monitoring and Assessment Program: Overall Evaluation*. National Academy Press, Washington, D.C.

Noon, B.R., T.A. Spies, and M.G. Raphael. 1999. Conceptual basis for designing an effectiveness monitoring program. Chapter 2 in The strategy and design of the effectiveness monitoring program for the Northwest Forest Plan. USDA Forest Service Gen. Tech. Rept. PNW-GTR-437.

Romesburg, H.C. 1981. Wildlife science: Gaining reliable knowledge. *Journal of Wildlife Management* 45:293–313.

Sauer, J.R., J.E. Hines, and J. Fallon. 2001. The North American Breeding Bird Survey, Results and Analysis 1966–2000. Version 2001.2, USGS Patuxent Wildlife Research Center, Laurel, MD.

Wade, P.R. 2000. Bayesian methods in conservation biology. *Conservation Biology* 14:1308–1316.

# 6 Factors to Consider When Designing the Monitoring Plan

The previous chapter provides a brief overview of the process of defining key questions, determining the monitoring design, and addressing issues of power and sample size. But there are other aspects of the monitoring plan that should be considered, especially those dealing with how you will analyze the data and how you will decide when a threshold or trigger point is reached. All too often data are collected because it "seems like a good idea" and then months or years later someone sits down to analyze the data and realizes that the resulting interpretation is highly limited and may not indeed address some key question. Hence, including aspects of the future analysis and interpretation of the data into the monitoring plan can help ensure that the data will be useful to people in the future.

Effective population monitoring is a critical aspect of adaptive management, and should be designed to test specific hypotheses that are relevant to management and policy decisions (Noss 1990). Unfortunately, monitoring often involves common mistakes such as overlooking explicit hypothesis testing or not paying attention to proper design and analysis (Hinds 1984). A properly constructed and reviewed sampling design will ensure accurate results, and allow users to make inferences to larger areas. In addition to the issues of statistical power and effect size estimation discussed in the previous chapter, there are several factors that should be addressed as a protocol is being developed. Before data collection begins, one must identify the limitations that will be imposed on the sampling techniques by the inherent variability of the animals and habitats being measured.

## USE OF EXISTING DATA TO INFORM SAMPLING DESIGN

Existing data available for the monitoring site or nearby areas can be exceptionally useful when designing the monitoring protocol. The resulting design can be much more robust and the ability to detect differences can be assessed prior to the onset of data collection. A few examples of uses of existing data are:

1. Estimating detectability and detection distances for mobile species (e.g., birds, fish, large mammals)

2. Estimating variance associated with potential indicators to aid in selection of the indicators most likely to detect effects with desired levels of power if they are present
3. Estimating variance to estimate sample sizes needed to achieve desired levels of power
4. Developing tradeoff scenarios to estimate the costs of the monitoring program versus the effect size that could be detected
5. Identifying appropriate levels of sampling effort through examination of variance stabilization
6. Maximizing efficiency during adaptive sampling tests
7. Estimating spatial and temporal patterns in habitat elements or populations

## DETECTABILITY

Since it is often logistically impossible to census an entire population of a species, monitoring generally relies upon a sample of that population. These data are often in the form of presence/not detected (e.g., species occurrence inventories), relative abundance (e.g., number of animals or their sign observed per unit time), or absolute abundance estimates (e.g., mark-recapture). Whether a sampling design collects an accurate sample that is representative of the population is dependent on the ability of the monitoring technique to detect the species. Detectability is defined as the probability that an individual will be observed (e.g., seen, heard, captured, etc.) in a particular habitat at a particular time, and is influenced by factors such as survey method, species, observer, habitat complexity, time of year, weather, density, and breeding phenology (Thomas 1996). These factors can result in missed individuals, multiple counting of individuals, recording an individual where none exists, or the overcounting of conspicuous species. Errors as a result of detectability problems are another source of variation in documenting population change.

The additional stochasticity associated with sampling can have two major consequences (Barker and Sauer 1995). First, efficiency of the monitoring technique in sampling the species is greatly compromised. Second, the sampling-induced variation can bias estimators of important population parameters. When a sample is collected and analyzed, those results are estimates of population parameters that are used in detecting population changes. Ignoring this error can reduce the likelihood of identifying critical trends in population data.

Population monitoring programs often have a number of different personnel responsible for collecting data. Several studies have suggested that observer error can be a significant source of variation in population data (Sauer et al. 1994; Barker and Sauer 1995; James et al. 1996; Thomas 1996). In fact, observer error has resulted in false population changes (Sauer et al. 1994). A popular approach to addressing detectability problems is by estimating a detection function (Burnham et al. 1980). Consequently, designing an effective and reliable monitoring protocol must emphasize standardized monitoring techniques that take into account species-specific biases, and proper training of data-collecting personnel.

# Factors to Consider When Designing the Monitoring Plan

## ESTIMATING DETECTION DISTANCES

When using some types of density estimators (see Chapter 7 for more details) such as variable width transects, variable circular plots, or mark-resight estimators, it is important to know what detection distances might be for the species of interest, and more importantly if detection distances vary from one vegetation type to another. Differences in detection distances do not present any particular problems when using these estimators, but if differences do occur among vegetation types, then simple indices to abundance (detection rates, resight rates) cannot be used reliably when comparing between or among vegetation types. In a very simple example, consider the detection distances for a salmonid in streams that have and have not been impacted by streamside activities leading to increased sediment loads. Or detection distances for olive-sided flycatchers in old-growth forests versus clearcuts. The function describing the detection decay curve will very likely differ between these two conditions and must be used to estimate abundance in an unbiased manner. Even estimates of abundance of patchily distributed plants may have different detection distances among vegetation types and this should be considered during design of the monitoring protocol, especially when distributing transects or points over an area to ensure complete, but not overlapping, coverage.

## ESTIMATING VARIANCE ASSOCIATED WITH INDICATORS

The ability to detect a difference between conditions, or a slope different from 0, will depend on the sampling error associated with the chosen indicator. Clearly, an indicator should be chosen that is meaningful to the question being asked, but there may very well be options to choose among. For instance, consider estimating the densities of a specific age class of fish from stream surveys versus electro-shocking mark-recapture estimates. Both techniques may produce similar numbers of observations, but if the stream survey work produces more consistent (less variable) estimates of abundance, then the ability to detect differences or trends would be greater (assuming similar sample sizes) with that technique than the mark-resight technique (or vice versa).

## ESTIMATING SAMPLE SIZE

The ability to detect a difference will increase with decreasing sampling error and increasing sample size. In a very simple example, consider $N = s^2 t^2 / d^2$, where $N$ = the estimated sample size, $s^2$ = the sample variance, $t$ = the t-statistic for a specified alpha level, and $d$ = the effect size or difference that you wish to detect. Sample sizes increase as (1) the variance increases, (2) the t-statistic increases (decreasing alpha level), or (3) the effect size decreases (you want to detect smaller differences). Knowing in advance what the variance might be for a sampling effort can allow the developer of the protocol to understand if it would even be possible to detect a difference given a limited amount of time and money. See Program MONITOR for more information on calculating sample sizes (Gibbs and Ramirez de Arellano 2007).

## Logistical Tradeoff Scenarios

Given a limited budget with which to work, a knowledge of the variance in a potential indicator can be useful in deciding what statistical power will be associated with the sampling design, at a given effect size and sample size. Conversely, knowledge of the variance can allow you to estimate the effect size that could be detected, given the logistics of funding, and a desired level of power. It should become quickly apparent that identifying indicators with low sampling error can mean considerable savings in sampling effort for intensive field studies.

## Variance Stabilization

You will need to make a decision regarding how many samples are sufficient to estimate variance for a monitoring protocol. Too few samples and the estimate may be unusually high and lead to an inability to detect differences. Too many samples and you are not being efficient with time and resources. One approach to understanding when sampling intensity is sufficient to provide a reasonable estimate of variance is to plot variance over number of samples (Figure 6.1). If you want to estimate the mean and variance for a set of values, one might ask how many data points would be needed to develop a reasonable estimate of variance and a mean. Variances and means fluctuate among just a few samples, but stabilize eventually; at some point, additional data are not likely to appreciably change the estimate. In the examples in Figure 6.1, variance stabilizes at about 10–15 samples, and the estimate of the mean stabilizes at about 15 samples. Clearly using fewer than 10 samples could lead to a poor estimate of the mean and variance. More than 30 samples are probably not necessary.

In Figure 6.2, estimates of dead wood from four 50-m transects at sample plots in young Douglas-fir stands show a high degree of variability over sample size, and

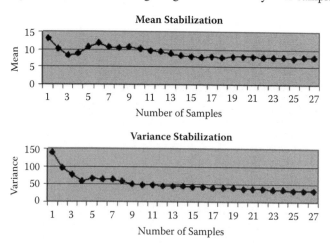

**FIGURE 6.1** Hypothetical estimates of variances and means based on number of samples in a population. Note that as the number of samples increases the fluctuations in the estimated mean and variance reach an asymptote.

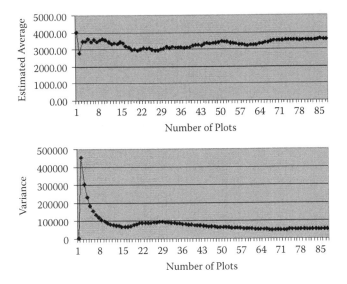

**FIGURE 6.2** Stabilization of the estimated mean (top) and variance (bottom) of the surface area per ha for coarse woody debris in young managed Douglas-fir stands in the Oregon Cascades. Note that for this habitat element, variances remain high compared to the mean, even with large sample sizes.

stabilization is observed only after a large number of plots are used in the analysis. Given such a large sample size needed to characterize the mean and variance, one would have to ask if some other indicator of this habitat element might be more efficiently characterized. For some habitat elements and populations that are highly patchy in nature, use of variance stabilization may be problematic because jumps in variance may occur as sampling efforts encounter new patches.

## SPATIAL PATTERNS

Populations and individuals can be dispersed throughout a habitat or a landscape in various patterns. Monitoring protocols must incorporate species-specific patterns in density and distributions. The three basic population spatial patterns are random, clumped, and regular (Figure 6.3; Curtis and Barnes 1988; Krebs 1989). Random dispersion is found in populations in which the spacing between individuals is irregular, and the presence of one individual does not directly affect the location of another individual. Clumped populations are characterized by patches or aggregations of individuals, and the probability of finding one individual increases with the presence of another individual. Regularly distributed populations have individuals that are distributed more or less evenly throughout an area, and the presence of one individual decreases the probability of finding another nearby individual.

The spatial distribution within populations is an important constraint on sampling designs. A number of factors influence the distribution of individuals across a habitat or landscape. Factors such as resource distribution (e.g., water, food, and cover) and habitat quality can affect the dispersion of individuals and populations

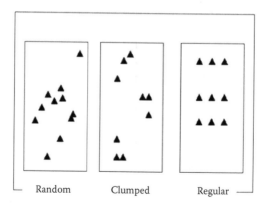

**FIGURE 6.3** The three basic spatial patterns found in populations are random, clumped, and regular.

(McComb 2007). A random distribution is often found in species that are dependent on ephemeral resources. Species that depend on temporary or seasonal resources may exhibit different types of distribution at different points in their life histories. In vertebrates, social behavior and territoriality can affect distribution. Highly territorial species tend to follow a regular distribution, while more gregarious and colonial nesting species tend to occur in clumps (Curtis and Barnes 1988; Newton 1998). In addition, the scale of observation is critical in determining the distribution of a population. Populations may appear regularly distributed at a fine scale, such as the site level, but may show a more random or clumped distribution throughout part or all of their geographic range.

Spatial patterns of organisms and populations can complicate sampling design, but there are a number of techniques to address these constraints. Nested quadrat sampling is often used to define a species-area curve for a community of organisms (Krebs 1989). The number of species will expectedly rise as the size of the quadrats increase, but a plateau in species number should eventually be reached within a patch type. This plateau is considered the minimum sampling area for that community. This technique has been used successfully for plant communities (Goldsmith and Harrison 1976). However, quadrats are not natural sampling units, and one must decide on a number of factors including size, shape, and number of quadrats. Refer to species area curves in the previous chapter to understand how numbers and sizes of plots can influence results.

A useful approach for determining the spatial distribution of a population is "plotless sampling." There are two general approaches to plotless sampling (Krebs 1989):

1. Select random organisms and measure distance to nearest neighbors
2. Select random points and measure distance from each point to nearest organisms

Plotless sampling has two major advantages. First, if the spatial pattern of the population is random, one can use plotless sampling to estimate the density of the population. Conversely, if the population density is known, then one can determine the

## TABLE 6.1
## Commonly Used Methods to Assess the Distribution of Individuals in a Population

| Methods | Data Types |
|---|---|
| Nearest-Neighbor Methods (Clark and Evans 1954) | Population density is known; spatial map of population is available |
| Byth and Ripley Procedure (Byth and Ripley 1980) | Sample of a population over a large area; independency of sample points |
| T-Square Sampling Procedure (Besag and Gleaves 1973) | Favored by field studies; random point assignments |
| Point-Quarter Method (Cottam and Curtis 1956) | Random point assignments; typically transect data |
| Change-in-Ratio Methods (Seber 1982) | Change in sex ratios data; population data |
| Catch-Effort Methods (Ricker 1975) | Exploited population data; catch per unit effort with time |

spatial pattern of that population. Both of these applications may prove valuable in the preliminary stages of monitoring. A first step is to estimate if a population has a random, clumped, or regular distribution; approaches for estimating these distributions are given in Table 6.1.

### TEMPORAL VARIATION

Spatial variation is a critical limitation of successful monitoring, but sample design must also consider temporal fluctuations in populations. A sampling design that does not emphasize a dynamic and repeatable monitoring scheme will not be able to accurately estimate populations or their response to management.

The temporal variation observed in populations is often the result of environmental variation (Akçakaya 1999). Environmental changes will have both direct and indirect effects on populations. Although we can often calculate probabilities regarding environmental fluctuations based on past records and averages (e.g., rainfall, flooding, etc.), it is often difficult, if not impossible, to predict when these changes might occur. In cases were population rates and demographics depend on environmental variables, population fluctuations are likely to occur (Akçakaya 1999). Sometimes these environmental changes can indirectly influence populations through predator densities, herbivory, sedimentation, nest predation, food resource availability, and cover. For example, a study on acorn production by oaks found that populations of white-footed mice, eastern chipmunks, and gray squirrels were significantly correlated with annual fluctuations in acorn crop (McShea 2000). In addition, these changes in small mammal populations were correlated to the nesting success of several ground-nesting birds. In eastern North America, Carolina wrens suffered a population crash following several severe winters in the mid 1970s (Figure 6.4; Barker and Sauer 1995; Sauer et al. 2001).

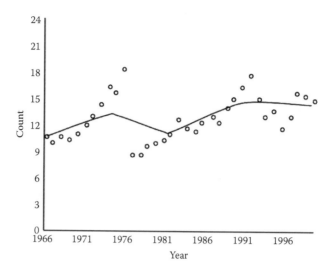

**FIGURE 6.4** Breeding bird survey data of Carolina wren populations in eastern United States, 1966–2001. (Redrafted from Sauer, J.R., J.E. Hines, and J. Fallon. 2007. The North American Breeding Bird Survey, Results and Analysis 1966–2006. Version 10.13.2007. USGS Patuxent Wildlife Research Center, Laurel, MD.)

Analysis and interpretation of long-term monitoring should recognize and evaluate the effects of these patterns on population data. Monitoring over extended time periods and careful documentation of environmental changes (including management actions) is the only effective method for determining the change and status of a population.

## COST

A proper sampling design emphasizes cost effectiveness, and improves the interpretation and presentation of results. Knowledge of the sample design promotes understanding and communication between project leaders and administrators, and further elucidates the critical goals of the program. Conversely, failure to consider a proper sampling design can reduce monitoring efficiency, inflate costs, and lead to inaccurate results. The inventorying and monitoring process consists of a hierarchy of decisions and events (see Chapter 5 and Jones 1986). Following the hierarchy of planning decisions, as well as constructing a reliable conceptual model, will allow managers to construct a cost-efficient sampling scheme. These decisions and goals will also dictate which type of inventory method is appropriate. Different monitoring techniques vary in their costs and intensity. For example, inventories relying on presence/not-detected data can be inexpensive because personnel with minimum training can sample fairly large areas. Comparatively, monitoring which relies on relative abundance or absolute abundance data (e.g., mark-recapture studies) demands highly skilled personnel and are generally more site intensive. Thus, a proper sampling design, constructed through a hierarchy of decisions and input, is critical to limiting costs and collecting relevant data.

## STRATIFICATION OF SAMPLES

Stratified random site selection can be an excellent method for limiting bias and collecting representative data. Although this topic will be discussed in greater detail in Chapter 11, considering stratification in your monitoring design is an important step. A stratum is an aggregation of mapping units that have similar abiotic or biotic characteristics (Jones 1986). Selecting appropriate strata is a critical step in habitat type and study site classification. In many cases, strata are based on existing vegetation and landforms. However, like many aspects of habitat assessment, strata selection can also be species specific. For example, if it is important to collect data on a particular species, a stratum can be created that encompasses several of the habitat components that are necessary for that species. Stratification of habitats can be based on dominant or subdominant vegetation types, topographical characteristics, soil composition, ichthyologic provinces, or management areas. However, monitoring effort and sampling are designed to be repeatable and creating strata based on vegetation can be problematic since these habitat boundaries may change over time (e.g., succession). If the characteristics used to create strata are variable over time, such as vegetation types, then it may be necessary to reevaluate strata classification, boundaries, and criteria every year.

## ADAPTIVE SAMPLING

Quite often during surveys of rare plants or animals, the species of interest is observed on very few of the sampling units, and units where the species is detected are in close proximity. Many common species also have a tendency to occur in population clusters because of dispersal mechanisms, behavior patterns (e.g., herding, colonialism), or habitat associations. Under these conditions, it is predictable that surveys conducted according to conventional designs will expend most of the sampling effort at locations where the species is unobserved. Adaptive sampling designs offer an approach to concentrate sampling units in the vicinity of population clusters, thus improving the efficiency of parameter estimation. The theoretical origins of adaptive sampling extend back at least 50 years (Wald 1947; Zacks 1969), and S.K Thompson has written extensively on the application of adaptive sampling designs to plant and wildlife surveys (Thompson and Ramsey 1983; Thompson 1990, 1992; Thompson and Seber 1996). Adaptive cluster sampling has been applied to surveys of rare trees (Roesch 1993), freshwater mussels (Box et al. 2002), and tailed frogs (Vesely and Stamp 2001), among other species and assemblages.

Adaptive cluster sampling refers to designs in which sample selection depends on the values of counts or other variables observed during the course of the survey. Initially, a probability procedure is used to select a set of sampling units in the study area. When any of the selected units satisfy some predetermined criterion, additional units are sampled in the vicinity of the qualifying unit. Sampling is extended until no further units satisfy the criterion. Typically the criterion is a count value of the target species.

For rare or highly aggregated populations, adaptive cluster sampling can greatly increase the precision of population size or density estimates when compared to a simple or stratified random design of equal cost (Thompson 1992). Adaptive cluster sampling can be applied to surveys conducted on quadrats, belt transects, and variable circular plots, among other types of sampling units. Thompson and Seber (1996) provide a comprehensive review of adaptive sampling designs and data analysis. There are two different considerations that may limit the applicability of adaptive sampling designs for some inventory and monitoring studies. First, adaptive sample selection may cause biases in conventional estimators for the population mean, variance, and total population size. Unbiased estimators have been described for simple and stratified adaptive sampling designs, but cannot be automatically computed with commonly used statistical software. Calculation of design-unbiased estimators for adaptive sampling designs requires a thorough understanding of statistics and the ability to program the estimators into one of the advanced statistical software packages (e.g., SAS®, S+®). A statistical consultant can aid in developing analytical methods appropriate for adaptive sampling, and provide similar assistance in performing data analysis. The second consideration in planning for adaptive sampling is the operational uncertainty surrounding the number of sampling units required for the survey. Unlike conventional designs in which the sample size is typically determined before starting fieldwork, there is no precise knowledge prior to data collection as to the number of units that will trigger secondary surveys. This uncertainty can be minimized by conducting pilot studies to estimate the mean and variance for sample sizes. Alternatively, survey protocols can limit the maximum number of sampling units visited in a study area, after making appropriate modifications to the population estimators.

## PEER REVIEW

Once a sampling design has been developed based on existing data or a pilot study, there is still some doubt that the data collected as part of the monitoring effort will adequately address all potential biases or inherent fluctuations in conditions over time. One step that can be extremely valuable in the development of a protocol is to draft the sampling design, providing the preliminary data analyses and rationale for indicators selected, sampling intensity, effect sizes, and power. This draft should be reviewed by both a statistician and an expert on the biology of the organism(s) and ecosystem(s). Recall that statistics are merely a tool to help make decisions. The design must be statistically sound, but it must also be biologically meaningful. External review of the design can help identify features of the sampling design that may limit the usefulness of the data in the analytical phase of the monitoring program. It is much better to be thorough in development of the design phase and minimize the risk of making errors now than to find out after a year or more of data collection that the results are biased, not independent, or too variable to make inferences.

## SUMMARY

In addition to developing clear objectives, having a random sample, and having adequate power in your design to detect trends, the monitoring plan should include a framework for analyzing and interpreting the data. Existing data from other studies or a pilot study may be particularly useful when estimating detectability, variance in potential indicators, estimating costs of monitoring, maximizing efficiency during adaptive sampling tests, and estimating spatial and temporal patterns in habitat elements or populations. These and other ancillary data may aid in the identification of strata as monitoring plans are designed. Stratified random site selection can be an excellent method for limiting bias and collecting representative data. For rare or highly aggregated populations, adaptive cluster sampling can greatly increase the precision of population size or density estimates when compared to a simple or stratified random design of equal cost. Finally, the monitoring plan should be subjected to rigorous peer review prior to implementation.

## REFERENCES

Akçakaya, H.R., M.A. Burgman, and L.R. Ginzburg. 1999. *Applied Population Ecology: Principles and Computer Exercises Using RAMAS EcoLab.* 2nd ed. Sinauer Associates, Sunderland, MA.

Barker, R.J., and J.R. Sauer. 1995. Statistical aspects of point count sampling. Pages 125–130 in *Monitoring Bird Populations by Point Counts.* C.J. Ralph, J.R. Sauer, and S. Droege (eds.). USDA Forest Service General Technical Report PSW-GTR-149.

Besag, J.E., and J.T. Gleaves. 1973. On the detection of spatial pattern in plant communities. *Bulletin of the International Statistical Institute* 45:153–158.

Box, J.B., R.M. Dorazio, and W.D. Liddell. 2002. Relationships between streambed substrate characteristics and freshwater mussels (Bivalvia: Unionidae) in coastal plain streams. *Journal of the North American Benthological Society* 21:253–260.

Burnham, K.P., D.R. Anderson, and J.L. Laake. 1980. Estimation of density from line transect sampling of biological populations. *Wildlife Monographs* 72:202 pp.

Byth, K., and B.D. Ripley. 1980. On sampling spatial patterns by distance methods. *Biometrics* 36:279–284.

Clark, P.J., and F. Evans. 1954. Distance to nearest neighbor as a measure of spatial relationships in populations. *Ecology* 35:445–453.

Cottam, G., and J.T. Curtis. 1956. The use of distance measures in phytosociological sampling. *Ecology* 37:451–460.

Curtis, H., and N.S. Barnes. 1988. *Biology.* 5th ed. Worth Publishers, New York. 1050 pp.

Gibbs, J.P., and P. Ramirez de Arellano. 2007. Program MONITOR: Estimating the Statistical Power of Ecological Monitoring Programs. Version 10.0. URL: www.esf.edu/efb/gibbs/monitor/monitor.htm

Goldsmith, F.B., and C.M. Harrison. 1976. Description and analysis of vegetation. Pages 85–157 in *Methods in Plant Ecology.* S.B. Chapman (ed.). Blackwell Scientific, Oxford, U.K.

Hinds, W.T. 1984. Towards monitoring of long-term trends in terrestrial ecosystems. *Environmental Conservation* 11:11–18.

James, F.C., C.E. McCulloch, and D.A. Wiedenfeld. 1996. New approaches to the analysis of population trends in land birds. *Ecology* 77:13–27.

Jones, K.B. 1986. Data types. Pages 11–28 in *Inventorying and Monitoring Wildlife Habitat.* A.Y. Cooperrider, R.J. Boyd, and H.R. Stuart (eds.). USDI Bureau of Land Management, Denver, CO.

Krebs, C.J. 1989. *Ecological Methodology.* Harper Collins, New York. 654 pp.

McComb, B.C. 2007. *Wildlife Habitat Management: Concepts and Applications in Forestry.* Taylor & Francis, CRC Press, Boca Raton, FL. 319 pp.

McShea, W.J. 2000. The influence of acorn crops on annual variation in rodent and bird populations. *Ecology* 81:228–238.

Newton, I. 1998. *Population Limitations in Birds*. Academic Press, New York.

Noss, R.F. 1990. Indicators for monitoring biodiversity: A hierarchal approach. *Conservation Biology* 4:355–364.

Ricker, W.E. 1975. Computation and interpretation of biological statistics of fish populations. Bulletin of the Fisheries Research Board of Canada. No. 191. 382 pp.

Roesch, F.A., Jr. 1993. Adaptive cluster sampling for forest inventories. *Forest Science* 39:655–669.

Sauer, J.R., J.E. Hines, and J. Fallon. 2007. The North American Breeding Bird Survey, Results and Analysis 1966–2006. Version 10.13.2007. USGS Patuxent Wildlife Research Center, Laurel, MD.

Sauer, J.R., J.D. Nichols, J.E. Hines, T. Boulinier, C.H. Flather, and W.L. Kendall. 2001. Regional patterns in proportion of bird species detected in the North American Breeding Bird Survey. Pages 293–296 in Wildlife, land, and people: Priorities for the 21st century. R. Field, R.J. Warren, H. Okarma, and P.R. Sievert (eds.). *Proceedings of the 2nd International Wildlife Congress*. The Wildlife Society, Bethesda, MD.

Sauer, J.R., B.G. Peterjohn, and W.A. Link. 1994. Observer differences in the North American Breeding Bird Survey. *Auk* 111:50–62.

Seber, G.A.F. 1982. *Estimation of Animal Abundance and Related Parameters*. Oxford University Press, New York. 654 pp.

Thomas, L. 1996. Monitoring long-term population change: Why are there so many analysis methods? *Ecology* 77:49–58.

Thompson, S.K. 1990. Adaptive cluster sampling. *Journal of the American Statistical Association* 85:1050–1059.

Thompson, S.K. 1992. *Sampling*. John Wiley & Sons, New York. 343 pp.

Thompson, S.K., and F.L. Ramsey. 1983. Adaptive sampling of animal populations. Technical Report 82. Dept. of Statistics, Oregon State University, Corvallis, OR.

Thompson, S.K., and G.A.F. Seber. 1996. *Adaptive Sampling*. John Wiley & Sons, New York. 270 pp.

Vesely, D.G., and J. Stamp. 2001. Report on surveys for stream-dwelling amphibians conducted on the Elliott State Forest, June–September 2001. Report prepared for the Oregon Department of Forestry. Pacific Wildlife Research, Corvallis, OR.

Wald, A. 1947. *Sequential Analysis*. John Wiley & Sons, New York. 230 pp.

Zacks, S. 1969. Bayes sequential designs of fixed size samples from finite populations. *Journal of the American Statistical Association* 64:1342–1349.

# 7 Putting Monitoring to Work on the Ground

To accurately monitor changes in habitat and populations, long-term monitoring programs must utilize a repeatable sampling scheme. Making inferences about a population from a sample taken over time depends on the ability of a monitoring plan to guide personnel to resample and revisit the same set of sample units (often points or plots) from one time period to the next in a standardized manner. Documentation of sample units throughout the study area and details for replicating sampling techniques are therefore critical. It is vitally important that there is proper documentation and storage of site locations and all notes pertinent to the sampling scheme for ensuring the long-term and repeatable success of monitoring plans. In this chapter we outline the steps that should be considered when establishing a monitoring scheme to ensure that it is consistently implemented over time.

## CREATING A STANDARDIZED SAMPLING SCHEME

### SELECTING SAMPLING UNITS

Monitoring plans should clearly identify the methods necessary for biologists to translate the conceptual sampling design into field practices that can be implemented even under challenging conditions. Methods for randomly selecting the locations or individual organisms to be included in the sample should be described in detail. Criteria or rules for establishing boundaries of the sampling area (extent) to match the spatial scale specified in the planning and design section should be identified. Also the procedures to stratify the sampling effort or to exclude certain regions (e.g., areas that are not habitat for species of concern) within the sampling frame should be described if these steps are required by the design.

### SIZE AND SHAPE OF SAMPLING UNITS

The size and shape of the sampling units have both logistical and statistical implications. Counting individual plants on plots delineated by a sampling frame was among the earliest approaches used by ecologists to estimate the frequency or density of plant populations. The technique may also be used for sedentary animals (e.g., mollusks, terrestrial salamanders). But researchers realized that count data obtained from plots are affected by the size and shape of the sampling unit. The optimum size and shape of a plot will differ according to the species, environmental conditions, and monitoring program objectives. Typically, the optimum plot configuration will be one that provides the greatest statistical precision (i.e., lowest standard error) for a given area sampled. Several investigators have developed approaches for determining the most appropriate plot size and shape for a particular population monitoring program (e.g., Hendricks 1956; Wiegert 1962). Krebs (1989) provides a useful review of standardized plot methods.

There are, however, many important general concepts to consider when determining plot size and shape. Square plots, also known as "quadrats," and circular plots have smaller boundary-to-interior ratios compared to rectangular shapes of equal area. Plots having an exaggerated length are sometimes referred to as "strip transects" or "belt transects." In some sampling conditions, it may be difficult for the surveyor to determine whether organisms occurring near the boundary of a plot are "in" or "out" of the plot and counting errors can result. In these circumstances, compact plot shapes are preferred. Boundary-to-interior ratios also decrease with increasing plot size, thus larger plots seemingly offer another approach for reducing counting errors. However, the tedious nature of counting organisms on a large plot under difficult field conditions may also cause surveyors to make mistakes. Counting errors are not the only factor to consider when determining plot size and shape. In heterogeneous habitats, count or abundance data collected on long plots often have been found to have lower statistical variance among plots than data from compact plots of the same total area (Krebs 1989).

While the usual notion of a plot is an area delineated by a frame or flagging, other techniques may be used to count organisms in a given area. Line transect and point transect sampling are specialized plot methods in which a search for the target organism is conducted along a narrow strip having a known area. See Chapter 8 for a more thorough discussion of application of transect sampling.

Population abundance also can be estimated by a variety of "plotless" monitoring methods that utilize measurements to describe the spacing of individuals in an area. These techniques are based on the assumption that the number of individuals in a population can be determined by measuring the average distance among individuals in the population or between individuals and randomly selected points. Distance methods have been commonly used for vegetation surveys, and are easily adapted to inventories of rare plants or other sessile organisms. The approach may also be useful for population studies of more mobile animal species by obtaining abundance estimates of their nests, dens, roosting sites, or scat piles. Indeed, sometimes the optimal option when considering your plot size and shape is to have no plots at all! Collecting data for monitoring programs based on distance methods may have some practical advantages over plots or transects:

1. Distance methods are not susceptible to counting errors that often occur near plot boundaries, thus may yield more accurate abundance estimates.
2. The time and effort to attain an adequate sample of distance measurements in an area often is less than that required to search for every target organism on a plot, thus increasing the efficiency of the monitoring program.

Field techniques vary depending upon the distance method selected for the monitoring program. All distance methods employ random selection procedures to choose points and compass bearings. Equipment requirements are minimal, usually only a compass, flagging, and a measuring device appropriate to the scale of the population and monitoring area are needed, but a detailed list of equipment used and descriptions of the protocol for using it must still be documented. Cottam and Curtis (1956) recommended a minimum of 20 measurements for estimating population density or abundance using the Point-Centered Quarter Method. However, data collection plans may be relatively complicated and sample size calculations may need to be performed in the field. To keep this from translating into biased results, a rigorous training program is recommended for personnel conducting the monitoring program. Two useful references for designing inventories based on distance methods are Seber (1982) and Bonham (1989).

## SELECTION OF SAMPLE SITES

A proper sampling design ensures that samples taken for a monitoring plan are representative of the population under study, and that any conclusions that are reached can be inferred and extrapolated to other areas and populations. The goal of the sampling design is to maximize efficiency by providing the best statistical estimates with the smallest amount of variance at the lowest cost (Krebs 1989). The sampling methods described below are three of the most popular methods for selecting sample sites when developing monitoring plans.

### SIMPLE RANDOM SAMPLING

Simple random sampling occurs when a random subset of sampling units are selected as samples from a population in such a way that every unit has an equal chance of being chosen (Krebs 1989). For example, a set of randomly located points is placed throughout a study area. At each point, information is collected on some aspect of the species of interest, such as its reproductive success. The data collected from this randomly generated set of points can be considered representative of the population within the study area. However, users often need large sample sizes with this approach because sampling schemes tend to be spatially unbalanced, and there is no attempt to reduce the effect of variability on estimates (Fancy 2000). Consequently, simple random sampling is generally not appropriate for large-scale monitoring because it is cost ineffective.

Nonetheless, randomization is essential in reducing bias and estimating the parameters of a population. In cases where it is important to have an adequate sample size from a limited area (e.g., watershed), samples can be distributed using a grid, cell design, or tessellation procedure (Stevens 1997). Most statistical analyses assume that sampling units were collected in a random fashion to reduce bias and maintain independence among samples (Krebs 1989). Although independence in ecological settings is difficult to ensure, randomization is an important aspect of any sampling design.

### SYSTEMATIC SAMPLING

Systematic sampling allows for simple and uniform sampling across an area, and is often conducted using a line transect or belt transect procedure. For example,

**FIGURE 7.1** Systematic sampling scheme using a random starting place and random compass bearing to establish a transect with systematically arranged sample points along the transect.

biologists often use point counts along preestablished transects to determine avian occupancy, density, and community composition across a habitat type. In this case, the starting point of the transect must be random, and point counts are placed at equal distances along the length of the transect (Figure 7.1).

Another common procedure is centric systematic area sampling. The study area is subdivided into equal squares and a sampling unit is taken from the center of each square (Krebs 1989). Systematic sampling provides an even coverage of the study area and is relatively cost efficient. However, if there is an environmental gradient (e.g., a moisture gradient, roads, fences, etc.) that happens to align with the orientation of a transect or grid, then estimates from this technique can be biased. In general, systematic sampling remains a popular method in monitoring programs and studies due to its ease of implementation and efficiency and is useful for sampling ecological data, but users must be aware of the potential for bias in some settings.

## STRATIFIED RANDOM SAMPLING

Stratified random sampling is a powerful technique for collecting reliable ecological data. This method consists of separating the population into subpopulations (strata) that do not overlap and are representative of an entire population (Krebs 1989). Strata are constructed based on criteria such as population density, habitat features, habitat quality, home range, or topography. The decision is often made using prior knowledge of the sampling situation in different areas. In some cases, however, what constitutes

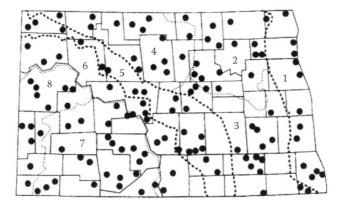

**FIGURE 7.2** Biotic regions of North Dakota and random quarter section sample units. Dotted lines designate biotic regions. (From Nelms, C.O., W.J. Bleier, D.L. Otis, and G.M. Linz. 1994. *American Midland Naturalist* 132:256–263. With permission.)

a stratum is not as clear, and it may be necessary to use preliminary data as a statistical basis for strata delineation. Iachan (1985) discussed several approaches for deciding strata boundaries. Once the strata are delineated, each stratum is sampled separately, and then samples are chosen randomly within the strata. In 1967, Stewart and Kantrud (1972) estimated populations of breeding birds in North Dakota by dividing the state into eight strata based on biotic characteristics (Figure 7.2). These biotic regions were relatively homogenous, and the number of sampling units in each stratum was proportional to the area of the biotic region. Sample units were selected randomly from each region. This design proved useful for other projects in the same area (Nelms et al. 1994). Stratified random sampling is favored in ecological studies for the following reasons (Cochran 1977):

1. Separate estimates for means and confidence intervals can be derived for each subpopulation allowing for comparisons among strata.
2. Cost per observation is typically reduced.
3. If strata are chosen well, then confidence intervals may be narrowed appreciably. This allows for a greater precision and confidence in the parameter estimates for the whole population.
4. Stratification may be administratively convenient if different organizational units are responsible for different parts of the sampling.

Although stratification represents an excellent sampling method for many monitoring plans, users should be aware of some of the inherent problems in creating strata. Recently, it has been suggested that designs based on the stratification of areas by "habitat types" derived from vegetation maps are not recommended (Fancy 2000). The reasoning is that habitat-type boundaries, especially those based on vegetation, can change over time, and this will cause problems for future sampling since strata are permanent classifications. A stratum is an area for the purpose of distributing a sample, and any changes will bias the sampling design. These changes in strata boundaries restrict the ability of biologists and managers to include new information in the sampling framework. Consequently, it is considered more appropriate to delineate special areas of interest based on physical features (e.g., terrain, geology, soil, topography, or ecoregions). However, many monitoring plans continue to base strata selection on habitat types due to their influence on animal abundance and distribution.

Yet this does not mean that we can't draw lessons from past and current monitoring designs. Monitoring agencies have different philosophies on sampling designs and framework. The National Park Service advocates sampling designs that emphasize areas of special interest, such as rare or declining habitat types (Fancy 2000). In many of their sampling designs, areas of special interest are sampled with higher frequencies using either stratification or a general approach of defining the units within the areas of interest and varying their selection probabilities (unequal probability approach). The USDA Forest Inventory and Analysis program's base grid is randomly located, and users can use this grid as a means of initial sampling. The North American Amphibian Monitoring Program uses a stratified random block design (Weir et al. 2005). Overall, although many monitoring programs differ in the specifics of their sampling framework, most do emphasize critical considerations such as random sampling, a grid or system that allows for the initial sampling of all areas, and documenting limitations to their sampling designs.

## LOGISTICS

Careful attention to the personnel, equipment, and permissions that are required for fieldwork is crucial. Incomplete data sets, unreliable measurements, and budget overruns are the common consequences of poorly implemented monitoring programs. Failure to prepare adequate safety plans for fieldwork may even jeopardize the welfare of the survey crew. The following are several steps that can help increase the logistical viability of a monitoring program:

1. Ensure consistent implementation of the monitoring plan among different units within an agency or among agencies conducting coordinated monitoring programs.
2. Maintain the scientific credibility of the project by standardizing materials and methods used during data collection and analysis, thus facilitating independent review and replication of the monitoring program.
3. Support the development of annual budgets and operation plans that biologists or contractors will be required to submit for the monitoring program to their unit supervisor.

### SAFETY PLAN

All federal and state and most industry organizations are required to prepare a safety plan. Gochfeld et al. (2006) provide an example of development of a health and safety plan for marine work, but the ideas and approaches are easily adaptable to other circumstances. Some biological monitoring programs may involve special procedures for hazards affecting the safety and welfare of personnel that are not typically covered in standard safety plans for natural resources organizations. Examples include:

1. Exposure to animal-borne diseases (e.g., rabies, hantavirus)
2. Risk of injury from handling wild animals (e.g., capturing large carnivores)
3. Risk of injury from special equipment or materials (e.g., electroshocker)
4. Hazardous activities (e.g., tree climbing, spelunking)

The U.S. National Center for Infectious Diseases provides fact sheets for many diseases for which biological technicians may be at risk (NCID 2003). Similar resources are available for a number of other safety considerations and should be sought out when applicable.

## Resources Needed

Proper equipment and adequate supplies are essential to the operation of monitoring programs. Your monitoring plan should therefore include a list of equipment and supplies needed for implementation. The list should provide a detailed description (including manufacturer and model number to allow future replacements), minimum functional specifications, and suppliers for specialized instruments and materials used in data collection and analysis. Any specialized laboratory or storage facilities that are required should also be described. Biologists in charge of implementing monitoring programs should include a checklist of equipment and supplies required for fieldwork in the operation plans and to survey crews. Monitoring plan developers should also identify special software packages necessary for performing data analysis or laboratory procedures. Data collected by different units within an agency conducting the same program may not be comparable if equipment and supplies are widely inconsistent. Developers of monitoring plans can minimize this possibility by ensuring the standardization of monitoring program materials.

## Permits

Obtaining the proper permits can be integral to the implementation or continuation of a monitoring program. Most states require surveyors to possess scientific collecting permits for studies involving the removal of rare plants or capturing native wildlife. Most studies of federally protected species such as migratory birds, endangered species, or CITES species also require obtaining specific permits from the U.S. Fish and Wildlife Service. Permits must likewise be acquired to use controlled substances or materials such as prohibited trap types (leg-hold traps in some states) and certain immobilizing agents (e.g., ketamine) and the syringes, dart guns, or other equipment used to administer the agents. When using radio transmitters, approved frequencies must be used as based on Federal Communications Commission regulations. Anytime that monitoring is conducted on private lands, written permission allowing entry should be acquired from the landowner. Failure to receive permission from a landowner can lead to dropping a plot from the sampling scheme in addition to creating ill will toward personnel involved in field sampling. A replacement technique that ensures randomization of plot locations should be developed. For instance, if a list of random plot locations is developed and permission is refused for one of the selected plot locations, then the next random plot in the list for that stratum should be selected. Nonetheless, if the reasons for refusing access rights are in some way related to the species of interest (a landowner does not want you to know that the species occurs on her/his land), then there is the potential for bias to creep into your sample.

## BIOLOGICAL STUDY ETHICS

If the monitoring is conducted in conjunction with a university, then the monitoring plan will need to be approved by the Institutional Animal Care and Use Committee (IACUC). Having an IACUC approval process that is well documented is required by universities but also should be used in other organizations as well. Capture, marking, and observation techniques may cause the subject animals to experience pain, permanent injuries, and increased mortality rates. In fact, some animal inventories and monitoring studies depend on lethal traps for the collection of voucher specimens or population data. The justification for such studies must balance the benefits of knowledge to be acquired with the welfare of individual animals and populations subjected to study methods. Most wildlife, fisheries, and zoological professional societies have adopted guidelines to assist field biologists in minimizing the adverse impacts on individual animals and populations (e.g., AFS, ASIH, and AIFRE 1987; Gaunt et al. 1997; American Society of Mammalogists 1998). Supervisors and field personnel should receive animal use training (required by most university scientists) and also become familiar with the standards for use of animals in field studies. In most cases, investigators must have explored alternative options to animal capture and handling and be prepared to justify any proposals to capture, restrain, harm, or kill an animal. The monitoring plan should ensure that methods used explicitly consider these standards.

## VOUCHER SPECIMENS

Techniques for the preparation of pressed plants for archiving in a herbarium, or preparation of animal skins and skulls for archiving in a museum, are provided in Anderson (1965) and Carter et al. (2007). The methods used to handle, prepare, and store plant or animal specimens collected in the field should be described. If laboratory analyses are required for the monitoring plan, the facility where the analyses will be conducted should be identified along with appropriate shipping methods. The museum or university collection that will ultimately house voucher specimens should also be identified. A lack of such considerations can physically fragment data and present obstacles to future analyses of specimens (e.g., subsequent genetic analysis).

## SCHEDULE AND COORDINATION PLAN

The schedule for sampling on a daily, weekly, monthly, and annual basis should be well documented to ensure consistency over the tenure of the monitoring plan. This schedule should include the major logistical activities, data collection, and fieldwork periods, and timing of analytical procedures. This will facilitate the estimation of reporting deadlines. It is particularly crucial that the monitoring plan describe factors that influence the appropriate season for conducting fieldwork. Examples of these factors include climatic or weather conditions, latitude or elevation, and animal activity patterns such as those relating to reproduction. The plan should also describe procedures that require coordination with other agencies and monitoring programs. Examples include the establishment of permanent monitoring plot monuments, determining radio frequencies to be used for monitoring program communication, and contracting arrangements. The deadlines for acquiring the necessary permits should likewise be noted.

## QUALIFICATIONS FOR PERSONNEL

One of the most important considerations in planning a monitoring program is to ensure that data collection and analytical procedures are performed by trained staff under the supervision of qualified biologists. Technological tools such as electronic data loggers improve the efficiency of field technicians, but these tools cannot compensate for the shortcomings of inexperienced or poorly trained personnel. The monitoring plan should specify the minimum qualifications and responsibilities of biologists, crew leaders, and crew members involved in conducting the monitoring program to ensure reliable and efficient data collection. Establishing written qualifications for personnel is particularly important for multiyear monitoring studies during which there is likely to be a significant amount of turnover among the monitoring program participants.

## SAMPLING UNIT MARKING AND MONUMENTS

Careful consideration should be given to selecting methods to mark sampling units and to install monuments at permanent plots. There are numerous marking systems available, and the final decision should be based on an assessment of the vulnerability of sampling units to vandalism and natural disturbance and the particular strengths and weaknesses of each system.

Plastic flagging and pin flags are inexpensive and suitable for temporary marking. However, plastic deteriorates in sunlight and may attract browsing by deer or cattle. Many large landowners reserve certain flagging colors for specific types of management activities; therefore, flagging guidelines should be reviewed prior to marking sampling sites. Polyvinylchloride (PVC) pipe and wood stakes are vulnerable to vandalism and can be lost during wildfire or floods. These materials should be used as short-term (<1 year) markers only and should not be used at all near locations frequently visited by humans or livestock.

Steel reinforcing rod (rebar) and T-posts are more durable materials for monumenting sampling units. They can be driven deeply into the ground to prevent all but the most determined vandals from removing them. Steel rebar and T-posts can also be relocated with a metal detector if they are buried during a flood or land management activities. If vandalism is not likely at the site, high-visibility paint or flagging can be used to make posts more detectable by field crews. Rebar can be fitted with a commercially available plastic cap or bent into a loop if the protruding end presents a hazard to cattle or humans. Electric or manually powered rock or masonry drills (available at rock climbing equipment retailers or masonry suppliers) can be used to insert a bolted marker into rock substrate when bedrock or boulders prevent the use of posts or stakes.

Trees can be marked with a paint ring to indicate a sampling point or as an aid for relocating an individual sample tree. Periodic repainting will be required at long-term monitoring sites. As with plastic flagging, tree marking guidelines vary by landowner and will be inappropriate on some sites. Trees can be provided a unique identifier by fastening two numbered tags (available at forestry supply companies) to the bole; one at breast height (1.5 m, 4.5 ft) and another at the base of the tree. Tags should be fastened with an aluminum nail to prevent rusting and loss of the tag. Two tags should be used in case the tree is harvested, then the basal tag will remain. Nails should be angled slightly downward and have a minimum of 7 cm (3 inches) of the shaft protruding from the tree to prevent diameter growth from damaging the tag. At permanent tree inventory plots, tree tags are customarily nailed to the aspect of the bole facing the plot center.

Whatever types of visual markers are used in the field, sampling unit locations should be recorded on a topographic map and with distances and compass bearings to permanent landmarks. Locations are typically determined using global positioning systems (GPS), geographic information systems (GIS), and other computer-based systems for data analysis, storage, and retrieval. GPS is a valuable tool for conducting field surveys and relocating sample sites. Using GPS, users can geo-reference sample site locations using latitude, longitude, or universal transverse mercator (UTM) coordinates. Along with GPS locations, careful notes taken on the location and description of sample sites can aid in relocation at future times. These notes should be reported in a consistent manner and held in office records. The details that must be recorded to ensure that sampling techniques can be replicated include equipment used, permits obtained, schedule for sampling, etc. Data management and documentation remains a critical component of project supervision.

However, it should not be assumed that the crew will always be able to relocate the sampling units using a GPS. Rugged terrain or dense canopy cover may prevent reception of the satellite signal, or the GPS may fail to operate once in the field. At long-term monitoring sites, landmark references should be recorded for every sampling unit. For short-term surveys and sites where sampling units are uniformly positioned along a transect or grid, landmark references only need to be recorded to the first unit. The crew can then rely on the site map and standardized spacing distances to navigate among sampling units at the site. Even this can take time, though—especially if sufficient time has elapsed between samples to allow vegetation to obscure permanent markers. Maps, UTM coordinates, and landmark references for sampling unit locations should be included as an appendix in the project report.

A number of other practices are helpful in maintaining consistent sampling units. For instance, the dimensions of plots, transects, or other sampling units should be described. Efficient techniques for positioning and measuring sampling units under field conditions should be identified. Providing a diagram or map to indicate the spacing and configuration of sampling units is also useful. For long-term monitoring projects, recommendations for marking and establishing monuments that are resistant to natural disturbances and vandalism should be provided. Elzinga et al. (2001) provided an excellent review of such techniques.

Putting Monitoring to Work on the Ground    125

## DOCUMENTING FIELD MONITORING PLANS

A comprehensive description of field methods for sampling the target species or habitat element is particularly important. The monitoring plan should address all of the following issues relevant to the population or habitat element being sampled:

1. Observational or capture techniques: Include a description of the equipment used and the rationale for the equipment chosen, pointing out the advantages and disadvantages of the chosen method with regards to precision and repeatability of the technique. Subtle details of techniques such as guidelines for trap placement, binoculars used, mensuration equipment used, and weather conditions may be very useful in reducing interannual variability in estimates.
2. Timing of sampling in terms of days, months, and years: Explain how the temporal framework for sampling interfaces with periods of activity, movement, growth, or reproduction for the species of interest or how it is associated with the function of the habitat element of interest. Point out the advantages and disadvantages of the proposed timing with regards to the precision and repeatability of estimating the index of interest.
3. Duration of sampling: Include a written explanation of why and how the sampling effort is adequate to develop a precise estimate over a period of time meaningful to the population of concern. Consider providing data from the literature or a pilot study that make it clear that the additional sampling effort beyond what is proposed is unlikely to add additional information (e.g., species detection curves).
4. Frequency of sampling: Document the periodicity of sampling that reflects the likelihood that the index to the population or habitat element is as precise as possible and that bias due to sampling at times when individuals are more or less detectable has been minimized. The advantages and disadvantages of the proposed sampling frequency should be pointed out relative to achieving precision and repeatability within the context of the annual and intergenerational changes in the populations. For instance, highly dynamic populations may need more frequent sampling than populations that are relatively stable and change slowly over time.
5. Data collection: Document exactly how data are to be collected, including reference to the significant digits with which data are recorded. Clearly state the taxonomic level expected, the degree of precision of the measurement, and the specific techniques used to acquire the datum.
6. Plant or animal marking techniques must be considered carefully. Any marking technique that introduces bias into estimates of survival or reproduction can lead to highly unreliable monitoring information, thus they must be considered carefully. References to support methods used to sample and mark plants and animals should be provided. In the case of radio transmitters, make it clear that transmitter mass should not exceed specific guidelines provided in the literature. Bands, ear tags, passive integrated transponder (PIT) tags, and other markers must not unduly modify

the organism's mobility, survival, reproductive potential, or other functions, which may result in an unreliable estimate of demographic parameters.
7. Use of equipment and materials should be precisely described. It is better to provide too much detail than too little regarding how equipment should be used, maintained, and stored.

The monitoring plan should receive peer review and be tested thoroughly before being implemented.

## QUALITY CONTROL AND QUALITY ASSURANCE

Data collection tasks that are vulnerable to observer error should include both a training program and compliance monitoring to ensure that the data are collected as accurately and precisely as possible. Data verification tests are highly recommended. For example, resurveys could be conducted by an independent examiner on a subset of sampling units to measure error rates. Monitoring plans should also establish criteria for acceptable levels of observer error, and describe remedial measures when data accuracy is not acceptable.

## CRITICAL AREAS FOR STANDARDIZATION

Projects designed to identify population distribution or abundance patterns across space and time must control for measurement biases and observer variability that could confound interpretation of the data analysis. This process is referred to as standardization and consists of those aspects of the data collection monitoring plan intended to ensure that observations and measurements are conducted using identical methods and under the same conditions across all sampling units included in the monitoring program. The primary objective of standardization is to make certain that the field techniques employed do not cause detection probabilities to vary among sampling units. Standardization is particularly critical for monitoring plans that utilize relative abundance indices, that is, counts of individuals per unit of time or effort. If the techniques do influence detection probabilities, then it cannot be assumed that counts or other field measurements accurately represent variation in the population parameter of interest (e.g., population size or population density). Formal parameter estimation procedures (e.g., sighting-probability models) utilize different pilot studies to permit analysis of detection probabilities according to age class, habitat, monitoring program personnel, and other factors. These different detection probabilities are then incorporated into the calculation of the estimator, instead of assuming that detection probabilities are fixed. Monitoring plans should consider the following factors when identifying critical areas of standardization.

### SEASON AND ELEVATION

Most plants occurring in temperate climates demonstrate a predictable pattern of growth, reproduction, and senescence in response to changing environmental conditions throughout the year along moisture gradients and elevational gradients.

Animals also often exhibit seasonal periodicity in habitat use and behavior patterns. These biological responses to the environment clearly influence detection probabilities in association with location, but may also strongly alter detection probabilities across different months of the year. Birds are an excellent example. Territorial male passerines sing regularly when establishing a territory and attracting a mate. During incubation and after hatching, however, singing drops off and indeed adults feeding young are not only quiet but cryptic. And when fledglings emerge from the nest then they often call but do not sing. Hence, sampling during the beginning, middle, and end of the nesting cycle can produce quite different estimates of abundance. Therefore, a monitoring plan should specify the appropriate season for sampling. If the monitoring plan is designed to be applied across a wide geographic range, it must indicate how sampling periods are to vary by region, latitude, or elevation.

### DIURNAL VARIABILITY

Activity patterns of wildlife and fish species often change in a relatively predictable manner throughout the 24-hour day. Using our example of bird sampling, males sing on territories most aggressively early in the morning, and the activity subsides during midday. On the other hand, some observational and capture techniques are insensitive to diurnal patterns of behavior because they rely on a trap or other device that is always ready to capture an animal or record a detection (e.g., screw traps for migratory fish, track plates for medium-sized carnivores). However, many field techniques are designed toward certain target behaviors that increase the detectability of some individuals in a population or their vulnerability to capture at particular periods throughout the day. Biologists developing monitoring plans should assess the significance of diurnal activity patterns on detection probabilities and, if necessary, specify the appropriate time of day to sample the target population.

### CLOTHING OBSERVERS WEAR WHILE MONITORING

Simply considering the clothing worn during field sampling can have an effect on the observations of some species. Standardizing field appearance and behaviors can be quite important. For instance, when sampling breeding birds, wearing red or orange can cause hummingbirds to be attracted to the observer, thereby biasing the estimates of abundance of this species. Wearing clothing that blends into the background to the degree possible may minimize such biases.

## BUDGETS

After a monitoring program is designed, a budget request must be made to ensure that the plan can be implemented as designed. Typically, this becomes an iterative process because the budget request for monitoring to achieve a certain statistical power to detect a given trend over time may exceed the capacity of the funding organization. The inability to procure adequate funding for a monitoring program often results in concessions being made to the effect size that can be detected by reducing the sample size, or changing the confidence level with which a trend can be detected.

Program MONITOR (Gibbs and Ramirez de Arellano 2007) provided a tool for assessing the costs associated with each sampling plot while simultaneously estimating the power associated with detecting a trend given a certain sample size. Hence, the implications of modifying the design on the funding levels needed to support the design can be explored prior to initiating the monitoring program. If funds are so limiting as to not provide an acceptable level of power, then changes can be made in the response variables, sampling techniques, or other design factors that may allow a revised form of the monitoring plan to be implemented. In a worst-case scenario, if the funding is simply not sufficient to provide an acceptable level of power to detect a trend, the monitoring program will likely have to be abandoned.

Budget planning is often broken into two components: fixed costs and variable costs. Fixed costs are costs which are not affected by the extent of the monitoring effort, and are typically those associated with equipment that is needed during the first year of the program, such as vehicles and capture or recording equipment. These items are necessary to purchase to begin the program, but can be used in subsequent years. It is important to estimate the life of this equipment so that there are funds for replacements as certain pieces cease to work adequately.

Variable costs are those that vary with the sampling intensity. Cumulative personnel expenses (salaries, benefits, and indirect costs) are often the largest portion of the budget, and these vary, depending on the number of sampling locations measured. In addition, it is important to calculate increased personnel costs over time due to raises and increased costs associated with benefits. Supplies are expendable items such as bait and flagging that must be replaced each year, so they also represent variable costs.

The greatest challenge to building and requesting a budget for a monitoring program is ensuring that funding will be available at the necessary time period (usually each year) over time. Given annual fluctuations in agency budgets, simply having the money allocated on a consistent basis becomes problematic. Most often, the program leader will have to make an effective argument regarding the importance of the information during each budget cycle. This should become easier over time as results are summarized and the value of the data increases. A related difficult decision for a program leader or an agency to make, however, is when to stop a monitoring program. The tendency is to try to keep it operational for as long as possible, but one must begin to ask what the return is on the investment. Monitoring programs that have been ongoing for years or decades can become institutionalized, and there can be considerable resistance to ending them. Yet this could come at the cost of taking funding from other more critical monitoring programs. It is therefore important to design monitoring programs with trigger points included in the monitoring plan (e.g., the population reaches a certain level of growth for a certain number of years) that clearly indicate the program's termination point. This will help facilitate the timely cessation of monitoring.

## SUMMARY

The logistics associated with implementing a monitoring plan may seem overwhelming but are integral to ensuring that the plan is implemented correctly and that the data will be useful. Adhering to random sampling monitoring plans, monumenting points, and using consistent plot designs from one time period to the next is critical. If stratification can reduce variance in estimates, then it should be used, but basing strata on vegetative characteristics should normally be avoided since vegetation structure and composition will likely change over time. Geographic or geologic features may be more useful and consistent strata.

Ensuring that personnel are trained and that the necessary state, federal, and institutional permits have been acquired takes considerable time. How and when these activities will be conducted should be included in planning documents. Where sampling occurs on private land, time must be allotted to determine landownership and contact land owners for access rights.

Monitoring plans should be developed and followed to ensure ethical treatment of animals, and standardization in timing, techniques, and locations over time. Personnel should be trained to follow these monitoring plans as closely as possible in order to reduce sampling error.

Finally, once the plan is designed and the costs associated with the logistics of the plan are estimated, a budget must be developed. Where budget constraints arise, then choices must be made regarding tradeoffs between data quality, response variables used, and statistical power to detect a trend. If these tradeoff analyses are not conducted before embarking on a monitoring program, then the end result could be data that are simply insufficient to detect a trend in the desired response variable, and funding and monitoring will likely end. On the other hand, collecting data consistently and estimating adequate power to detect trends has the potential to create a positive feedback loop in which acquiring funding for continuation of the program becomes easier over time as the value of the data increases. Including a trigger point in the monitoring plan that dictates when the monitoring should end is important, however, to avoid the monitoring program from becoming institutionalized and funded for its own sake.

## REFERENCES

American Fisheries Society (AFS), American Society of Ichthyologists and Herpetologists (ASIH), and American Institute of Fishery Research Biologists (AIFRE). 1987. Guidelines for the use of fishes in field research. Approved Sept 1987, Winston-Salem, NC. Published March–April 1988. *Fisheries* 13(2):16–23.

American Society of Mammalogists, Animal Care and Use Committee. 1998. Guidelines for the capture, handling, and care of mammals as approved by the American Society of Mammalogists. *Journal of Mammalogy* 79:1416–1431.

Anderson, R.M. 1965. Methods of collecting and preserving vertebrate animals. *National Museum of Canada Bulletin* 69:1–199.

Bonham, C.D. 1989. *Measurements for Terrestrial Vegetation*. John Wiley & Sons, New York. 338 pp.

Carter, R., C.T. Bryson, and S.J. Darbyshire. 2007. Preparation and use of voucher specimens for documenting research in weed science. *Weed Technology* 21:1101–1108.

Cochran, W.G. 1977. *Sampling Techniques*. 3rd ed. John Wiley & Sons, New York. 428 pp.

Cottam, G., and J.T. Curtis. 1956. The use of distance measures in phytosociological sampling. *Ecology* 37:451–460.

Elzinga, C.L., D.W. Salzer, J.W. Willoughby, and J.P. Gibbs. 2001. *Monitoring Plant and Animal Populations*. Blackwell Science, Malden, MA. 368 pp.

Fancy, S.G. 2000. Guidance for the design of sampling schemes for inventory and monitoring of biological resources in National Parks. Unpublished report dated March 24, 2000 from the National Park Service Inventory and Monitoring Program. 19 pp.

Gaunt, A.S., L.W. Oring, K.P. Able, D.W. Anderson, L.F. Baptista, J.C. Barlow, and J.C. Wingfield. 1997. *Guidelines to the Use of Wild Birds in Research*. A.S. Gaunt and L.W. Orning (eds.). Ornithological Council, Washington, D.C.

Gibbs, J.P., and P. Ramirez de Arellano. 2007. Program MONITOR: Estimating the Statistical Power of Ecological Monitoring Programs. Version 10.0. ESF State University of New York, Syracuse.

Gochfeld, M., C.D. Volz, J. Burger, S. Jewett, C.W. Powers, and B. Friedlander. 2006. Developing a health and safety plan for hazardous field work in remote areas. *Journal of Occupational and Environmental Hygiene* 3(12):671–683.

Hendricks, W.A. 1956. *The Mathematical Theory of Sampling*. Scarecrow Press, New Brunswick, NJ. 364 pp.

Iachan, R. 1985. Plane sampling. *Statistics and Probability Letters* 50:151–159.

Krebs, C.J. 1989. *Ecological Methodology*. Harper Collins, New York. 654 pp.

National Center for Infectious Diseases (NCID). 2003. NCID publications available online. U.S. Centers for Disease Control and Prevention, National Center for Infectious Diseases publications. http://www.cdc.gov/ncidod/publicat.htm

Nelms, C.O., W.J. Bleier, D.L. Otis, and G.M. Linz. 1994. Population estimates of breeding blackbirds in North Dakota, 1967, 1981–1982 and 1990. *American Midland Naturalist* 132:256–263.

Seber, G.A.F. 1982. *The Estimation of Animal Abundance and Related Parameters*. 2nd ed. Edward Arnold Publ., London. 676 pp.

Stevens, D.L., Jr. 1997. Variable density grid-based sampling designs for continuous spatial populations. *Environmetrics* 8:167–195.

Stewart, R.E., and H.A. Kantrud. 1972. Vegetation of prairie potholes, North Dakota, in relation to quality of water and other environmental factors. U.S. Geol. Surv. Prof. Pap. 585-D. 36 pp.

Weir, L.A., J.A. Royle, P. Nanjappa, and R.E. Jung. 2005. Modeling anuran detection and site occupancy on North American Amphibian Monitoring Program (NAAMP) routes in Maryland. *Journal of Herpetology* 39:627–639.

Wiegert, R.G. 1962. The selection of optimum quadrat size for sampling the standing crop of grasses and forbs. *Ecology* 43:125–129.

# 8 Field Techniques for Population Sampling and Estimation

Field techniques refer to the standardized methods employed to select, count, measure, capture, mark, and observe individuals sampled from the target population for the purpose of collecting data required to achieve study objectives. The term also includes methods used to collect voucher specimens, tissue samples, and habitat data. The choice of field techniques to use for a particular species or population is influenced by five major factors:

1. Data needed to achieve inventory and monitoring objectives
2. Spatial extent and duration of the project
3. Life history and population characteristics
4. Terrain and vegetation in the study area
5. Budget constraints

## DATA REQUIREMENTS

The types of data required to achieve inventory or monitoring objectives should be the primary consideration in selecting field techniques. Four categories of data collection are discussed below along with some suggestions for electing appropriate field techniques for each.

### OCCURRENCE AND DISTRIBUTION DATA

For some population studies, simply determining whether a species is present in an area is sufficient for conducting the planned data analysis. For example, biologists attempting to conserve a threatened salamander may need to monitor the extent of the species' range and degree of population fragmentation on a land ownership. One hypothetical approach is to map all streams in which the salamander is known to be present, as well as additional streams that may qualify as the habitat type for the species in the region. To monitor changes in salamander distribution, data collection could consist of a survey along randomly selected reaches in each of the streams to determine if at least one individual (or some alternative characteristic such as an egg mass) is present. Using only a list that includes the stream reach (i.e., the unique identifier), the survey year, and an occupancy indicator variable, a biologist could prepare a time series of maps displaying all of the streams by year and distinguish the subset

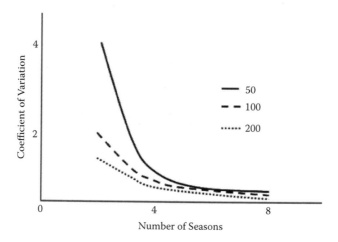

**FIGURE 8.1** Simulation-based coefficient of variation for estimated trend in occupancy (on the logistic scale) where 50, 100, or 200 landscapes are each surveyed 3 times per season, for multiple seasons. (Redrafted from MacKenzie, D.I., and J.A. Royle. 2005. *Journal of Applied Ecology* 42:1105–1114.) Estimates of occupancy can be facilitated by use of computer programs such as PRESENCE (MacKenzie et al. 2003).

of streams that were known to be occupied by the salamander. Such an approach could support a qualitative assessment of changes in the species distribution pattern, thereby attaining the program's objectives, and generate new hypotheses as to the cause of the observed changes.

It is far easier to determine if there is at least one individual of the target species on a sampling unit than it is to count all of the individuals. Determining with confidence that a species is not present on a sampling unit also requires more intensive sampling than collecting count or frequency data because it is so difficult to dismiss the possibility that an individual eluded detection. Probability of occurrence can be estimated using approaches such as those described by MacKenzie and Royale (2005). MacKenzie (2005) offered an excellent overview for managers of the trade-off between number of units sampled per year and the number of years (or other unit of time) for which the study is to be conducted. The variation in the estimated trend in occupancy decreases as the number of years of data collection increases (Figure 8.1). A similar level of precision can be achieved by surveying more units over fewer years versus surveying fewer units over a longer period.

## POPULATION SIZE AND DENSITY

National policy on threatened and endangered species is ultimately directed toward efforts to increase or maintain the total number of individuals of the species within their natural geographic range (Suckling and Taylor 2006). Total population size and effective population size (i.e., the number of breeding individuals in a population; Lande and Barrowclough 1987) most directly indicate the degree of species endangerment and effectiveness of conservation policies and practices. Population size or

# Field Techniques for Population Sampling and Estimation

more accurately density per unit area is usually used as the basis for trend analyses because changes in density integrate changes in natural mortality, exploitation, and habitat quality. In some circumstances, it may be feasible to conduct a census of all individuals of a particular species in an area to determine the population density. Typically, however, population size and density parameters are estimated using statistical analyses based on only a sample of population members. Population densities of plants and sessile animals can be estimated from counts taken on plots or data describing the spacing between individuals (i.e., distance methods) and are relatively straightforward. Population analyses for many animal species must account for animal response to capture or observation, observer biases, and different detection probabilities among subpopulations. Pilot studies are usually required to collect the data necessary to address these factors in the analysis. Furthermore, mark-recapture studies, catch-per-unit effort surveys, and other estimation methods require multiple visits to sampling units (Pradel 1996). These considerations increase the complexity and cost of studies designed for population parameter estimation.

## Abundance Indices

The goals and objectives of some biological inventories and monitoring studies can be met with indices of population density or abundance, rather than population estimators. The difference between estimators and indices is that the former yield absolute values of population density while the latter provide relative measures of density that can be used to compare indices to populations among places or times. Indices are founded on the assumption that index values are closely associated with values of a population parameter, although the precise relationship between the index and parameter usually is not quantified. Examples of abundance or density indices are plant canopy cover, numbers of individuals captured per 1,000 trap nights, and counts of individuals observed during a standardized unit of time, among many others. From a data collection perspective, density indices often require less sampling intensity and complexity than population estimation procedures. However, population indices are not comparable among different studies unless field techniques are strictly standardized. Furthermore, the assumption that an abundance index closely approximates population density is rarely tested (Seber 1982).

## Fitness Data

For rare or declining populations, estimates of survival in each life stage as well as reproductive rates are required. These data not only provide useful trigger points for estimating rates of decline (lambda) they also allow trigger points for removal of a species from a threatened or other legal status. Collecting these sorts of data is often labor intensive and expensive. In a study on northern spotted owls, for instance, millions of dollars have been spent collecting these types of data (Lint 2001). This is not particularly surprising as the types of data that would be necessary to understand the population dynamics of a bird are numerous and complicated to generate. Nest densities, clutch sizes, hatching rates, fledging rates, and survival rates to maturity and survival rates as reproductive adults would be a minimum data set. New approaches to estimating individual contributions to population growth and changes in distributions of quantitative traits and alleles include genetic analyses, which can lead to even more detailed understanding of the potential for a population to adapt to variations in environmental factors (Pelletier et al. 2009).

## Research Studies

Studies of habitat relationships or cause-and-effect responses require coordinated sampling of the target population and environmental measurements or stressors to which the population may respond. Data collection efforts tend to be complex, requiring multiple sampling protocols for the target population, study site attributes, and landscape pattern metrics. The funding required to conduct research studies typically limits their application to species or populations in greatest need of management planning, such as those listed as threatened or endangered. Manipulative studies are often carried out to generate the necessary data, but when these focus on a threatened species, ethical questions regarding the conduct of the experiment placing the species at even greater risk, at least locally, often emerge. Hence, it is often monitoring of both environmental conditions and aspects of population density or fitness that is used to assess associations in trends between population parameters and environmental parameters.

## SPATIAL EXTENT

Clearly the scope of inference will influence the type of sampling technique used. Breeding bird atlas techniques commonly use large grids placed over entire states to assess the occurrence of species in a grid cell. Such approaches and those of the Breeding Bird Survey (Sauer et al. 2008) can be conducted through volunteer efforts. On the other hand, monitoring the trends in reproductive rates of northern spotted owls, northern goshawks, or grizzly bears over their geographic ranges requires a huge budget to collect the level of population data over large areas needed to understand trends. Great care must be taken when deciding what technique to use because both budgets and sample size requirements enter into logistics. Indeed, it is often the trade-off between more detailed data and the cost of producing those data that drives decisions regarding monitoring designs for species at risk.

## FREQUENTLY USED TECHNIQUES FOR SAMPLING ANIMALS

The array of techniques available to sample animals is vast and summarized elsewhere in techniques manuals (e.g., Bookhout 1994). We summarize a few examples of commonly used techniques, but strongly suggest that those of you developing monitoring plans do a more complete literature search on sampling of the species that are of most concern in your monitoring program. We first provide a brief overview of techniques used to sample vertebrates and then point out which techniques are commonly used among various taxonomic groups.

# Field Techniques for Population Sampling and Estimation 135

## AQUATIC ORGANISMS

Some aquatic organisms can and have been monitored using techniques that are essentially identical to those used for terrestrial vertebrates. For instance, in Brazil, arapaima have been monitored using a point count technique that counts individuals as they surface for aerial breathing (Castello et al. 2009). Point counts were more logistically and economically feasible, were determined to more accurately represent population changes over time, and led to more effective management, but a conventional mark-recapture technique was also attempted with the same fish species (Castello et al. 2009).

Yet cases such as the arapaima are uncommon because this species is detected when surfacing for aerial breathing, has a low enough population density in a small enough area to be counted effectively, and possesses certain subtle visual and acoustic characteristics that allow for the identification of individuals (Castello et al. 2009). Most techniques used to sample aquatic organisms are conceptually similar to those used to sample terrestrial organisms. But constraints placed on observers of dealing with sampling in or on the water and at various water depths require that many techniques be more specialized. There are a variety of techniques commonly used to sample fish and aquatic amphibians as well as aquatic invertebrates (Slack et al. 1973). A systematic assessment of stream reaches using either snorkel surveys (Hankin and Reeves 1988) or electrofishing equipment is commonly used in shallow streams and rivers (Cunjak et al. 1988).

In estuaries and large rivers, quantitative studies are often confounded by the high variability of fish populations and the high efficiency of fish sampling gear (Poizat and Baran 1997). In light of this, Poizat and Baran (1997) undertook a study assessing the efficacy of surveying fishermen, compared with a scientist-managed gill-net sampling approach, and determined that combining both approaches is the best way to increase confidence that observed trends are real. In other words, if both sets of survey data suggest the same trend, it is safer to infer that the trends are real than if the data sets suggest different trends or there exists but one type of data.

Manta tows, which are comparable to line-transect methods but must account for uniquely marine conditions such as turbidity of water, tides, and sea-condition characteristics, are often utilized to monitor general characteristics of coral reefs and their associated populations. The technique, which has been employed in both scientist-run and community-based programs, consists of towing a snorkeler trained to observe certain variables behind a boat at constant speed along a predetermined stretch of reef (Bass and Miller 1996; Uychiaoco et al. 2005). In one study along the Great Barrier Reef, where manta tows have been employed since the 1970s, the sampled line is broken up into zones that take 2 minutes to sample, and every 2 minutes the boat stops for the observer to record data on an aquatic data sheet (Bass and Miller 1996). In these surveys, data often include counts of conspicuous species, such as giant clams, or of entire assemblages, such as carnivorous and herbivorous fish, but the technique is also used for monitoring habitat (Bass and Miller 1996; Uychiaoco et al. 2005). Indeed, observations of suites of variables designed to inform practitioners about the state of coral reefs over time—such as reef slope, dominant benthic form, dominant hard coral genus, and structural complexity of coral—are also commonly recorded (Bass and Miller 1996; Uychiaoco et al. 2005).

Welsh (1987) proposed a habitat-based approach for amphibians in small headwater streams, and time-constrained and area-constrained approaches have also been used for headwater species (Hossack et al. 2006). Pond-breeding species or species that inhabit deeper water are often sampled using minnow traps, nets, or call counts of vocalizing frogs and toads (Kolozsvary and Swihart 1999; Crouch and Paton 2002).

Tracking of individual animals through tags, passive integrated transponders (PIT tags), and similar techniques is expensive but provides information on animal movements and estimates of population size and survival. Such approaches have been used with species of high interest such as coho salmon in the Pacific Northwest (Wigington et al. 2006).

## TERRESTRIAL AND SEMI-AQUATIC ORGANISMS

The diversity of forms, sizes, and life histories among terrestrial vertebrates has led to the development of hundreds of field techniques designed for different species and survey conditions. Table 8.1 lists the most widely used field techniques for collecting wildlife data, but it is by no means an exhaustive list of all inventory and monitoring methods. Techniques are separated into observational, capture, and marking methods and by the mode by which data are collected. A comprehensive review of all the different field techniques for terrestrial and semiaquatic organisms is a separate book in itself (see Bookhout 1994). Here we provide a brief overview of some of the commonly used techniques.

For certain species and conditions, it may be feasible to determine a count of individual members of the population on quadrats (sample plots) randomly or systematically positioned in the study area. Searches can be conducted on foot, all-terrain vehicles, or airplane depending on the scale and circumstances of the survey. Quadrat sampling is commonly used for plants and habitat elements, but with animals quadrat sampling poses some challenges. If animals are mobile during the sampling period, then there needs to be some reasonable assurance that an individual is not double-counted at multiple quadrats as it moves. Size, spacing, and mobility of the organisms must all be considered.

Point counts are perhaps the most extensively used technique for measuring bird abundance and diversity in temperate forests and on rangelands but have also been used to estimate abundance of other diurnal species such as squirrels. Variations in the technique have been described for different species and to meet different data needs (Verner and Ritter 1985; Verner 1988; Ralph et al. 1995; Huff et al. 2000). Ralph et al. (1995) provided a collection of papers examining sample size adequacy, bird detectability, observer bias, and comparisons among techniques.

Spot mapping, also referred to as territory mapping, often is used to estimate avian population densities by locating singing males during a number of visits to the study area and delineating territory boundaries. The technique is further described in Ralph et al. (1995). Nest searches can be used to assess reproductive success in an avian population by monitoring the survival of eggs and nestlings over the course of

# TABLE 8.1
# Field Techniques for Inventory and Monitoring Studies of Terrestrial and Semiaquatic Vertebrates

| Data Type | Mode | Technique | Target Groups of Species | References |
|---|---|---|---|---|
| Observational | Direct | Quadrats; fixed-area plots | Sessile or relatively immobile organisms | Bonham 1989 |
| | | Avian point counts | Bird species that sing or call on territories | Ralph et al. 1995 |
| | | Spot mapping and nest searches | Territorial bird species | Ralph et al. 1993 |
| | | Line transect | Large mammals, birds | Anderson et al. 1979 |
| | | Call playback response | Wolves, ground squirrels, raptors, woodpeckers | Ogutu and Dublin 1998 |
| | | Standardized visual searches | Large herbivores, | Cook and Jacobson 1979 |
| | | Census | Cave-dwelling bats; large herbivores | Thomas and West 1989 |
| | Animal Sign | Foot track surveys | Medium-large mammals | Wilson and Delahay 2001 |
| | | Pellet & scat counts | Medium-large mammals | Fuller 1991 |
| | | Food cache searches | Large carnivores; beaver | Easter-Pilcher 1990 |
| | | Structures (e.g., dens, nests) | Arboreal mammals; fossorial mammals; bears | Healy and Welsh 1992 |
| | Remote Sensing | Track plates | Medium-large mammals | Wilson and Delahay 2001 |
| | | Photo and video stations | Medium-large mammals | Moruzzi et al. 2002 |
| | | Ultrasonic detectors | Bats | Thomas and West 1989 |
| | | Audio monitoring | Frogs | Crouch and Paton 2002 |
| | | Hair traps | Small-medium mammals, large carnivores | McDaniel et al. 2000 |
| | | Radio telemetry | Limited by animal body size | USGS 1997 |
| | | GPS telemetry | Limited by animal body | Girard et al. 2002 |
| | | Marine radar | Bats, migrating birds | Harmata et al. 1999 |
| | | Harmonic radar | Bats, amphibians, reptiles | Pellet et al. 2006 |

*continued*

## TABLE 8.1 (continued)
## Field Techniques for Inventory and Monitoring Studies of Terrestrial and Semiaquatic Vertebrates

| Data Type | Mode | Technique | Target Groups of Species | References |
|---|---|---|---|---|
| Capture | Passive | Pitfalls | Salamanders, lizards, small mammals | Mengak and Guynn 1987; Enge 2001 |
| | | Snap traps | Small mammals | Mengak and Guynn 1987 |
| | | Box traps | Small-medium mammals | Powell and Proulx 2003 |
| | | Funnel-type traps | Snakes, turtles | Enge 2001 |
| | | Leg-hold and snares | Large mammals | Bookhout 1994 |
| | | Mist nets | | Kuenzi and Morrison 1998 |
| | Active | Drives to an enclosure | Medium-large mammals with predictable flight response | deCalesta and Witmer 1990 |
| | | Cannon nets | Medium-large mammals | Bookhout 1994 |
| | | Immobilizing agents | Large mammals | Bookhout 1994 |
| | | Hand capture | Salamanders | Kolozsvary and Swihart 1999 |
| Marking | | Tags | Birds/mammals | Nietfeld et al. 1994 |
| | | Mutilation | Small mammals | Wood and Slade 1990 |
| | | Pigments | Small mammals | Lemen and Freeman 1985 |
| | | Collars and bands | Birds/mammals | Nietfeld et al. 1994 |

a breeding season. Both techniques are labor intensive and are not commonly used for inventories, but the information gained from these methods (i.e., territory densities, productivity) may be better indicators of population trends and habitat quality than simply counts of individuals.

Line transect and point transect sampling are specialized plot methods in which a search for the target organism is conducted along a narrow strip having a known area. Rarely can it be assumed that all animals are detected along the transect. However, if the probability of detection can be predicted from the distance between the animal and the centerline of the transect, then a detection function can be used to estimate population density. The approach can be adapted to surveys conducted by foot, snorkeling, and ground or air vehicles. Buckland et al. (1993) provided a complete, though highly technical, introduction to line transect and point transect methods. The approach has been widely applied to surveys of vertebrates, including desert tortoise (Anderson et al. 2001), marbled murrelets (Madsen et al. 1999),

# Field Techniques for Population Sampling and Estimation

**FIGURE 8.2** Ensatina salamander detected on a time-constrained sample in managed forests of the Oregon Cascades.

songbirds in oak–pine woodlands (Verner and Ritter 1985), and mule deer (White et al. 1989).

Audio recordings of animal vocalizations have been used to elicit calls and displays from species otherwise difficult to detect. The technique has been applied in studies of blue grouse (Stirling and Bendell 1966), northern spotted owls (Forsman 1988), ground squirrels (Lishak 1977), and others. The number of responses by the target species elicited by the recording is tallied during a prescribed interval and provides a population density index.

Standardized visual searches refer to techniques used to determine species occurrence, species richness, or relative density values, where sampling effort is standardized by space or time. Examples include road counts for large mammals (Rudran et al. 1996), raptor migration counts (Hussell 1981), and visual encounter surveys for terrestrial amphibians (Crump and Scott 1994) (Figure 8.2). Some visual search techniques do not necessarily equalize the amount of survey effort among sampling units. Instead, animal counts or species detected are standardized during analysis by dividing the number of observations by a unit of area or time. Variability among observers and environmental conditions may be significant sources of error unassociated with the sampling technique and should be assessed prior to data collection to minimize biases and improve precision.

Under certain circumstances, it may be possible to effectively observe all individuals in the target population. In such cases, population size can be determined directly from the count of individuals; no statistical procedures are necessary. Accurate counts on individuals depend upon a natural tendency of population members to aggregate, at least during predictable periods (e.g., cave-roosting bats). Furthermore, all locations where the individuals aggregate must be known in the study area, and there must be adequate surveyors available to make simultaneous counts at all locations.

In many cases, the target population is highly cryptic or too wary to be observed directly and budget constraints prevent the investigator from utilizing capture methods for data collection. In these situations, it is often possible to infer the presence of a species or determine an index value for population density by observing

**FIGURE 8.3** Virginia opossum attracted to a hair snare baited with apples. Baiting can attract animals and may increase the likelihood of detecting a species but may also give a biased index to abundance.

animal signs. Signs are tracks, scat piles, fecal pellets, scent-marking posts, or animal constructions (e.g., arboreal nests, beaver lodges, burrow openings) that can be accurately identified as evidence of a particular species. If searches for such evidence are conducted on standardized transects or quadrats, then observations may provide a reliable index to population density. Data analyses are similar to those for direct observations. Davis and Winstead (1980) and Wemmer et al. (1996) provide an overview of methods based on animal sign.

Elusive species can be sampled using remote sensing devices positioned across the study area. Track plates (Zielinski and Stauffer 1996) and hair traps (Scotts and Craig 1988; McDaniel et al. 2000) are inexpensive and suitable for determining the occurrence and distribution of rare mammals in the study area. Ultrasonic detectors can be used to monitor bat populations (Kunz et al. 1996); however, it is not always possible to reliably distinguish among all bat species. Remote camera stations with data-loggers (Cutler and Swann 1999; Moruzzi et al. 2002) not only detect occurrence of the species, but also may yield information about sex, age, and activity patterns of individuals. Baits at a track plate or camera station can be used to increase the probability of detecting a cryptic or rare species, but can also bias any estimates made or disproportionately attract common omnivorous species, such as Virginia opossum or northern raccoon (Figure 8.3).

Radio-telemetry has been used for many years to collect data on wildlife and fish movement, home range size, and habitat selection (Figure 8.4). Transmitters weighing <1.0 g are now commercial available, making it possible to track all but the smallest vertebrates. Tracking systems utilizing global positioning system (GPS) satellites permit monitoring of animal locations in real-time without requiring surveyors to

# Field Techniques for Population Sampling and Estimation

**FIGURE 8.4** Cheetah with a radio collar at Moremi Wildlife Reserve, Botswana. (Photo by Nancy McGarigal. Used with permission.)

determine radio signal directions in the field and are becoming more reasonably priced. A collection of abstracts on wildlife telemetry methods (USGS 1997) provides a useful introduction to the topic for terrestrial studies. Marine band radar has been used to count migratory birds at observation points (Harmata et al. 1999) and monitor activity patterns of marbled murrelets (Burger 2001).

Finally, genetic analyses of tissue collected from hair traps, scat, or other tissues have led to an explosion of approaches to assess populations, dispersal, and evolutionary patterns (Haig 1998; Mills et al. 2000).

## LIFE HISTORY AND POPULATION CHARACTERISTICS

Certain techniques are more commonly used with some taxonomic groups than others. In this section we provide guidance as to the types of techniques that you might consider depending on the species included in your monitoring program.

### AMPHIBIANS AND REPTILES

The small size, cryptic nature, and fossorial habits of salamanders make data collection particularly difficult. Many terrestrial amphibians move only short distances and are not susceptible to passive capture techniques. These species are usually sampled using visual searches and hand capture techniques with sampling effort standardized by area (Jaeger and Inger 1994; Bailey et al. 2004) or a time constraint (Crump and Scott 1994) (Figure 8.2). Species that migrate between aquatic breeding ponds and terrestrial habitat types may be susceptible to pitfall traps with drift fences (Corn 1994). For pond-breeding species, egg masses are often more detectable than adults of the same species, making egg masses more suitable for population monitoring studies. Shaffer et al. (1994) and Olson et al. (1997) provided an excellent introduction to techniques for amphibian inventories in ponds.

Cover-board surveys have been widely adopted for estimating the relative abundance of amphibian and reptile populations in different habitat types (Grant et al.

**FIGURE 8.5** A gopher snake and a garter snake uncovered from under a cover board used to sample reptiles in an Oregon oak savannah.

1992; Harpole and Haas 1999; Engelstoft and Ovaskake 2000). Cover boards are objects such as boards or metal roofing that provide daytime cover for animals and, when lifted, reveal a sample of the animals in the area that use it (Figure 8.5). Hence, it is a sample of a plot for those species that seek cover.

## BIRDS

As a group, many species of birds are sampled using distance sampling because they are so mobile and vocal (Rosenstock et al. 2002, and see Buckland et al. 1993 for an overview of distance sampling). Consequently, there are two primary means of detecting many species of birds, increasing the likelihood that they may be detected, especially during the breeding season when males are often territorial. Diurnal species sampled in grasslands, marshes, or other rather uniform vegetation conditions often can be sampled using transects (e.g., Ribic and Sample 2001). Samples taken in areas where rugged terrain or other factors prevent the use of transects rely on point counts (Buckland 1987). Other commonly used techniques include spot mapping to understand territory densities (e.g., Dobkin and Rich 1998), nest searches to understand nest densities, and variations on the Mayfield method of calculating nest success (Johnson and Shaffer 1990). Nocturnal species present additional challenges, but point counts for owls during their mating season can be effective especially if call-back recordings are used to elicit responses (Hardy and Morrison 2000). But call-back recordings can introduce biases when recordings are played in areas having different vegetative structure or topography. Finally, banding and band returns can be used to estimate longevity and age structures of populations (Pollock and Raveling 1982).

## MAMMALS

Small mammals are often sampled using traps of various forms (Figure 8.6). Live traps are often used as they are suitable for mark-recapture estimates of population size, or live or kill traps can be used to estimate catch per unit effort estimates of relative abundance. When any kind of trap is used, animal welfare guidelines should be reviewed and followed.

# Field Techniques for Population Sampling and Estimation    143

**FIGURE 8.6** Live traps attached to trees are particularly effective at sampling flying squirrels and other arboreal species. (Photo of flying squirrel by Cheron Ferland, used with permission; photo of woman checking trap by Laura Navarrette, used with permission.)

Detection probabilities are influenced by the number of trap-nights (the number of traps multiplied by the number of nights sampled). Capture rates are easily influenced by factors such as density-dependent intraspecific interactions, weather, and habitat, so capture probabilities must be calculated to allow an unbiased estimate of relative abundance (Menkens and Anderson 1988). This is particularly important when assessing trends over time during circumstances where weather and other conditions affecting catchability vary from year to year.

Larger mammals, especially those that form social aggregations or occur as clustered populations, may be more effectively surveyed by observational techniques than capture methods. Aerial surveys for ungulates are often conducted using distance sampling procedures, but again observability among vegetation types, weather conditions, and topographies needs to be considered to ensure unbiased estimates of abundance (Pollock and Kendall 1987).

Occurrence and indices of abundance for mid-size mammals are often addressed using remotely activated cameras, scent stations, track counts or track plates, or spotlight surveys (Figure 8.3; Gese 2001). Each of these techniques has advantages of being reasonably low cost and effective at detecting certain species (see Gese 2001), but estimation of population size is usually not possible.

Bats present a unique sampling problem. Using ultrasonic recordings of their feeding calls can help in distinguishing some species occurring in an area (O'Farrell et al. 1999), but estimates of abundance are often conducted at roost or maternity sites.

## EFFECTS OF TERRAIN AND VEGETATION

Imagine that you are counting all the birds that you can see or hear along a transect that extends from the center of a forested patch to the center of an adjacent field. You wish to determine if the relative abundance of birds differs between the two habitat types. You detect 28 birds in the forest and 34 in the field. Are there more birds in the field than in the forest? Perhaps. But probably not. Your ability to detect birds in the forest is hampered by the decreased visibility in the forest compared to a more open field. Hence, if you corrected for the differences in detectability as a function of the vegetation type (using techniques common in distance sampling), then the number of birds detected per unit area might be much greater in the forest than in the field. Consequently, simply raw counts of animals without considering the distance from the observer to the animal are often biased and should not be used as response variables in monitoring programs. This is particularly important where vegetation structure is likely to change over the course of the monitoring time frame.

On human time scales, terrain does not (usually) change appreciably from one time period to another, but observations must still be standardized by their detectability in various settings.

## MERITS AND LIMITATIONS OF INDICES COMPARED TO ESTIMATORS

Indices are often used to assess changes in populations over time based on an assumption that some aspect of detection is related to actual animal density. For instance, the following are a few examples of indices to abundance:

- Track counts (Conner et al. 1983)
- Pellet group counts (Fuller 1991)
- Capture rates (Cole et al. 1998)
- Detection rates of singing male birds (McGarigal and McComb 1999)
- Relative dominance (plant populations)
- Counts of squirrel leaf nests (Healy and Welsh 1992)
- Counts of beaver lodges or caches (Easter-Pilcher 1990)

These examples do not provide estimates of populations; rather, an index is measured with the assumption that the index is related to the population or its fitness in a known manner and that observed changes in the index measurement over time will reflect changes in population according to this relationship. Assumptions such as the more tracks seen, the more individuals present, or the more male birds heard singing, the more birds reproducing at the site are often made. But the reliability of these assumptions is brought into question, and indeed the opportunity for bias associated with indices to abundance is quite high. For instance, track counts may be related to animal abundance or to animal activity levels, but on the other hand, they could be related to the characteristics of the substrate, the weather, or any combination of these options. Likewise, counts of singing male birds may represent trends in abundance of territorial males, but if some males do not attract a mate, then numbers of singing males may not indicate abundance of nesting females nor reproductive output. Capture rates of animals over space and time can be related to animal abundance or to their vulnerability to capture in different areas of habitat quality. Consequently, although indices to abundance are often used because of logistical constraints, considerable caution must be exercised when interpreting the results. Indeed, it is often useful to conduct a pilot study that will allow you to state with a known level of certainty what the relationship is between the index and the actual population (or fitness) for the species being monitored. To do so requires an estimate of the population.

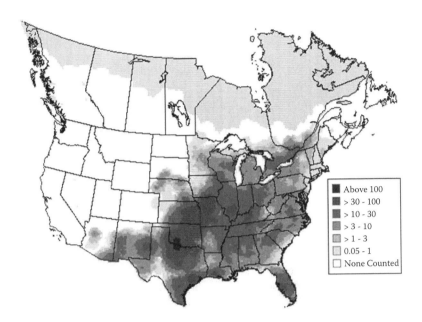

**FIGURE 2.2** Abundance map for the eastern meadowlark based on Breeding Bird Survey data collected between the summers of 1994–2003. (From Sauer, J.R., J.E. Hines, and J. Fallon. 2007. The North American Breeding Bird Survey, Results and Analysis 1966–2006. Version 10.13.2007. U.S. Geological Survey Patuxent Wildlife Research Center, Laurel, MD.)

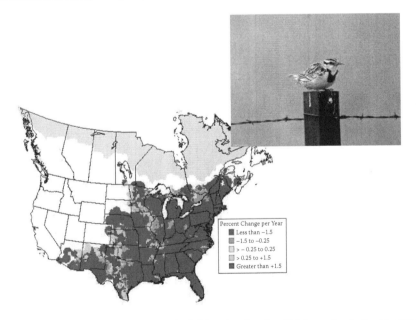

**FIGURE 2.3** Trend map for the eastern meadowlark based on Breeding Bird Survey trend estimates collected between the summers of 1966–2003. (From Sauer, J.R., J.E. Hines, and J. Fallon. 2007. The North American Breeding Bird Survey, Results and Analysis 1966–2006. Version 10.13.2007. U.S. Geological Survey Patuxent Wildlife Research Center, Laurel, MD; photo inset by Laura Erickson is used with permission.)

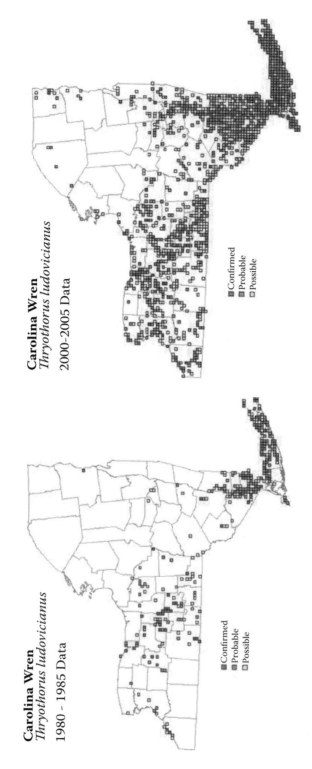

**FIGURE 2.9** Changes in the distribution of the Carolina wren between two statewide atlases conducted in 1980–1985 and 2000–2005. This species has shown one of the most dramatic increases in occupancy of any species recorded during the atlas project.

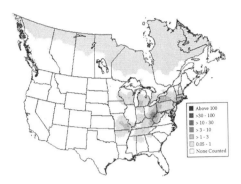

**FIGURE 4.5** Distribution of blue-winged warblers in the United States and Canada. (From Adapted from Sauer, J.R., J.E. Hines, and J. Fallon. 2006. The North American Breeding Bird Survey, Results and Analysis 1966–2006. Version 6.2.2006. USGS Patuxent Wildlife Research Center, Laurel, MD.)

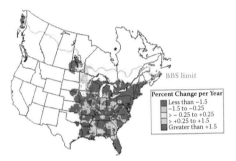

**FIGURE 5.7** Changes in abundance of yellow-billed cuckoos over time varies from one portion of its geographic range to another. (From Sauer, J.R., J.E. Hines, and J. Fallon. 2001. The North American Breeding Bird Survey, Results and Analysis 1966–2000. Version 2001.2, USGS Patuxent Wildlife Research Center, Laurel, MD.) Care should be given to ensure that the data relate to the scope of inference of the monitoring plan that is being developed. Local trends may be informative, while regional trends may not; if your area of interest were in Louisiana, then national trends would be misleading.

**FIGURE 5.8** Predicted changes in abundance of eastern towhees over its geographic range, 1966–1996. (From Sauer, J.R., J.E. Hines, and J. Fallon. 2001. The North American Breeding Bird Survey, Results and Analysis 1966–2000. Version 2001.2, USGS Patuxent Wildlife Research Center, Laurel, MD; photo inset by Laura Erickson is used with permission.)

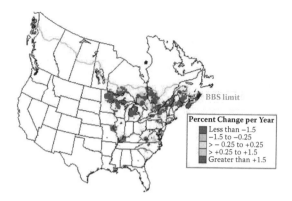

**FIGURE 5.10** Predicted changes in American woodcock abundance over the species range. (From Sauer, J.R., J.E. Hines, and J. Fallon. 2001. The North American Breeding Bird Survey, Results and Analysis 1966–2000. Version 2001.2, USGS Patuxent Wildlife Research Center, Laurel, MD.)

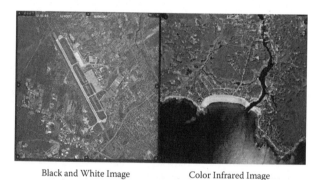

Black and White Image        Color Infrared Image

**FIGURE 9.3** Examples of panchromatic (black and white) and color infrared aerial photographs.

**FIGURE 9.4** San Francisco Bay area, California. This Band 3,2,1 image shows the spring runoff from the Sierras and other neighboring mountains into the bay and out into the Pacific. (Image from Landsat 7 Project, NASA.)

Field Techniques for Population Sampling and Estimation 145

Estimators provide additional information to the user and can help address some of the biases inherent in many indices. Fortunately, there are a number of tools available to estimate abundances that are based on sampling theory and can result in known levels of confidence placed around the estimates. For instance, distance estimators provide a mechanism for estimating abundance of organisms from points or transects where detectability might differ among vegetative types or stream reaches. Population estimators are generally available as free software:

http://nhsbig.inhs.uiuc.edu/wes/density_estimation.html
http://nhsbig.inhs.uiuc.edu/wes/populations.html
http://www.ruwpa.st-and.ac.uk/distance/

A variety of information is used as the basis for estimates of populations (capture-recapture and band returns), some of which can also provide estimates of survival and reproduction. The overwhelming advantage of using population estimators is that estimates of abundance, survival, and age class distribution can be made with estimates of confidence. Clearly with replicate sampling of independent sites, indices to abundance can be calculated with confidence intervals, but there is still doubt regarding the assumption of unbiased association between the index and the population characteristic of interest. Consequently, it is important when designing monitoring protocols to ensure that estimators are considered, and if it is not logistically possible to use estimators as a response variable in the monitoring program, then the selected index should be justified relative to its known association with the population characteristic of interest. If this cannot be done, then both the assumptions and the implications of violating those assumptions should be clearly stated.

## ESTIMATING COMMUNITY STRUCTURE

Although much of this book focuses on monitoring species and the environmental conditions in which they live, at times managers may be concerned with maintaining or developing conditions that promote diverse or functional communities, or be able to detect declines in functional diversity in the face of environmental stressors. Diversity metrics have evolved over decades and provide a means of comparing complexity between or among places or times. Unfortunately, most diversity metrics also bury information on individual species responses within one or a few numbers. To understand what is really happening within a community, a diversity metric must be deconstructed to see changes in individual species or populations. The devil is indeed in the detail. Nonetheless, diversity metrics are still used as a guide to community structure and function. Typically, a diversity metric consists of an estimate of the number of species in a unit of space and time (species richness) and the distribution of individuals among those species (evenness). In most evenness metrics, the maximum achievable value is 1.00 (an equal number of individuals per unit area represented among all species in the community; i.e., no species dominates the community). But consider the example in Table 8.2. The hypothetical forest and grassland have the same diversity and evenness. So are they the same community? Do they function in a similar manner? Obviously not. In fact, the degree to which

## TABLE 8.2
### Example of Estimates of Species Diversity and Community Similarity for Three Hypothetical Communities

|  | Community | | |
| --- | --- | --- | --- |
|  | Forest | Savannah | Grassland |
| Black-capped chickadee | 10 | 4 | 0 |
| Song sparrow | 2 | 4 | 0 |
| American robin | 4 | 3 | 0 |
| Grasshopper sparrow | 0 | 3 | 4 |
| Marsh wren | 0 | 1 | 2 |
| Savannah sparrow | 0 | 1 | 10 |
| **Evaluation** | | | |
| Species richness | 3 | 6 | 3 |
| Number of individuals | 16 | 16 | 16 |
| Simpson's evenness | 0.797 | 0.956 | 0.797 |
| Simpson's species diversity[a] | 0.567 | 0.850 | 0.567 |
| **Community Similarity (%)[b]** | | | |
| Forest–savannah | 0.563 | | |
| Forest–grassland | 0.000 | | |
| Savannah–grassland | 0.313 | | |

[a] White 1986.
[b] Itow 1991.

the two communities are similar in species representation and distribution of individuals among species in common between the two communities indicates that the percent similarity between the two communities is 0. In fact, each community has more in common with a more species-rich and more even community represented in a savannah than in the forest or grassland. So is the savannah better because it is more species rich and more diverse? No, it is just different. In short, these metrics of community structure can be very difficult to interpret without digging into the details that comprise the metrics.

Sampling communities in a manner that produces unbiased estimates of community structure also can be problematic. Species richness assessments may be as simple as developing a list of species detected in an area using standardized techniques. But some species are likely more easily detected than others. Some are active at different times of the day, and some move more than others. So estimates of richness are confounded by differences in detectability. When diversity metrics are calculated using these data, additional complications arise because estimates of abundance for each species in the community must be unbiased and based on the same unit of space and time. Again, differential detectability of species and biases resulting from movement arise, making evenness estimates confounded between actual differences in abundance among species and differences in detection among species. Surveys to determine the presence of a species in an area typically require less sampling intensity than fieldwork necessary to collect other population statistics.

Another metric that can be calculated from data collected within a monitoring framework deals with unequal sampling efforts when trying to estimate the number of species in an area. Because rare species are often less likely to be detected than common species, we can estimate the number of species based on rarefaction curves so that we can compare the number of species found in two areas when the sampling effort differed (Simberloff 1972).

## ESTIMATING BIOTIC INTEGRITY

Karr (1981) developed and index to biotic integrity (IBI) that was designed to be used to compare aspects of fish communities among sites in a standardized manner that reflected the water quality of the site with regards to its capacity to support fish communities. It included aspects of species richness, fish health, and a number of other parameters. Researchers have since adapted this technique for other taxa including aquatic macroinvertebrates and birds (Bryce et al. 2002). Although the general structure of IBIs is similar to that developed by Karr (1981), each IBI is typically crafted to a reference condition representive of the ecoregion within which a sampled area occurs. Hence, an IBI developed for the mixed mesophytic forest of the Appalachians may have a similar structure to that of one in the Cascade Range mountains in Oregon, but the parameters measured would be based on a very different set of reference conditions. For that reason IBIs are not typically used in terrestrial monitoring protocols, and like diversity indices, they can bury information that must be extracted by deconstructing the index.

## STANDARDIZATION AND PROTOCOL REVIEW

Regardless of the approach taken, certain aspects of a sampling design should be standardized to minimize bias in the resulting data. (See Chapter 7 for a more detailed overview of standardization techniques.) At the very least, specific attention should be paid to consistently sampling the same location over time, sampling during the same season and time of day, and using the same equipment from one sampling period to the next.

In addition to consistency in sampling techniques, locations, and times, most of the approaches described in this chapter will need approval by an Institutional Animal Care and Use Committee (IACUC) or comparable review board at universities and agencies. Standards of care for animals when conducting research are rigorous and should be applied to monitoring programs as well (see Laber et al. 2007 and Hafner 2007 for a perspective on limitations to such reviews).

## BUDGET CONSTRAINTS

The choice of which sampling technique to use when presented with options and a set of goals and objectives is as often driven by constraints of time and money as by the ideal technique to acquire the data needed. For instance, while information on survival of juveniles in the face of climate change may be a primary consideration for a long-lived species, a suitable surrogate might be an index to the abundance of young of that year and adult animals in the population. Since budget decisions are often made based on the social seriousness of the issue, and since social values change over time, so do budgets. Consequently, it often is wise to develop a monitoring program using techniques that are robust to concerns regarding bias and precision and that are cost effective, even if they do not result in the ideal level of fitness data that would be needed to answer demographic questions more definitively. More detailed demographic information may be collected in a supplementary manner if the level of social concern reaches a point where it is mandated.

## SUMMARY

There is a huge array of methods available to detect organisms, enumerate them, and assess their fitness of organisms. Matching the appropriate approach with the goals and objectives of the study is a key first step in designing a monitoring program. The data needed to achieve inventory and monitoring objectives are often the driving factor in deciding which techniques will be used in a monitoring program. In addition the scale of the monitoring program may also influence the methods used. For narrowly focused programs on highly valued species, detailed measurements of fitness may be appropriate. As the scale of the project over space and time increases, then estimates of abundance or occurrence may be the most feasible approach.

Furthermore, different taxa vary in their propensity to be sampled using different methods. Whereas sessile animals can be sampled using quadrat sampling, wide-ranging species may more easily be sampled using cameras or hair traps that may provide information on occurrence or individual movements. In a similar way, simply the logistics associated with sampling in inhospitable (to humans) terrain limits the choices of sampling strategies available. Finally, budgetary and logistic restrictions often mean that the ideal sampling system may not be feasible because the time or money is not available to implement the technique to adequately meet goals and objectives.

Amid all of the decisions regarding which technique to employ, you should continually be aware of the need to standardize approaches over space and time, and seek to minimize any biases in your sampling methods.

## REFERENCES

Anderson, D.R., K.P. Burnham, B.C. Lubow, L. Thomas, P.S. Corn, P.A. Medica, and R.W. Marlow. 2001. Field trials of line transect methods applied to estimation of desert tortoise abundance. *Journal of Wildlife Management* 65:583–597.

Anderson, D.R., J.L. Laake, B.R. Cran, and K.P. Burnham. 1979. Guidelines for line transect sampling of biological populations. *Journal of Wildlife Management* 43:70–78.

Bailey, L.L., T.R. Simons, and K.H. Pollock. 2004. Estimating site occupancy and species detection probability parameters for terrestrial salamanders. *Ecological Applications* 14:692–702.

Bass, D.K., and I.R. Miller. 1996. Crown-of-thorns starfish and coral surveys using the manta tow and scuba search techniques. Long-term monitoring of the Great Barrier Reef standard operational procedure number 1. Australian Institute of Marine Science, Townsville.

Bonham, C.D. 1989. *Measurements of Terrestrial Vegetation*. John Wiley & Sons, New York. 338 pp.

Bookhout, T.A. (ed.). 1994. *Research and Management Techniques for Wildlife and Habitats*. The Wildlife Society, Bethesda, MD. 740 pp.

Bryce, S.A., R.M. Hughes, and P.R. Kaufman. 2002. Development of a bird integrity index: Using bird assemblages as indicators of riparian condition. *Environmental Management* 30:294–310.

Buckland, S.T. 1987. On the variable circular plot method of estimating animal density. *Biometrics* 43:363–384.

Buckland, S.T., D.R. Anderson, K.P. Burnham, and J.L. Laake. 1993. *Distance Sampling: Estimating Abundance of Biological Populations*. Chapman and Hall, New York. 446 pp.

Burger, A.E. 2001. Using radar to estimate populations and assess habitat associations of marbled murrelets. *Journal of Wildlife Management* 65:696–715.

Castello, L., J.P. Viana, G. Watkins, M. Pinedo-Vasquez, and V.A. Luzadis. 2009. Lessons from integrating fishers of arapaima in small-scale fisheries management at the Mamirauá Reserve, Amazon. *Environmental Management* 43:197–209.

Cole, E.C., W.C. McComb, M. Newton, J.P. Leeming, and C.L. Chambers. 1998. Response of small mammals to clearcutting, burning, and glyphosate application in the Oregon coast range. *Journal of Wildlife Management* 62:1207–1216.

Conner, M.C., R.F. Labisky, and D.R. Progulske, Jr. 1983. Scent-station indices as measures of population abundance for bobcats, raccoons, grey foxes and opossums. *Wildlife Society Bulletin* 11:146–152.

Cook, R.D., and J.O. Jacobson. 1979. A design for estimating visibility bias in aerial surveys. *Biometrics* 35:735–742.

Corn, P.S. 1994. Straight line drift fences and pitfall traps. Pages 109–117 in *Measuring and Monitoring Biological Diversity: Standard Methods for Amphibians*. W.R. Heyer, M.A. Donnelly, R.W. McDiarmid, L.C. Hayek, and M.S. Foster (eds.). Smithsonian Institution Press, Washington, D.C.

Crouch, W.B., and P.W.C. Paton. 2002. Assessing the use of call surveys to monitor breeding anurans in Rhode Island. *Journal of Herpetology* 36:185–192.

Crump, M.L., and N.J. Scott, Jr. 1994. Visual encounter surveys. Pages 84–92 in *Measuring and Monitoring Biological Diversity: Standard Methods for Amphibians*. W.R. Heyer, M.A. Donnelly, R.W. McDiarmid, L.C. Hayek, and M.S. Foster (eds.). Smithsonian Institution Press, Washington, D.C.

Cunjak, R.A., R.G. Randall, and E.M.P. Chadwick. 1988. Snorkeling versus electrofishing: A comparison of census techniques in Atlantic salmon rivers. *Naturaliste Canadien* 115:89–93.

Cutler, T., and D.E. Swann. 1999. Using remote photography in wildlife ecology: A review. *Wildlife Society Bulletin* 27:571–581.

Davis, D.E., and R.L. Winstead. 1980. Estimating the numbers of wildlife populations. Pages 221–245 in *Wildlife Management Techniques Manual*. S.D. Schemnitz (ed.). The Wildlife Society, Boston, MA.

deCalesta, D.S., and G.W. Witmer. 1990. Drive-line census for deer within fenced enclosures. USDA Forest Service Research Paper NE-643. 4 pp.

Dobkin, D.S., and A.C. Rich. 1998. Comparison of line-transect, spotmap, and point-count surveys for birds in riparian areas of the Great Basin. *Journal of Field Ornitholology* 69:430–443.

Easter-Pilcher, A. 1990. Cache size as an index to beaver colony size in northwestern Montana. *Wildlife Society Bulletin* 18:110–113.

Enge, K.M. 2001. The pitfalls of pitfall traps. *Journal of Herpetology* 35:467–478.

Engelstoft, C., and K.E. Ovaskake. 2000. Artificial cover objects as a method for sampling snakes (*Contia tenuis* and *Thamnophis* spp.) in British Columbia. *Northwestern Naturalist* 81:35–43.

Forsman, E.D. 1988. A survey of spotted owls in young forests in the northern Coast Range of Oregon. *Murrelet* 69:65–68.

Fuller, T.K. 1991. Do pellet counts index white-tailed deer numbers and population change? *Journal of Wildlife Management* 55:393–396.

Gese, E.M. 2001. Monitoring of Terrestrial Carnivore Populations. Pages 373–396 in *Carnivore Conservation*. J.L. Gittleman et al. (eds.). Cambridge University Press, Cambridge, U.K.

Girard, I., J.P. Ouellet, R. Courtois, C. Dussault, and L. Breton. 2002. Effects of sampling effort based on GPS telemetry on home-range size estimations. *Journal of Wildlife Management* 66:1290–1300.

Grant, B.W., A.D. Tucker, J.E. Lovich, A.T. Mills, P.M. Dixon, and J.W. Gibbons. 1992. The use of coverboards in estimating patterns of reptile and amphibian abundance. Pages 379–403 in *Wildlife 2001*. R. Seigel and N. Scott (eds.). Elsevier Science, London, U.K.

Hafner, M.S. 2007. Field research in mammalogy: An enterprise in peril. *Journal of Mammalogy* 88:1119–1128.

Haig, S. 1998. Molecular contributions to conservation. *Ecology* 79:413–425.

Hankin, D.G., and G.H. Reeves. 1988. Estimating total fish abundance and total habitat area in small streams based on visual estimation methods. *Canadian Journal of Fisheries and Aquatic Sciences* 45:834–844.

Hardy, P.C., and M.L. Morrison. 2000. Factors affecting the detection of elf owls and western screech-owls. *Wildlife Society Bulletin* 28:333–343.

Harmata, A.R., K.M. Podruzny, J.R. Zelenak, and M.L. Morrison. 1999. Using marine surveillance radar to study bird movements and impact assessment. *Wildlife Society Bulletin* 27:44–52.

Harpole, D.N., and C.A. Haas. 1999. Effects of seven silvicultural treatments on terrestrial salamanders. *Forest Ecology and Management* 114:349–356.

Healy, W.M., and C.J.E. Welsh. 1992. Evaluating line transects to monitor grey squirrel populations. *Wildlife Society Bulletin* 20:83–90.

Hossack, B.R., P.S. Corn, and D.B. Fagre. 2006. Divergent patterns of abundance and age-class structure of headwater stream tadpoles in burned and unburned watersheds. *Canadian Journal of Zoology* 84:1482–1488.

Huff, M.H., K.A. Bettinger, H.L. Ferguson, M.J. Brown, and B. Altman. 2000. A habitat-based point-count protocol for terrestrial birds, emphasizing Washington and Oregon. United States Forest Service General Technical Report PNW-GTR-501.

Hussell, D.J.T. 1981. The use of migration counts for monitoring bird population levels. Pages 92–102 in *Estimating Numbers of Terrestrial Birds*. C.J. Ralph and J.M. Scott (eds.). *Studies in Avian Biology* no. 6.

Itow, S. 1991. Species turnover and diversity patterns along an evergreen broad-leaved forest coenocline. *Journal of Vegetation Science* 2:477–484.

Jaeger, R.G., and R.F. Inger. 1994. Quadrat sampling. Pages 97–102 in *Measuring and Monitoring Biological Diversity: Standard Methods for Amphibians*. W.R. Heyer, M.A. Donnelly, R.W. McDiarmid, L.C. Hayek, and M.S. Foster (eds.). Smithsonian Institution Press, Washington, D.C.

Johnson, D.H., and T.L. Shaffer. 1990. Estimating nest success: When Mayfield wins. *Auk* 107:595–600.

Karr, J.R. 1981. Assessment of biotic integrity using fish communities. *Fisheries* 6(6):21–27.

Kolozsvary, M.B., and R.K. Swihart. 1999. Habitat fragmentation and the distribution of amphibians: Patch and landscape correlates in farmland. *Canadian Journal of Zoology* 77:1288–1299.

Kuenzi, A.J., and M.L. Morrison. 1998. Detection of bats by mist-nets and ultrasonic sensors. *Wildlife Society Bulletin* 26:307–311.

Kunz, T.H., D.W. Thomas, G.C. Richards, C.R. Tidemann, E.D. Pierson, and P.A. Racey. 1996. Observational techniques for bats. Pages 105–114 in *Measuring and Monitoring Biological Diversity*. D.E. Wilson, F.R. Cole, J.D. Nichols, R. Rudran, and M.S. Foster (eds.). Smithsonian Institution Press, Washington, D.C.

Laber, K., B.W. Kennedy, and L. Young. 2007. Field studies and the IACUC: Protocol review, oversight, and occupational health and safety considerations. *Lab Animal* 36:27–33.

Lande, R., and G.F. Barrowclough. 1987. Effective population size, genetic variation, and their use in population management. Pages 87–123 in *Viable Populations for Conservation*. M.E. Soulé (ed.). Cambridge University Press, Cambridge, U.K. 204 pp.

Lemen, C.A., and P.W. Freeman. 1985. Tracking mammals with fluorescent pigments: A new technique. *Journal of Mammalogy* 66:134–136.

Lint, J. 2001. Northern spotted owl effectiveness monitoring plan under the Northwest Forest Plan: Annual summary report 2000. Northwest Forest Plan Interagency Monitoring Program, Regional Ecosystem Office, Portland, OR.

Lishak, R.S. 1977. Censusing 13-lined ground squirrel with adult and young alarm calls. *Journal of Wildlife Management* 41:755–759.

MacKenzie, D.I. 2005. Was it there? Dealing with imperfect detection for species presence/absence data. *Australia and New Zealand Journal of Statistics* 47:65–74.

MacKenzie, D.I., J.D. Nichols, J.E. Hines, M.G. Knutson, and A.D. Franklin. 2003. Estimating site occupancy, colonization and local extinction probabilities when a species is not detected with certainty. *Ecology* 84:2200–2207.

MacKenzie, D.I., and J.A. Royle. 2005. Designing efficient occupancy studies: General advice and tips on allocation of survey effort. *Journal of Applied Ecology* 42:1105–1114.

Madsen, S., D. Evans, T. Hamer, P. Henson, S. Miller, S.K. Nelson, D. Roby, and M Stapanian. 1999. Marbled murrelet effectiveness monitoring plan for the Northwest Forest Plan. USDA Forest Service General Technical Report PNW-GTR-439. 51 pp.

McDaniel, G.W., K.S. McKelvey, J.R. Squires, and L.F. Ruggiero. 2000. Efficacy of lures and hair snares to detect lynx. *Wildlife Society Bulletin* 28:119–123.

McGarigal, K., and W.C. McComb. 1999. Forest fragmentation and breeding bird communities in the Oregon Coast Range. Chapter 13 in *Forest Fragmentation: Wildlife and Management Implications*. J.P. Rochelle, L.A. Lehman, and J. Wisniewski (eds.). Brill Press, Netherlands.

Mengak, M.T., and D.C. Guynn. 1987. Pitfalls and snap traps for sampling small mammals and herpetofauna. *American Midland Naturalist* 118:284–288.

Menkens, G.E., and S.H. Anderson. 1988. Estimation of small-mammal population size. *Ecology* 69:1952–1959.

Mills, L.S., J.J. Citta, K.P. Lair, M.K. Schwartz, and D.A. Tallmon. 2000. Estimating animal abundance using noninvasive DNA sampling: Promise and pitfalls. *Ecological Applications* 10:283–294.

Moruzzi, T.L., T.K. Fuller, R.M. DeGraaf, R.T. Brooks, and W.J. Li. 2002. Assessing remotely triggered cameras for surveying carnivore distribution. *Wildlife Society Bulletin* 30:380–386.

Nietfeld, M.T., M.W. Barrett, and N. Silvy. 1994. Wildlife marking techniques. Pages 140–168 in *Research and Management Techniques for Wildlife and Habitats*. T.A. Bookhout (ed.). The Wildlife Society, Bethesda, MD.

O'Farrell, M.J., B.W. Miller, and W.L. Gannno. 1999. Qualitative identification of free-flying bats using the Anabat detector. *Journal of Mammalogy* 80:11–23.

Ogutu, J.O., and H.T. Dublin. 1998. The response of lions and spotted hyaenas to sound playbacks as a technique for estimating population size. *African Journal of Ecology* 36:83–95.

Olson, D.H., W.P. Leonard, and R.B. Bury. 1997. Sampling amphibians in lentic habitats: Methods and approaches for the Pacific Northwest. Society for Northwestern Vertebrate Biology, Olympia, WA.

Pellet, J., L. Rechsteiner, A.K. Skrivervik, J.F. Zürcher, and N. Perrin. 2006. Use of harmonic direction finder to study the terrestrial habitats of the European tree frog (*Hyla arborea*). *Amphibia-Reptilia* 27:138–142.

Pelletier, F., D. Reale, J. Watters, E.H. Boakes, and D. Garant. 2009. Value of captive populations for quantitative genetics research. *Trends in Ecology and Evolution* 24:263–270.

Poizat, B., and E. Baran. 1997. Fishermen's knowledge as background information in tropical fish ecology: A quantitative comparison with fish sampling results. *Environmental Biology of Fishes* 50(4):435–449.

Pollock, K.H., and W.L. Kendall. 1987. Visibility in aerial surveys: A review of estimation procedures. *Journal of Wildlife Management* 51:502–510.

Pollock, K.H., and D.G. Raveling. 1982. Assumptions of modern band-recovery models, with emphasis on heterogeneous survival rates. *Journal of Wildlife Management* 46:88–98.

Powell, R.A., and G. Proulx. 2003. Trapping and marking terrestrial mammals for research: Integrating ethics, performance criteria, techniques, and common sense. *ILAR Journal* 44:259–276.

Pradel, R. 1996. Utilization of capture-mark-recapture for the study of recruitment and population growth rates. *Biometrics* 52:703–709.

Ralph, C.J., S. Droege, and J.R. Sauer. 1995. Managing and monitoring birds using point counts: Standards and applications. Pages 161–168 in *Monitoring Bird Populations by Point Counts*. C.J. Ralph, J.R. Sauer, and S. Droege (eds.). USDA Forest Service General Technical Report PSW-GTR-149.

Ralph, C.J., G.R. Geupel, P. Pyle, T.E. Martin, and D.F. DeSante. 1993. Handbook of field methods for monitoring landbirds. USDA Forest Service General Technical Report PSW-GTR-144. 41 pp.

Ribic, C.A., and D.W. Sample. 2001. Associations of grassland birds with landscape factors in southern Wisconsin. *American Midland Naturalist* 146:105–121.

Rosenstock, S.S., D.R. Anderson, K.M. Giesen, T. Leukering, and M.F. Carter. 2002. Landbird counting techniques: Current practices and an alternative. *Auk* 119:46–53.

Rudran, R, T.H. Kunz, S.C. Jarman, and A.P. Smith. 1996. Observational techniques for nonvolant mammals. Pages 81–114 in *Measuring and Monitoring Biological Diversity—Standard Methods for Mammals*. D.E. Wilson, F.R. Cole, J.D. Nichols, R. Rudran, and M.S. Foster (eds.). Smithsonian Institution Press, Washington, D.C.

Sauer, J.R., J.E. Hines, and J. Fallon. 2008. The North American Breeding Bird Survey, Results and Analysis 1966–2007. Version 5.15.2008. USGS Patuxent Wildlife Research Center, Laurel, MD.

Scotts, D.J., and S.A. Craig. 1988. Improved hair-sampling tube for the detection of rare mammals. *Australian Wildlife Research* 15:469–472.

Seber, G.A.F. 1982. *The Estimation of Animal Abundance*. 2nd ed. Griffin Publishers, London. 654 pp.

Shaffer, H.B., R.A. Alford, B.D. Woodward, S.J. Richards, R.G. Altig, and C. Gascon. 1994. Quantitative sampling of amphibian larvae. Pages 130–141 in *Measuring and Monitoring Biological Diversity: Standard Methods for Amphibians*. W.R. Heyer, M.A. Donnelly, R.W. McDiarmid, L.C. Hayek, and M.S. Foster (eds.). Smithsonian Institution Press, Washington, D.C.

Simberloff, D.S. 1972. Properties of the rarefaction diversity measurement. *American Naturalist* 106:414–418.

Simberloff, D.S. 1978. Use of rarefaction and related methods in ecology. Pages 150–165 in *Biological Data in Water Pollution Assessment: Quantitative and Statistical Analyses*. K.L. Dickson, J. Cairns, Jr., and R.J. Livingston (eds.). American Society for Testing and Materials STP 652, Philadelphia, PA.

Slack, K.V., R.C. Averett, P.E. Greeson, and R.G. Lipscomb. 1973. Methods for collection and analysis of aquatic biological and microbiological samples: U.S. Geological Survey Techniques of Water-Resources Investigations. Report Number 05-A4.

Stirling, D.F., and J.F. Bendell. 1966. Census of blue grouse with recorded calls of the female. *Journal of Wildlife Management* 30:184–187.

Suckling, K., and M. Taylor. 2006. Critical habitat and recovery. Pages 50–67 in D.D. Goble, J.M. Scott, and F.W. Davis. *The Endangered Species Act at 30: Vol. 1: Renewing the Conservation Promise*. Island Press, Washington, D.C.

Thomas, D.W., and S.D. West. 1989. Sampling methods for bats. USDA Forest Service General Technical Report PNW 243. 20 pp.

U.S. Geological Survey (USGS). 1997. Forum on Wildlife Telemetry: Innovations, Evaluations, and Research Needs; September 21–23, 1997, Snowmass Village, CO. Program and Abstracts. U.S. Geological Survey and The Wildlife Society.

Uychiaoco, A.J., H.O. Arceo, S.J. Green, M.T. De la Cruz, P.A. Gaite, and P.M. Aliño. 2005. Monitoring and evaluation of reef protected areas by local fishers in the Philippines: Tightening the adaptive management cycle. *Biodiversity and Conservation* 14:2775–2794.

Verner, J. 1988. Optimizing the duration of point counts for monitoring trends in bird populations. USDA Forest Service Research Note PSW-395.

Verner, J., and L.V. Ritter. 1985. A comparison of transects and point counts in oak–pine woodlands of California. *Condor* 87:47–68.

Welsh, H.H., Jr. 1987. Monitoring herpetofauna in woodland habitats of northwestern California and southwestern Oregon: A comprehensive strategy. Pages 203–213 in *Multiple-Use Management of California's Hardwood Resources*. T.R. Plumb and N.H. Pillsbury (eds.). USDA Forest Service General Technical Report PSW-100.

Wemmer, C., T.H. Kunz, G. Lundie-Jenkins, and W. McShea. 1996. Mammalian sign. Pages 157–176 in *Measuring and Monitoring Biological Diversity—Standard Methods for Mammals*. D.E. Wilson, F.R. Cole, J.D. Nichols, R. Rudran, and M.S. Foster (eds.). Smithsonian Institution Press, Washington, D.C.

White, G.C., R.M. Bartmann, L.H. Carpenter, and R.A. Garrott. 1989. Evaluation of aerial line transects for estimating mule deer densities. *Journal of Wildlife Management* 53:625–635.

White, M.J. 1986. Segregation and diversity measures in population distribution. *Population Index* 52:198–221.

Wigington, P.J., J.L. Ebersole, M.E. Colvin, S.G. Leibowitz, B. Miller, B. Hansen, H. Lavigne, D. White, J.P. Baker, M.R. Church, J.R. Brooks, M.A. Cairns, and J.E. Compton. 2006. Coho salmon dependence on intermittent streams. *Frontiers in Ecology and the Environment* 4:513–518.

Wilson, G.J., and R.J. Delahay. 2001. A review of methods to estimate the abundance of terrestrial carnivores using field signs and observation. *Wildlife Research* 28:151–164.

Wood, M.D., and N.A. Slade. 1990. Comparison of ear-tagging and toe clipping in prairie voles, *Microtus ochrogaster*. *Journal of Mammalogy* 71:252–255.

Zielinski, W.J., and H.B. Stauffer. 1996. Monitoring *Martes* populations in California: Survey design and power analysis. *Ecological Applications* 6:1254–1267.

# 9 Techniques for Sampling Habitat

We define habitat as the resources necessary to support a population over space and through time (McComb 2007). Species depend on a number of environmental resources within a specific unit of space for their long-term persistence; these resources comprise a species' habitat. All of these resources play an important role in the quality of habitat, and any fluctuation in these resources can affect a species at both the individual and population level (Pulliam 1988). Legislative mandates in the United States make it legally necessary to monitor habitats, as well as populations of species that are designated as management indicator or endangered species (e.g., National Environmental Policy Act, Endangered Species Act, National Forest Management Act). Repeated inventories or long-term monitoring programs are excellent approaches to determine the interactions and dynamics of habitat use by a broad suite of species (Jones 1986). A properly designed monitoring protocol is appropriate for understanding and evaluating the long-term dependency of selected species on various habitat components. Management that modifies or reduces habitat quality can change the overall fitness of an organism, and consequently affect the growth of the population as a whole (McComb 2007). Consequently, managers will want to identify and incorporate the interactions between habitat and population change in order to make informed management decisions through an adaptive management process (Barrett and Salwasser 1982).

Resource availability is dynamic for most, if not all, species. Consequently, the challenge when developing a monitoring protocol is assessing whether changes in occurrence, abundance, or fitness in a population are independent from or related to changes in habitat availability and quality (Cody 1985). To assess the responses of animals or plants to changes in habitat quality, monitoring protocols must measure and document appropriate habitat elements. Many of these attributes could be incorporated into existing monitoring efforts associated with vegetation changes (e.g., USDA Forest Service Forest Inventory and Analysis efforts). A number of techniques exist for measuring site-level habitat characteristics, composition, and structural conditions. For example, point sampling and line sampling (e.g., line-intercept method) are two excellent methods for collecting data on a number of habitat characteristics such as percentage of plant cover, substrate type, and horizontal or vertical complexity (Jones 1986; Krebs 1989). Plot and plotless sampling (e.g., point-center-quarter method) are more intensive techniques for determining the density, composition, and biomass of various habitat components in an area (Krebs 1989). These data can be used to characterize the composition and frequency of habitat elements for an area. Furthermore, if these sampling points are permanent, then it is possible to document changes in habitat elements over time. This is especially important, since the carrying capacity of a habitat is dynamic and depends on a number of resources that can fluctuate due to natural (e.g., food, hurricanes, disease, etc.) and/or anthropogenic (e.g., silviculture, water management) disturbances. These disturbances can occur on a coarse scale, affecting hundreds of hectares and multiple patches within a landscape, or at a finer scale, affecting only a small area within a larger patch type. How a population responds to disturbances and the resultant environmental changes depends on the species' habitat requirements. Whether or not a monitoring protocol can determine causes of population change will be a direct result of its ability to monitor that species' habitat.

It is important to differentiate between habitat, which is species specific, and a habitat type, which is a biophysical classification of the environment that can be useful for understanding possible effects on a suite of species. Habitat types are often used as surrogates for a species habitat, but this approach must be viewed with caution. Habitat for one species rarely if ever represents habitat for another species given the multiple factors involved in defining a species' habitat (Johnson 1980).

## SELECTING AN APPROPRIATE SCALE

Knowing how a species selects habitat can provide clues as to which habitat elements to measure at what scale to ensure independence among samples and to more accurately characterize habitat for a species. Habitat selection is a set of complex behaviors that individuals in a population have developed over time to ensure fitness. These behaviors are often innate and have evolved to allow populations to be resilient to the variations in conditions that occur over time (Wecker 1963). Additionally, these behaviors are frequently evolved so that each species selects habitat in a manner that allows it to reduce competition for resources with other species. So the evolutionary selection pressures on each species, both abiotic and biotic, have led species to develop distinct strategies for survival and have thereby linked habitat selection and population dynamics.

Some species are habitat generalists and can use a broad suite of food and cover resources. These species tend to be highly adaptable and occur in a wide variety of environmental conditions. Other species are habitat specialists. These species are adapted to survive by capitalizing on the use of a narrow set of resources, which they are adapted to use more efficiently than most other species. Consider where you might find spring salamanders in the eastern United States or torrent salamanders in the western part of the country. Both species occur in clear, cold, headwater streams, and they tend to be most abundant where predatory pressure from fish is minimal or nonexistent. Changes over time in the occurrence and abundance of both species are of interest to managers due to the concern that forest management activities that reduce canopy cover and raise stream temperatures could threaten populations of these species (Lowe and Bolger 2002; Vesely and McComb 2002). Clearly, though, habitat generalists and specialists are simply two ends of a spectrum of species' strategies for survival in landscapes faced with variable climates, soils, disturbances, competitors, and predators.

## HIERARCHICAL SELECTION

Many studies have been conducted to assess habitat selection by wildlife species. Johnson (1980) suggested that many species select habitat at four levels and called these levels first-, second-, third-, and fourth-order selection (Figure 9.1). First-order selection is selection of the geographic range. The geographic range defines, quite literally, where in the world this species can be found. In our example from Figure 9.1, pileated woodpeckers are found in forests throughout eastern and western North America. Considering two extremes underscores the significance of understanding first-order selection: Figure 9.2 shows geographic range maps for two species, Weller's salamander, found in spruce forests above 1,500 m (5,000 ft) in the southern Appalachians, and black-capped chickadees found throughout the northern United States and southern Canada. These two species have widely divergent geographic ranges, and any monitoring protocols developed must consider these differences.

Consider the importance of populations of a species at the center versus the periphery of its geographic range. Populations at the periphery may be in lower-quality habitat if either biotic or abiotic factors are limiting its distribution. However, recall that environments are not static. They are constantly changing. Climate changes and earthquakes change the topography, and some species arrive while others leave. It is those populations at the periphery of their geographic range that are on the front line of these changes. Hence, although it may be tempting to think of these peripheral populations as somewhat expendable, they may be critical to population maintenance as large-scale changes in habitat availability occur. Given the contemporary

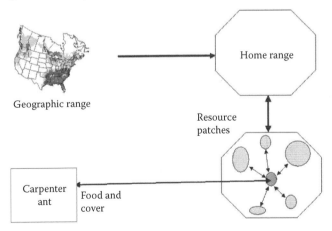

**FIGURE 9.1** Hierarchical habitat selection as described by Johnson (1980). This generalized concept is illustrated using pileated woodpeckers as an example. (Range map from USGS Biological Resources Division.)

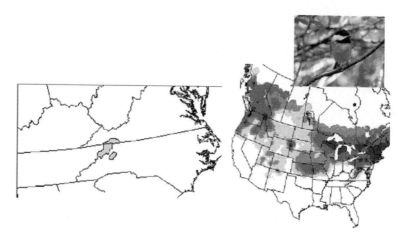

**FIGURE 9.2** Range maps for a geographically restricted species, Weller's salamander (left) and a cosmopolitan species, black-capped chickadee (right). (Maps from USGS Biological Resources Division; photo inset by Laura Erickson is used with permission.)

rate of climate change, these peripheral populations may be even more important over the next few hundred years (McCarty 2001).

Although Johnson (1980) did not describe metapopulation distribution as a selection level, it is important to realize that within the geographic range, populations often are distributed among smaller, interacting subpopulations that contribute to overall population persistence. In this structure, defined as a metapopulation, the growth, extinction, and recolonization of and by subpopulations is an adaptation to habitat quality changes caused by iterative disturbance and regrowth. The distribution of the subpopulations is important to consider during development of a monitoring plan to ensure not only that the subpopulations are included in the monitoring framework, but that suitable habitat patches not currently occupied, as well as the intervening matrix, are also included.

Johnson (1980) described second-order selection as the establishment of a home range: the area that an individual or pair of individuals uses to acquire the resources that it needs to survive and reproduce. Not all species have established home ranges, but most do. Individuals which have nests, roosts, hibernacula, or other places central to their daily activities move in an area around that central place to acquire food, use cover, drink water, and raise young. Home ranges are not the same as territories. A territory is the space, usually around a nest, that an individual or pair defends from other individuals of the same species and occasionally other individuals of other species. Territories may be congruent with a home range, smaller (if just a nest site is defended), or may not be present at all.

Species with larger body mass need more energy to support that mass than smaller-bodied species. Herbivores tend to have smaller home ranges than carnivores of the same size, both because energy available to herbivores is more abundant and because with each increase in trophic level, there is a decrease in energy availability. Home range sizes also vary among individuals within a species. Generally speaking, if food resources are less abundant or more widely distributed, home range size increases. But within a species the home range size has an upper limit that is governed by balancing energy input from food with energy loss by movement among food patches. For instance, Thompson and Colgan (1987) reported larger home ranges for American marten during years of low prey availability than in years of high prey availability. In this example, to ensure that observations of American marten are independent from one another, sample sites should be distributed in a manner that no individual marten is likely to visit more than one site. Because the marten can be difficult to detect in some settings, however, subsamples within sample sites may be used to ensure that the species is detected if it is present.

Third-order selection is the use of patches within a home range where resources are available to meet an individual's needs. Biologists often can delineate a home range based upon observed daily or seasonal movements of individuals going about their business of feeding, resting, and raising young. But this area is not used in its entirety. Rather, there are some places within the home range that are used intensively, and other parts of the home range that are rarely used (Samuel et al. 1985). Selection of these patches is assumed to represent the ability of the individual to effectively find and use resources that will allow it to survive and reproduce. Sampling in these patches can influence the detectability of a species to the extent that sampling says more about the habitat elements and resources important to the species of interest rather than a species' density or abundance.

Fourth-order selection is the selection of specific food and cover resources acquired from the patches used by the individual within its home range. Given the choice among available foods, a species should most often select those foods that will confer the greatest energy or nutrients to the individual. Which food (or nest site) to select is often a tradeoff among availability, digestibility, and risk of predation (Holmes and Schultz 1988). Factors that influence the selection of specific food and cover resources most often tend to be related to energetic gains and costs, but there are exceptions. The need for certain nutrients at certain times of the year can have little to do with energetics and much to do with survival and fitness. For instance, band-tailed pigeons seek a sodium source at mineral springs to supplement their diet during the nesting season (Sanders and Jarvis 2000).

Collectively, these levels of habitat selection influence the fitness of individuals, populations, and species. Habitat quality is dependent on not only the food and cover resources but also the number of individuals in that patch. Many individuals in one patch means that there are fewer resources per individual. Habitat quality and habitat selection is density dependent. Hence, sampling of habitat elements must consider the likely effects of population density on the probability of detecting a resource. Sampling browse in a forest with high deer densities will result in an estimate of browse availability only in relation to that population size, and not to the potential of the forest to provide browse biomass.

## REMOTELY SENSED DATA

Some information on habitat availability for selected species can be efficiently collected over large areas using remotely sensed data. In the very simplest form, maps allow users to better define their sampling framework by locating and delineating important areas and habitat types. Users can classify areas based on a variety of physiographic characteristics, and are able to document site-specific information, such as winter roosts, vernal pools, and breeding grounds. Data for maps can be based on a number of sources including aerial photographs, satellite imagery, orthophotos, or planimetric maps (Kerr 1986).

## AERIAL PHOTOGRAPHY

Monitoring and inventory programs often use aerial photography for creating maps. Generally, these programs use aerial photos from scales of 1/2,000 to 1/60,000 for the purposes of classification and planning (Kerr 1986). Aerial photographs should not be considered as maps because they contain inherent degrees of radial and topographical distortions. Thus, aerial photographs must be corrected for geometric and directional inaccuracies. However, many geographic information systems (GISs) are capable of correcting these problems. Orthophoto maps are commonly used for large-scale monitoring projects (covering 8,000–400,000 ha). These maps are corrected aerial photos with topographic information superimposed. If available, these types of data can be very useful during design of a sampling scheme. Aerial photography is usually available in black and white, true color, and color infrared (Figure 9.3).

Color-infrared aerial photography is useful for identifying riparian vegetation and seasonally dominant vegetation (Cuplin et al. 1985; Kerr 1986). The primary use of color infrared photographs is for analyzing and classifying vegetation types. This is because healthy green vegetation is a very strong reflector of infrared radiation and appears bright red on color infrared photographs. Oftentimes, aerial photography is taken at different times in the year, allowing for seasonal comparisons of vegetation types (i.e., leaf-on and leaf-off photography). Yearly differences can be documented to describe changes in habitat area and composition over time.

Black and white image        Color infrared image

**FIGURE 9.3** (*A color version of this figure follows page 144.*) Examples of panchromatic (black and white) and color infrared aerial photographs.

Although aerial photography can be essential in creating maps for habitat types and area classification, the accuracy of these maps depends on photo interpretation by humans. Misclassification of habitat types is one common result of human error during photo interpretation. Consequently, although aerial photography is an indispensable tool for creating a sampling framework over large geographic areas, the accuracy, reliability, and usefulness of the map is influenced by the expertise of those responsible for interpreting and classifying the photography.

## SATELLITE IMAGERY

Satellite imagery has become a popular method for describing gross vegetation types over large geographic areas (see Figure 9.4). The majority of satellite imagery data comes from the Landsat program. The Landsat program refers to a series of satellites put into orbit around the earth to collect environmental data about the earth's surface (Richards and Jia 1999). The U.S. Department of Interior and NASA initiated this program under the name Earth Resources Technology Satellites. Spatially, these data consist of discrete picture elements, referred to as pixels, and are radiometrically classified into discrete brightness levels (Richards and Jia 1999). This reflected light is recorded in different wavelength bands, the number of which depends on the type of sensor carried by the satellite. For instance, Landsat 7 was launched in April 1999 and carries the Enhanced Thematic Mapper Plus (ETM+), which produced data in seven specific bands and one panchromatic band (Richards and Jia 1999).

The area each pixel covers on the ground is referred to as that image's resolution. The pixel size for Landsat image data is typically 30 m². Using the digital

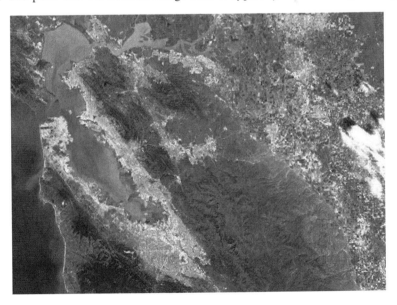

**FIGURE 9.4** (*A color version of this figure follows page 144.*) San Francisco Bay area, California. This Band 3,2,1 image shows the spring runoff from the Sierras and other neighboring mountains into the bay and out into the Pacific. (Image from Landsat 7 Project, NASA.)

reflectance data, these images can be initially classified by humans, but later extrapolated through computer classification, to the rest of the image. Satellite imagery produces digital data that carry the distinctive advantage of being able to be processed by computers that can distinguish small differences in color in a consistent, easily repeatable way.

There are, however, several drawbacks to using satellite imagery. In general, satellite imagery does not provide the detail that is found in aerial photography (Kerr 1986). Computer interpretation and classification often has difficulty in separating vegetation species or areas of the same reflectance. This may pose a problem if one is attempting to classify areas based on a species of plant. The scale of image resolution may not be sufficient to address the needs of the monitoring protocol. Although satellite imagery has proved useful for regional analyses (see Scott et al. 1993), the level of spatial detail and resolution may not be appropriate for small-scale monitoring programs.

## Accuracy Assessment and Ground-Truthing

The accuracy of photo interpretation and satellite imagery depends on ground-truthing and subsequent accuracy assessment. The process of classifying such broad ranges of environmental features into specific, and often simplified, land use and land cover classes leads to inevitable errors in classification. Having a well-distributed grid of Global Positioning System (GPS) field surveys is a useful method for rectifying such problems. This process, *ground-truthing*, is an absolute necessity for any map, whether or not the map is based on satellite imagery or aerial photography. However, this method can be relatively expensive if there are a large number of ground points across an extensive geographic area. Recently, aerial videography has become a useful method for assessing the accuracy of Landsat imagery. This method consists of airborne video data systems that tag each frame of video with geographic coordinates from a GPS, providing a cost- and time-effective method for obtaining data on vegetative communities over large geographic areas (Graham 1993). This method can be combined with ground-truthing procedures, but allows for a decreased dependency on ground-based field surveys. These methods of accuracy assessment, and others like them, allow for rigorous statistical analyses to produce acceptably accurate land use and land cover maps. It is important to stress that to legitimately use and create accurate maps based on satellite imagery or aerial photography, monitoring protocols must incorporate ground-truthing and accuracy assessment.

## Vegetation Classification Schemes

Vegetation classification is a common use of map data. Although vegetation can change over time and may not be a particularly useful means of stratifying samples, vegetation structure and composition are often highly related to a species occurrence or abundance. Biologists and managers often use a hierarchal approach to stratify vegetation types based on dominant and subdominant vegetation, landform, soil composition, or other factors that they deem pertinent (Kerr 1986). Once a classification criterion is determined, the landscape can be separated into discrete units so that

# Techniques for Sampling Habitat

samples can be distributed into an adequate number of habitat types that represent strata in a stratified random sample. Vegetation stratification can yield important information about the use of different types by a species. The Resource Inventory Committee of British Columbia has outlined a good approach to vegetation stratification (Robinson et al. 1996):

1. Delineate the project area boundary.
2. Conduct a literature review of the habitat requirements of the focal species. If there is enough available and accurate information on the habitat requirements of the species, then it may be possible to identify those vegetation components that relate to habitat quality. However, caution should be used when relying on habitat associations from previous studies since many studies may not be applicable to the region of study.
3. Develop a system of habitat stratification that you expect will coincide with species habitat requirements.
4. Use maps, aerial photographs, or satellite imagery to review and select sample units that are reflective of the study area.
5. Evaluate the availability of each habitat strata within the study area. If there are any habitat types that are inaccessible, then no inferences can be made to similar habitat types throughout the project area.

## CONSISTENT DOCUMENTATION OF SAMPLE SITES

To accurately monitor changes in habitat and populations, long-term monitoring protocols must emphasize a repeatable sampling scheme. Making inferences about a population from a sample taken over time depends on the ability of a protocol to resample and revisit the same set of points from one year to the next. Monitoring protocols should incorporate consistent documentation of sample sites throughout the study area. This typically involves the use of GPSs, GISs, and other computer-based systems for data analysis, storage, and retrieval. GPS is a valuable tool for conducting field surveys and relocating sample sites. Using GPS, researchers can geo-reference sample site locations using latitude, longitude, or UTM coordinates. Along with GPS, careful notes should be taken on the location and description of sample sites. These notes should be reported in a consistent manner and held in office records. Data management and documentation remains a critical component of project supervision. It is vitally important that there is proper documentation and storage of sample site locations for ensuring the long-term and repeatable success of monitoring protocols.

## GROUND MEASUREMENTS OF HABITAT ELEMENTS

Oftentimes a monitoring framework will be based on remotely sensed data as a first coarse-scale approach to documenting change followed by more detailed ground-based sampling. Changes in resource availability for species often are predicted using measurements of habitat elements and habitat types. Habitat measurements include abiotic and biotic features such as rock, soil, elevation, vegetation types, snags, ground cover, and litter. These data apply to a specific habitat type, which is defined as an area that is characterized by consistent abiotic and biotic features (Jones 1986). To assess habitat availability, a number of measurements are documented at various scales and include (Jones 1986):

- Size of sites (area)
- Location and position of habitat sites, including distance between habitat sites and heterogeneity within an area
- Edge influences created by habitat size interfaces and ecotones
- Temporal availability of habitat sites

These data are collected over various scales; however, the relevant scale at which to collect and interpret these data will be defined by the organism (see previous section).

Information that relates to the abiotic or biotic influences on the species is considered a habitat element or component (Jones 1986). These elements can be highly significant measurements in terms of the distribution, fitness, and stability of a population throughout their range. Habitat elements are characterized both spatially and temporally, and can represent an entire habitat type or a specific area occupied by the species. Examples of abiotic and biotic data that are typically included are described by a number of authors (Cuplin et al. 1985; Cooperrider 1986; Jones 1986) (Table 9.1):

## TABLE 9.1
## Examples of Habitat Elements That Should be Considered When Identifying Vegetation Types and Sampling Designs

| Habitat Feature | Habitat Elements |
|---|---|
| Landform and physical features | Average elevation, slope, and aspect |
| | Average surface rock cover, talus and rock outcrop abundance and size |
| | Soil depth and composition |
| | Vertical and horizontal heterogeneity of soils and rocks |
| Vegetation-derived features | Average leaf and litter cover per unit area |
| | Snag density and size |
| | Dead and down woody debris |
| | Average height, density, and composition of vegetation types |
| | Vertical and horizontal heterogeneity of litter, snags, and live vegetation |
| Aquatic physical features | Streamflow pattern (riffles, pools, glides, and cascades) |
| | Streamflow volume |
| | Stream width, depth, and gradient |
| | Water temperature |
| | Turbidity |
| | Surface acreage |
| | Stream channel stability |

# Techniques for Sampling Habitat

The measurement and interpretation of habitat elements is central to understanding current conditions and desired future conditions in patches and landscapes. In the subsequent sections we describe a few simple techniques for measuring the availability of key habitat elements, but more comprehensive information on field sampling of habitat elements can be found in James and Shugart (1970), Hays et al. (1981), Noon (1981), and Bookhout (1994).

## METHODS FOR GROUND-BASED SAMPLING OF HABITAT ELEMENTS

Some habitat elements are particularly important to many species depending on their size, distribution, and abundance. These include: percent cover, height, density, and biomass of trees, shrubs, grasses, forbs, and dead wood. Other habitat elements are associated with only a few species, such as stream gradients (e.g., beaver; Allen 1983) and tree basal area (e.g., downy woodpecker; Schroeder 1982). The following constitutes suggestions for an exercise that will train you in some of the more common habitat sampling methods:

Visit two areas with very different management histories such as a recent clearcut and an unmanaged forest. First, compare habitat elements between the two stand types. Then, choose a particular organism and, using life history information, a habitat suitability index model, and a geographic information system, assess the habitat quality of each site for that species.

### RANDOM SAMPLING

Probably the most important part of sampling habitat is to sample randomly within the area of interest (stand, watershed, stream system, etc.). Systematic or subjective sampling can introduce bias into your estimates and lead to erroneous conclusions. In this example you will be sampling two forest stands, but it could be wetlands, prairies, or any other unit of land or water. Within your stand you should collect a random sample of data describing the habitat elements. For the purposes of this exercise, you will collect data from three or more randomly located points in each stand.

1. Using a random numbers table (nearly all statistics books have these) first select a three-digit number that is a compass bearing (in degrees) that will lead you into the stand. If the number that you select does not lead you into your stand then select another number until you have a bearing that will work.
2. Select another three-digit number that is a distance in meters. Using your compass to establish the bearing and either a 30-m tape measure or pacing, walk the randomly selected distance in the direction of the assigned bearing and establish a sample point. You will collect habitat data at this point. Once you have completed collecting data at this point, you repeat the process of random number selection three or more times in this stand and then three or more times in the second stand.

## Vegetation Sampling

### Measuring Density

One of the most common habitat elements that you will measure is density of items, usually trees, snags, logs, shrubs, or other plants. Density is simply a count of the elements over a specified area. Counting individual plants on plots delineated by a sampling frame is the most commonly used approach to estimate the frequency or density of plant populations. Measurements of plant biomass also can be obtained from the dry weights of vegetation clipped on plots having standardized heights, as well as width and length. Standardized photographic methods (i.e., photoplots) can be used to capture a visual record of a sampling unit that is useful for monitoring studies (Elzinga et al. 1998).

Data collected from plots are affected by the size and shape of the sampling unit. Square plots, known as "quadrats," and circular plots have smaller boundary-interior ratios compared to rectangular shapes of equal area. In heterogeneous habitats, data collected on long plots often have been found to have lower statistical variance than data from compact plots of the same total area (Krebs 1989).

The optimum size and shape of a plot will differ according to plant life-form, environmental conditions, and survey objectives. Typically, the optimum plot configuration will be one that provides the greatest statistical precision (i.e., lowest standard error) for a given area sampled. Several investigators have developed approaches for determining the most appropriate plot size and shape for a particular population survey (Hendricks 1956; Wiegert 1962). Krebs (1989) provided a useful review of standardized plot methods.

When estimating the density of trees, you usually will count all the trees in a circular plot, usually 0.04 ha (0.1 acre) in size. Saplings and tall shrubs are usually measured in a 0.004-ha (0.01-acre) plot. Small shrubs and tree seedlings are usually measured in a 0.0004-ha (0.001-acre) plot.

1. From plot center, measure out in each cardinal direction (N, E, S, W) 11.3 m (37.2 ft) (the radius of a 0.04-ha [0.1-acre] plot). Mark these places with flagging.
2. Using a diameter tape or a Biltmore stick, measure the diameter at 1.3 m (4.5 ft) above ground of all live trees in the plot that are >15 cm (6 in.) in diameter at breast height (dbh) and record the species of each tree. Repeat this procedure for all dead trees >15 cm dbh. Expand this sample to a hectare (or acre) by multiplying the estimates by the proper conversion factor, in this case using 25/1 converts our number to a per hectare estimate (10/1 will generate a per acre estimate). This procedure, with the appropriate conversion factors, can be repeated for smaller plot sizes to estimate more tediously sampled habitat components such as seedling numbers.

### Estimating Percent Cover

Use your four 11.3-m (37.2-ft) radii as transects, by walking along each and stopping at five equidistant points to estimate canopy cover. There are a number of techniques available to estimate canopy cover including moosehorns (Garrison 1949) and densitometers (Lemmon 1957). A simple approach to estimating cover is to estimate the presence or absence of vegetation using a sighting tube (a piece of PVC pipe with crosshairs) (James and Shugart 1970). At each of the 20 points on your transects, look directly up and see if the crosshairs intersect vegetation (if so, record a 1) or sky (if so, record a 0). Repeat this at each of the five points on each of the four transects. Tally the number of "1's" recorded from these points. Divide by 20 to estimate the proportion of the plot covered by vegetation. A similar approach to estimating cover can be used by looking down rather than up, but estimates of ground covered by grasses and forbs often use somewhat different techniques. One method of measuring ground cover for a plant species is to calculate the percentage of points in contact with the target species from a standardized arrangement of points on the sampling unit. The technique utilizes marks at fixed intervals along a tape, or pins held along a linear frame positioned above the sampling unit (i.e., point frame). Points are particularly useful for estimating the cover of short grasses, dense patches of forbs, and other low plant species. Bonham (1983) reviews several types of simple apparati to standardize point observations and greatly improve the precision of cover measurements.

Ground and shrub cover can also be measured using a fiberglass measuring tape stretched tautly between two end poles. The surveyor proceeds along the tape, recording each point where the tape first intercepts the vertical plane of any aboveground plant part of the target species, and then the point where interception with the target species is interrupted. Cover is estimated as a percentage of the total length of tape intercepting the target species. Interception points can be reliably observed along the tape to approximately eye level; a staff or contractor's level held vertically and moved along the tape by the surveyor can extend the effective height of measurements to 4 m or more. The position of the end poles should be permanently monumented to ensure re-measurements are reliably positioned.

## Estimating Tree Height

Use a clinometer with a percent scale (look through the viewfinder and you should see two scales, with units given on them if you look straight up or straight down).

1. Walk 30 m (100 ft) from the base of the tree or other object that you wish to measure.
2. Looking through the viewfinder, align the horizontal line in the viewfinder with the top of the tree. Record the number on the percent scale (top).
3. Looking through the viewfinder, align the horizontal line in the viewfinder with the base of the tree. Record the number on the percent scale (bottom).
4. If the top number is positive and the bottom number is negative (<0) then add the absolute values of these two numbers together to estimate height in feet.
5. If the top number is positive and the bottom number is also positive (>0), then subtract the absolute value of the bottom number from the top number to estimate height in feet.

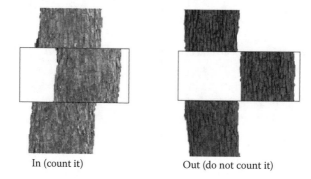

In (count it)  Out (do not count it)

**FIGURE 9.5** When using a wedge prism you have two images to compare—the one you see through the prism and the one above or below the prism. If they overlap you count the tree as an "in" tree. If the images do not overlap then the tree is not counted. (Image from Jesse Caputo; used with permission.)

### Estimating Basal Area

Basal area is the cross-sectional area of all woody stems at 1.3 m (4.5 ft) above ground. It is a measure of tree dominance on a site. The higher the basal area, the greater the dominance by trees. There are two ways to estimate basal area. First, you can use your estimates of dbh from your sample of trees (see estimating density, above) to calculate the area of each stem ($A = 3.1416*r^2$, where r = dbh/2). By summing the areas on a 0.04-ha (0.1-acre) plot and then multiplying the total by 25, you arrive at an estimate of basal area per hectare (multiply by 10 to estimate basal area per acre).

Alternatively, you can use a wedge prism (Figure 9.5) using a plotless method for estimating basal area. Holding the prism over the plot center (do not stand at the center), look at a tree through the prism. If the image that you see through the prism is connected to the image of the tree outside of the prism, then tally the tree and record its species. If the image that you see through the prism is disconnected from the image outside the prism, then do not record the tree. Moving in a circle around the prism that you continue to hold over plot center, record all trees that have the prism image connected to the image outside of the prism regardless of whether they fall in the 0.04-ha (0.1-acre) plot or not. Tally the number of trees that were recorded. Generally you will use a 10-factor prism, that is, each tallied tree represents 10 other trees per acre. Multiply the number of trees tallied by 10; this estimates the basal area in square feet per acre for this site. Other factor prisms do exist, however, and may be more appropriate for your survey, depending on the specific characteristics of the habitat being sampled (Wensel et al. 1980).

Another class of "plotless" techniques utilizes measurements to describe the spacing of individuals in an area to estimate population density. Commonly referred to as "distance methods," these techniques are founded on the assumption that number of individuals in a population can be determined by measuring the average distance among individuals, or between individuals and randomly selected points. Two of the most widely used techniques include the point-centered quarter method (Cottam and

# Techniques for Sampling Habitat

Curtis 1956) and the wandering-quarter method (Cantana 1963). Distance methods have been commonly used for vegetation surveys, and are easily adapted to inventories of rare plants or other sessile organisms. Collecting data for surveys based on distance methods may have some practical advantages over plots or transects. First, distance methods are not susceptible to counting errors that often occur near plot boundaries, thus may yield more accurate density estimates. Secondly, the time and effort to attain an adequate sample of distance measurements in an area often is less than that required to search for every target organism on a plot, thus increasing the efficiency of the survey. Cottam and Curtis (1956) recommended a minimum of 20 measurements for estimating population density or abundance using the point-centered quarter method. However, it is always prudent to assess the adequacy of the sample size empirically using widely available formulas (Bonham 1989).

## Sampling Dead Wood

Dead wood has been recognized as a habitat element important to many species (McComb 2007). There are a number of methods for sampling dead wood, but the most comprehensive uses line intercept sampling to generate estimates of a variety of dead wood characteristics (Harmon and Sexton 1996). The line transects often radiate out from a plot center, and the diameter of logs on which they fall is measured at the point of intersection. Usually an estimate of decay class (1–5) is also made (McComb 2007). Diameters of elliptically shaped logs should be measured in two dimensions and then averaged to get a single value. These measurements are then used. Volume, projected area, and surface area of small and large CWD can be calculated using the following equations (Marshall et al. 2000, 2003):

- Volume/ha (m³/ha) = ($\pi^2$/(8 * transect length)) * $\Sigma$(log diameters²)
- Total projected area/ha (m²/ha) = ((50 * $\pi$)/transect length) * $\Sigma$(log diameters)
- Total surface area (m²/ha) = ((50 * $\pi^2$)/transect length) * $\Sigma$(log diameters)

where transect length is in meters and log diameter is in centimeters.

## Estimating Biomass

Biomass of vegetation is usually estimated to provide information on winter food availability for herbivores. A number of species, including deer, moose, elk, and hare, depend most on these browse resources after the vegetation senesces. Herbivores usually will only eat woody growth resulting from the most recent growing season, and during winter, this includes the twigs and buds, but not leaves.

Within a 1.1-m (3.7-ft) radius plot, use clippers to clip all of the twigs within the plot that have resulted from the most recent growing season. Remove and discard the leaves and place the twigs in a bag. Return to the lab and weigh the bag with the twigs. Remove the twigs and weigh the empty bag. Subtract the bag weight from the bag + twigs weight to estimate biomass per 0.0004-ha (0.001-acre) plot. Multiply this number by 2,500 to estimate biomass (kg) per hectare (or 1,000 to generate a per acre estimate). Because the proportion of the food biomass composed of water can vary from one period of the growing season to the next and from one year to the next, usually the samples are oven dried and then weighed to standardize estimates of biomass.

**TABLE 9.2**

**Comparison of Average and Range of Habitat Elements Between Clearcut (With a Legacy of Living and Dead Trees) and Uncut Forests, Cadwell Forest, Pelham, Massachusetts**

|  | Clearcut | Mature Forest |
|---|---|---|
| Trees > 15 cm/ha | 3 (0–6) | 308 (234–412) |
| Snags > 15 cm/ha | 1 (0–2) | 22 (4–43) |
| Basal area/ha | 2.4 (0–3) | 16 (12–18) |
| Canopy cover (%) | 4 (0–7) | 95 (90–100) |
| Canopy height (m) | 23 (18–34) | 27 (23–33) |
| Browse (kg/ha) | 1,234 (554–2,600) | 387 (122–788) |

## USING ESTIMATES OF HABITAT ELEMENTS TO ASSESS HABITAT AVAILABILITY

The aggregate data on habitat element abundance, sizes, and distribution can now be used to estimate habitat availability. Using Table 9.2 and the downy woodpecker as an example (if you are carrying out this exercise, you can use your own numbers and species information from your field samples and research), consider how you would interpret these data. DeGraaf and Yamasaki (2001) described habitat for downy woodpeckers as "... woodlands with living and dead trees from 25–60 cm dbh; some dead or living trees must be greater than 15 cm dbh for nesting." Although both sites contain trees and snags of sufficient size, the canopy cover data in Table 9.2 would suggest that the clearcut is not functioning as a woodland. Thus, we would probably not consider it suitable habitat for downy woodpeckers (though they certainly do use snags in openings at times). Similar descriptions of habitat are available for many species in North America and Europe; this type of subjective assessment of habitat availability can be easily accomplished.

## USING ESTIMATES OF HABITAT ELEMENTS TO ASSESS HABITAT SUITABILITY

In addition to using your data to understand if a site might be used by a species, habitat suitability index models have been developed to understand if some sites might provide more suitable habitat than others (e.g., Schroeder 1982). Very few of these models have been validated, especially not using fitness as a response variable. Nonetheless, they do represent hypotheses based on the assumption that there is a positive relationship between the index and habitat carrying capacity. In the case of the downy woodpecker, its habitat suitability is based on two indices: tree basal area (Figure 9.6) and density of snags >15 cm dbh (Figure 9.7). Considering first the uncut stand (Table 9.2), note that there is an average of 16 $m^2$/ha of basal area and 22 snags per ha (8.8 per 0.4 ha). The corresponding suitability index score for each variable is 1.0 and the overall habitat suitability is calculated (in this case) as the minimum of the two values. Hence this should be very good habitat for downy woodpeckers.

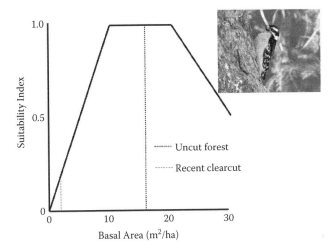

**FIGURE 9.6** Habitat suitability relationship for downy woodpeckers for one of two suitability indices: basal area. (Redrafted from Schroeder, R.L. 1982. Habitat suitability index models: Downy woodpecker. USDI Fish and Wildlife Service FWS/OBS-82/10.38; photo inset by Laura Erickson is used with permission.)

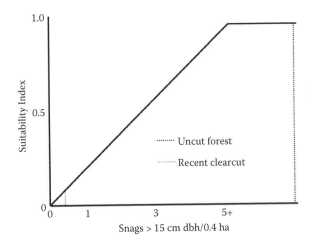

**FIGURE 9.7** Habitat suitability relationship for downy woodpeckers for one of two suitability indices: snag density. (Redrafted from Schroeder, R.L. 1982. Habitat suitability index models: Downy woodpecker. USDI Fish and Wildlife Service FWS/OBS-82/10.38.)

In the recent clearcut, however, the suitability index for snags is approximately 0.1 and for basal area is approximately 0.2. Hence, the overall suitability in the recent clearcut for this species is 0.1; according to this index, it is both unsuitable in general and much less suitable relative to the uncut stand. Further, in the recent clearcut, snag density, which has the lowest index score, is the factor most limiting habitat quality for downy woodpeckers. The best way to use these types of models is to compare the habitat suitability of one site to another rather than as definitive data pertinent to species abundance, density, or fitness. It is also not justifiable to use habitat suitability scores as the basis for value judgments; indeed, if we were to use this technique to assess habitat suitability for snowshoe hares then we might find the recent clearcut to be much better habitat.

## ASSESSING THE DISTRIBUTION OF HABITAT ACROSS THE LANDSCAPE

It is often important to know if patches are suitable habitat for a species and how they are arranged on a landscape. In Figure 9.8, a 490-ha forest has been broken into habitat types based on overstory cover and stand structure. Field samples were taken at 117 points distributed across the forest, and habitat elements were sampled at each point. Habitat suitability index values are then calculated at each point and extrapolated to the habitat types as portrayed in this figure to illustrate how habitat availability for a species can be displayed over a landscape. A different pattern would emerge for other species using this same approach, and these would have to be overlain on patches used as the basis for management. In addition to synthesizing monitoring information to display changes in spatial patterns, these types of maps can guide planning for management activities in order to achieve habitat patterns leading to a desired future condition for the landscape.

## LINKING INVENTORY DATA TO SATELLITE IMAGERY AND GIS

Ohmann and Gregory (2002) described a technique called the *gradient nearest neighbor* (GNN) approach to populate raster cells in satellite imagery with ground plot data (Figure 9.9). They used direct gradient analysis and nearest-neighbor

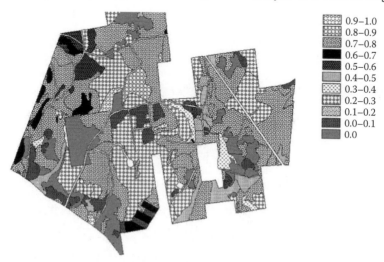

**FIGURE 9.8** An example of a mosaic of habitat patches of varying suitability based on extrapolation of ground inventory data to digitized patches in Cadwell Memorial Forest, Pelham, Massachusetts.

# Techniques for Sampling Habitat

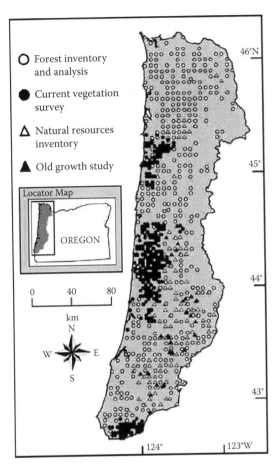

**FIGURE 9.9** A distribution of ground plots is used as the basis for imputing ground-based estimates of structure and composition to unmeasured locations using a GNN approach. (From Ohmann, J.L., and M.J. Gregory. 2002. *Canadian Journal of Forest Research* 32:725–741. Used with permission.)

imputation to ascribe detailed ground attributes of vegetation to each pixel in a digital landscape map (Figure 9.10). The gradient nearest-neighbor method integrates vegetation measurements from regional grids of field plots, mapped environmental data, and Landsat Thematic Mapper (TM) imagery (Ohmann and Gregory 2002). The accuracy of maps created with this approach is normally questioned, but at the regional level, mapped predictions represent the variability in the sample data very well and predict area by vegetation type accurately in many forests where the approach has been tested. For instance, prediction accuracy for tree species occurrence and several measures of vegetation structure and composition was good to moderate in the Oregon Coast Range (Ohmann and Gregory 2002). Vegetation maps produced with the gradient nearest-neighbor method are often used for regional-level monitoring, planning, and policy analysis. Their utility decreases as they are used over smaller spatial scales.

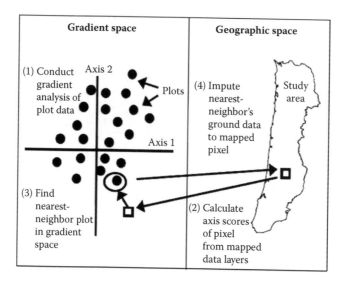

**FIGURE 9.10** Steps in the gradient nearest-neighbor method. (From Ohmann, J.L., and M.J. Gregory. 2002. *Canadian Journal of Forest Research* 32:725–741. Used with permission.)

Results of the maps generated using the GNN approach can then be reclassified in a nearly infinite number of ways to represent habitat availability for various species over large areas (Figure 9.11). With repeated monitoring of ground plots and synchronous acquisition of satellite image data, new projections of habitat availability can be developed over time to understand trajectories of change. In addition, by linking the ground plot data to vegetation growth models, changes in vegetation structure and composition and resulting habitat can be projected into the future under various assumptions regarding land use, natural disturbances, or other environmental stresses (Spies et al. 2007).

## MEASURING LANDSCAPE STRUCTURE AND CHANGE

In addition to assessing habitat elements at various spatial scales over space and time, the pattern of patches across landscapes can be monitored. The assumption often made by landscape ecologists is that pattern influences process and processes create patterns. Consequently, knowledge of aspects of landscape patterns such as characteristics of patches (size, shape, edges, etc.) and among patches (richness, diversity, arrangement) can be informative when inferring how processes might be influenced by changes in landscape pattern. One often-used tool used to generate measurements of landscape pattern is FRAGSTATS (McGarigal et al. 2002). FRAGSTATS quantifies the extent and configuration of patches in a landscape. The user must define the basis for the analysis, including the extent (outer bounds) and grain (the finest level of detail) in the landscape. All patches in a landscape must be classified in a manner that is logical to the objectives of the analysis (e.g., related to habitat for a species or group of species). FRAGSTATS computes three groups of metrics. For a given landscape mosaic, FRAGSTATS computes several metrics for (1) each patch in the

# Techniques for Sampling Habitat

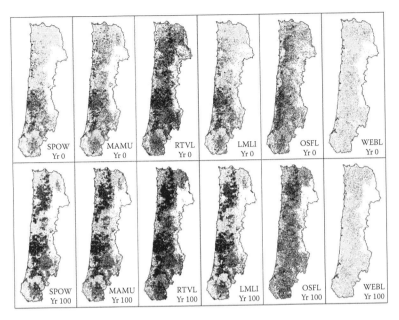

**FIGURE 9.11** An example of applying habitat suitability models to an ecoregion using GNN-based data from the Oregon Coast Range and projecting habitat change 100 years into the future for northern spotted owls (SPOW), marbled murrelets (MAMU), red tree voles (RTVL), low mobility lichens (LMLI), olive-sided flycatchers (OSFL), and western bluebirds (WEBL). (From Spies, T.A. et al. 2007. *Ecological Applications* 17:48–65. Used with permission.)

mosaic, (2) each patch type (class) in the mosaic, and (3) the landscape mosaic as a whole (McGarigal et al. 2002). These metrics can easily become the basis for long-term monitoring to assess changes in landscape patterns and allow inference to changes in landscape processes.

## SUMMARY

Effectively sampling habitat within a monitoring framework requires that the key habitat elements for the species are defined. Habitat is a species-specific concept, so unless monitoring is focused on habitat types (a biophysical classification of the environment), then a set of habitat elements must be identified as important to one of four levels of habitat selection: geographic range, home range, resource patches, or habitat resources. Sampling these habitat elements must consider these scales of habitat selection when defining the grain and extent of a monitoring framework for the species of interest. Often remotely sensed data are used as a first assessment of habitat patches, but ground plot data are usually stratified by patch types identified from remotely sensed data. Ground plot data can provide more detailed information on the availability of habitat elements when extrapolated to specific patch types, or can be used to populate a GIS layer using gradient nearest-neighbor approaches. Remotely sensed data, ground plot data, and GIS layers interpreted as habitat for the species of interest can all be used to monitor changes over time in habitat availability for a species.

Similarly, changes in the pattern of landscape structure can be assessed over time to understand how changes in pattern might influence key ecological processes on the landscape. FRAGSTATS is a commonly used tool to conduct such assessments.

## REFERENCES

Allen, A.W. 1983. Habitat suitability index models: Beaver. USDI Fish and Wildlife Service FWS/OBS-82/10.30 (revised). Washington, D.C.

Barrett, R.H., and H. Salwasser. 1982. Adaptive management of timber and wildlife habitat using DYNAST and wildlife-habitat relationships models. Pages 350–365 in *Proceedings of the Western Association of Fish and Game Agencies.*

Bonham, C.D.1983. Range vegetation classification. *Rangelands* 5:19–21.

Bonham, C.D. 1989. *Measurements of Terrestrial Vegetation.* John Wiley & Sons, New York.

Bookhout, T.A. 1994. *Research and Management Techniques for Wildlife and Habitats.* 5th ed. The Wildlife Society, Bethesda, MD.

Cantana, H.J. 1963. The wandering quarter method of estimating population density. *Ecology* 44:349–360.

Cody, M.L. 1985. *Habitat Selection in Birds.* Academic Press, New York.

Cooperrider, A.Y., R.J. Boyd, and H.R. Stuart, eds. 1986. *Inventory and Monitoring of Wildlife Habitat.* USDI BLM, Service Center, Denver, CO. 858 pp.

Cottam, G., and J.T. Curtis. 1956. The use of distance measures in phytosociological sampling. *Ecology* 37:451–460.

Cuplin, P., W.S. Platts, O. Casey, and R. Masiton. 1985. A comparison of riparian area ground data with large scale airphoto interpretation. Pages 67–68 in *Riparian Ecosystems and Their Management: Reconciling Conflicting Uses: Proceedings, First North American Riparian Conference.* R.R. Johnson, C.D. Ziebell, D.R. Patton, P.F. Folliott, and R.H. Hamre, tech. coords. USDA Forest Service General Technical Report RM-120.

DeGraaf, R.M., and M. Yamasaki. 2001. *New England Wildlife: Habitat, Natural History, and Distribution.* Univ. Press of New England, Hanover, NH.

Elzinga, C.L., D.W. Salzer, and J.W. Willoughby. 1998. Measuring and monitoring plant populations. Bureau of Land Management, Technical Reference 1730–1, Denver, CO.

Garrison, G.A. 1949. Uses and modifications for the "moosehorn" crown closure estimator. *Journal of Forestry* 47:733–735.

Graham, L.A. 1993. Airborne video for near real-time vegetation mapping. *Journal of Forestry* 90(12):16–21.

Harmon, M.E., and J. Sexton. 1996. Guidelines for measurements of woody detritus in forest ecosystems. U.S. LTER Network Office, University of Washington, College of Forest Resources. U.S. LTER Publication 20.

Hays, R.L., C. Summers, and W. Seitz. 1981. Estimating wildlife habitat variables. USDI Fish and Wildlife Service FWS/OBS-81/47.

Hendricks, W.A. 1956. *The Mathematical Theory of Sampling.* The Scarecrow Press, New Brunswick, NJ. 364 pp.

Holmes, R.T., and J.C. Schultz. 1988. Food availability for forest birds: Effects of prey distribution and abundance on bird foraging. *Canadian Journal of Zoology* 66:720–728.

James, F.C., and H.H. Shugart, Jr. 1970. A quantitative method of habitat description. *American Birds* 24:727–736.

Johnson, D.H. 1980. The comparison of usage and availability measurements for evaluating resource preference. *Ecology* 61:65–71.

Jones, K.B. 1986. Data types. Pages 11–28 in *Inventory and Monitoring of Wildlife Habitat*. A.Y. Cooperrider, R.J. Boyd, and H.R. Stuart (eds.). USDI BLM, Service Center, Denver, CO. 858 pp.

Kerr, R.M. 1986. Habitat mapping. Pages 49–72 in *Inventory and Monitoring of Wildlife Habitat*. A.Y. Cooperrider, J.B. Raymond, and H.R. Stuart (eds.). USDI BLM, Service Center. Denver, CO. 858 pp.

Krebs, C.J. 1989. *Ecological Methodology*. Harper Collins, New York. 654 pp.

Lemmon, P.E. 1957. A new instrument for measuring forest canopy overstory density. *Journal of Forestry* 55:667–668.

Lowe, W.H., and D.T. Bolger. 2002. Local and landscape-scale predictors of salamander abundance in New Hampshire headwater streams. *Conservation Biology* 16:183–193.

Marshall, P.L., G. Davis, and V.M. LeMay. 2000. Using line intersect sampling for coarse woody debris. BC Forest Service, Forest Research Technical Report TR-003. 34 pp.

Marshall, P.L., G. Davis, and S.W. Taylor. 2003. Using line intersect sampling for coarse, woody debris: Practitioners' questions addressed. BC Forest Service, Forest Research Extension Note EN-012. 10 pp.

McCarty, J.P. 2001. Ecological consequences of recent climate change. *Conservation Biology* 15:320–331.

McComb, B.C. 2007. *Wildlife Habitat Management: Concepts and Applications in Forestry*. Taylor & Francis, CRC Press, Boca Raton, FL. 319 pp.

McGarigal, K., S.A. Cushman, M.C. Neel, and E. Ene. 2002. FRAGSTATS: Spatial Pattern Analysis Program for Categorical Maps. Computer software program produced by the authors at the University of Massachusetts, Amherst.

Noon, B.R. 1981. Techniques for sampling avian habitats. Pages 41–52 in *The Use of Multivariate Statistics in Studies of Wildlife Habitat*. D.E. Capen (ed.). USDA Forest Service General Technical Report RM-87.

Ohmann, J.L., and M.J. Gregory. 2002. Predictive mapping of forest composition and structure with direct gradient analysis and nearest neighbor imputation in coastal Oregon, USA. *Canadian Journal of Forest Research* 32:725–741.

Pulliam, H.R. 1988. Sources, sinks, and population regulation. *American Naturalist* 132:652–661.

Richards, J.A., and X. Jia. 1999. Interpretation of hyperspectral image data. Pages 313–337 in *Remote Sensing Digital Image Analysis: An Introduction*. Springer, New York.

Robinson, C.L.K., D.E. Hay, J. Booth, and J. Truscott. 1996. Standard methods for sampling resources and habitats in Coastal Subtidal Regions of British Columbia: Part 2—Review of Sampling with Preliminary Recommendations. Canadian Technical Report of Fisheries and Aquatic Sciences 2119. Fisheries and Oceans Canada.

Samuel, M.D., D.J. Pierce, and E.O. Garton. 1985. Identifying areas of concentrated use within the home range. *Journal of Animal Ecology* 54:711–719.

Sanders, T.A., and R.L. Jarvis. 2000. Do band-tailed pigeons seek a calcium supplement at mineral sites? *Condor* 102:855–863.

Schroeder, R.L. 1982. Habitat suitability index models: Downy woodpecker. USDI Fish and Wildlife Service FWS/OBS-82/10.38.

Scott, J.M., F. Davis, B. Csuti, R. Noss, B. Butterfield, C. Groves, H. Anderson, S. Caicco, F. D'Erchia, T.C. Edwards, Jr., J. Ullirnan, and R.G. Wright. 1993. Gap analysis: A geographic approach to protection of biological diversity. *Wildlife Monographs* 123.

Spies, T.A., B.C. McComb, R. Kennedy, M. McGrath, K. Olsen, and R.J. Pabst. 2007. Potential effects of forest policies on biodiversity in a multi-ownership province. *Ecological Applications* 17:48–65.

Thompson, I.D., and P.W. Colgan. 1987. Numerical responses of marten to a food shortage in north-central Ontario. *Journal of Wildlife Management* 51:824–835.

Vesely, D., and W.C. McComb. 2002. Terrestrial salamander abundance and amphibian species richness in headwater riparian buffer strips, Oregon Coast Range. *Forest Science* 48:291–298.

Wecker, S.C. 1963. The role of early experience in habitat selection by the prairie deer mouse, *Peromyscus maniculatus bairdi*. *Ecological Monographs* 33:307–325.

Wensel, L.C., J. Levitan, and K. Barber. 1980. Selection of basal area factor in point sampling. *Journal of Forestry* 78:83–84.

Wiegert, R.G. 1962. The selection of optimum quadrat size for sampling the standing crop of grasses and forbs. *Ecology* 43:125–129.

# 10 Database Management

For many monitoring projects, data management is often considered a nuisance and of less importance than sampling design, objective setting, and data collection and analysis. Yet a proper database management system is a critical component of any monitoring plan and should be considered early in the planning process. In many ways, such a system serves as the ship's log of a monitoring mission and should detail every step of data collection, storage, and dissemination. Sound data management is so vital because a monitoring project adapts and changes over time and as such, so might the data. Furthermore, because most monitoring projects are conducted over many years and include the inevitable changes in staff, data collection and methodologies, land ownership and accessibility, and shifting technologies, improper data management can fail to document these changes and undermine the entire monitoring initiative. In addition, because online data dissemination and digital archives are becoming increasingly popular (if not necessary), data management serves as a much needed blueprint of instructions for future users of the data who might not have been involved in any aspect of the original monitoring plan.

## THE BASICS OF DATABASE MANAGEMENT

The data generated from monitoring programs are often complex and the protocols used to generate these data can change and adapt over time. Consequently, the system used to describe these data, and the methods used to collect them, must be comprehensive, detailed, and flexible to changes. In a perfect world, monitoring data are collected and often entered into a database (as opposed to being stored in a filing cabinet). A comprehensive database should include six basic descriptors of the data that detail how they were collected, measured, estimated, and managed. Ultimately, these basic descriptors ensure the long-term success of a monitoring effort because they describe the details of data collection and storage.

The six essential descriptors are: *what* (the type of organism), *how many* (units of observation for individual organisms or colonies, presence/absence, detection/non-detection, relative abundance, distance measurements), *where* (the geographic location at which the organism was recorded and what coordinate system was referenced), *when* (the date and time of the recording event), *how* (what sort of record is represented and other details of data-collection protocols; e.g., 5-minute point counts, mist-netting, clover trap, etc.), and *who* (the person responsible for collecting the data). Each of these components represents an important aspect of data collection that facilitates future use. For example, information on *how* a recording event was made allows someone separate from the data collection to properly account for variation in effort and detection probability, deal with data from multiple protocols, and determine whether the data are from multiple species or single-taxon records.

## THE GENERAL STRUCTURE OF A MONITORING DATABASE

Unfortunately, there is no "one size fits all" solution to the basic structure of a monitoring database. Monitoring programs are diverse and so are the data they collect. There are, however, several basic and standardized templates that can be used when creating a monitoring database (Huettmann 2005; Jan 2006). As an example, the Darwin Core is a simple data standard that is commonly used for occurrence data (specimens, observations, etc., of living organisms) (Bisby 2000). The Darwin Core standard specifies several database components including record-level elements (e.g., record identifier), taxonomic elements (e.g., scientific name), locality elements (e.g., place name), and biological elements (e.g., life stage). Jan (2006) provided another excellent example of a functional structure for an observational database. Using the terminology of Jan (2006), biological sampling information relates to field site visits, and each of these visits is considered a *Gathering*. Each Gathering event is to be described by the occurrence and/or abundance of a species and additional site information including site name, the period of time, the name of the collector, the method of collection, and geography. The geography field could indicate country codes using the International Organization for Standardization (ISO) standards (www.iso.org), and it should have an attribute detailing whether this information is currently valid because political boundaries and names change over time (e.g., new countries form, their names can change) (Jan 2006). Geospatial data are stored under the heading of GatheringSite and include coordinate data (e.g., latitude and longitude, altitude), gazetteer data (e.g., political or administrative units), and geoecological classifications (e.g., geomorphological types). It is important that this field allows for high-resolution georeferencing for subsequent integration with a GIS (e.g., using five significant digits for latitude and longitude coordinates). The *Unit* field includes organisms observed in the field, herbarium specimens, field data, taxonomic identifications, or descriptive data. An *Identifications* field details the species' common name, species' scientific name, and a species code (using the Integrated Taxonomic Information System [ITIS; www.itis.gov]) to a Unit (specimen, observation, etc.). Identifications can then be connected to a taxon database using a TaxIdRef field. The organization of any monitoring database should have these necessary information fields (although field names may vary) and will likely require the use of a digital database for storage and manipulation.

## DIGITAL DATABASES

Digital databases are now considered an invaluable and commonly used tool for storing data generated from monitoring programs. Even in remote field sites, researchers are using mobile Global Positioning System (GPS) and Personal Digital Assistant (PDA) units to record georeferenced census tracks and species observations (Traviani et al. 2007) (Figure 10.1). Using any laptop computer, these data can then be quickly integrated into database management software such as CyberTracker (cybertracker.org), Microsoft® Excel (office.microsoft.com/en-us/excel), or Microsoft® Access (office.microsoft.com/en-us/access). By using a digital database, researchers gain the ability to georeference census points for later integration into a GIS, such as ArcGIS

# Database Management

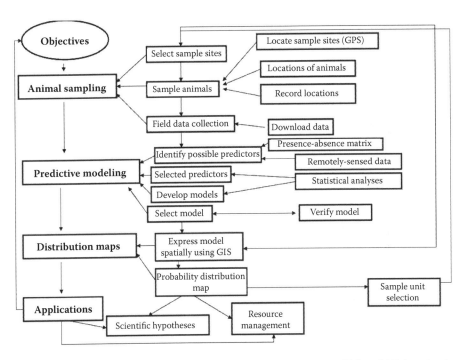

**FIGURE 10.1** A flowchart of an integrated framework for using GPS and PDA technology for collecting monitoring information. By collecting monitoring data in a real-time, digital format these data can be used for more sophisticated purposes such as species distribution mapping. Tasks highlighted in bold indicate the places in which advanced methods can provide increased accuracy. (Redrafted from Traviani, A., et al. 2007. *Diversity and Distributions*, 13.)

(esri.com/software/arcgis/), allowing for additional analytical options such as predictive species distribution modeling (Figure 10.1). Traviani et al. (2007) provided an excellent review and application of a field-based database framework for using digitally stored data to subsequently map animal distributions in remote regions. A key advantage to recording data into a digital database during the collection event itself is the ability to develop and maintain multiple databases. Digital databases also increase the capacity to integrate data into online data management programs and thereby to access data at later dates.

In addition to this, database managers often use online and digital databases because they can be readily linked to other databases for greater functionality. Connecting multiple databases results in a relational database management system (RDMS) (Figure 10.2), which allows for queries to be made among multiple databases. In light of these developments, the structure and framework of many large monitoring databases are increasingly sophisticated and data on demographic rates, abundance, and species occurrences can be linked with other geographic information stored in ancillary databases. For example, a standard relational database can consist of subtables of data that are connected through a common record ID number (Figure 10.2). A Main table normally contains information on the sampling units, the units used for data

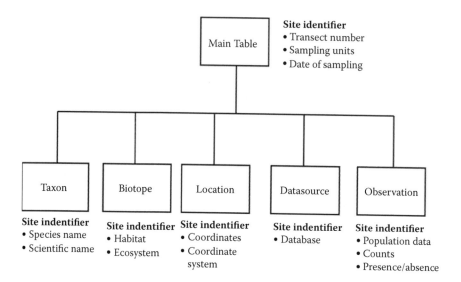

**FIGURE 10.2** A relational database management system combines data from several different databases. These databases are typically linked by a standard unique identifier—in this case, a site identity number—which allows a user to extract data from multiple data sets.

presentation, the years of the study, and notes on sampling design. Other frequently used tables include a Taxon table (information about the organism sampled in each data set; see ITIS [www.itis.org] for globally accepted species names), a Biotope table (habitat of the organism), a Location table (geographical details of the monitoring site), a Datasource table (reference to the original source of the data), and the actual Observation table (original population data). In this case, a relational database and a common record identifier enable the user to perform multiple queries based on species, taxonomic group, habitats, areas, latitudes, or countries. This is particularly powerful because a user can query a unique identifier that refers to a specific study site and then extract data on that site from multiple data tables. In practice, developing, maintaining, and retrieving data from an RDMS often requires knowledge of Structured Query Language (SQL; http://en.wikipedia.org/wiki/SQL), a widely used database filter language that is specifically designed for management, query, and use of RDMS. SQL is a standardized language with a huge user community that is recognized both by the American National Standards Institute (ANSI) and International Organization for Standardization (ISO; iso.org/iso/home.htm). It is implemented in many popular relational database management systems including Informix (ibm.com/software/data/informix), Oracle (oracle.com/index.html), SQL Server (microsoft.com/sql/default.mspx), MySQL (mysql.com), and PostreSQL (postresql.org).

## DATA FORMS

All data collectors should use a standard data form that is approved by stakeholders in the monitoring program (Table 10.1). Copies of these data forms should be included as an appendix to the planning document. The appendix should also provide a data

## TABLE 10.1
## Example of a Field Data Sheet Used in Association With a Project Designed to Monitor the Occurrence and Number of Detections of Birds in Agricultural Lands in the Willamette Valley of Oregon

Date _____  Sample Point _____  Landscape _____  Observer _____
Weather _____                     Time Begin _____  Time End _____

| Obs. Num. | Species | Number | Distance (m) | Repeat (Y/N) | Behavior | Patch Type |
|---|---|---|---|---|---|---|
| 1 | | | | | | |
| 2 | | | | | | |
| 3 | | | | | | |
| 4 | | | | | | |
| 5 | | | | | | |
| 6 | | | | | | |
| 7 | | | | | | |
| 8 | | | | | | |
| 9 | | | | | | |
| 10 | | | | | | |
| 11 | | | | | | |
| 12 | | | | | | |
| 13 | | | | | | |
| 14 | | | | | | |
| 15 | | | | | | |
| 16 | | | | | | |
| 17 | | | | | | |

*Note:* The inset photograph is used not only to locate sample points but also to plot locations of bird observations.
*Species:* Use 4-letter code.
*Number:* Number of individuals.
*Repeat:* Enter "Y" only for repeat observations of the same bird.
*Behavior:* F(eeding), R(esting), O (flyover).
*Patch type:* P(lowed), G(rass), W(Grassed waterway), S(hrubby), T(reed).

format sheet that identifies the data type, unit of measurement, and the valid range of values for each field of the data collection form. The data format sheet should also identify all codes and abbreviations that may be used in the form.

## DATA STORAGE

For many monitoring programs, hardcopy data are still collected despite the increasing availability of digital formats. In general, there are three primary obstacles to the shift of hardcopies to digital formats. First, significant amounts of (historical) hardcopy data remain to be digitized (e.g., in archives, libraries, and filing cabinets). Second, although technological advances are making the collecting of digital data in the field more feasible (Traviani et al. 2007), many field data are still collected in field notebooks when field conditions are difficult and the field-site is remote. Third, many digital data sets are still getting printed as hardcopy for cultural and logistical reasons.

In many cases, even when data are compiled digitally, hardcopies are collected as an important backup for many monitoring programs, or such data are retained and maintained as critical sources of information for legacy programs that have been running for decades. Yet despite this, and although other examples of obstacles associated with the transition to digital documentation doubtless exist, there are many reasons why monitoring data should be collected and stored in a digital format (with necessary backup systems). The advantages of using digital field data collection methods include immediate data availability (e.g., real-time online data entry), lack of labor-intense data key-in sessions afterwards, and automated metadata and processing. Given these benefits, a more universal use of digital data collection would be ideal, but many universities, governments, NGOs, and agencies are hesitant to embrace current technologies. Reasons for reluctance to go entirely digital are varied but typically include a lack of computational training and insufficient infrastructure for using and storing digital data.

## METADATA

Databases resulting from biological inventories and monitoring studies can benefit other scientific research efforts and facilitate species conservation programs for many decades. The usefulness of a database for these purposes, however, is determined not only by the rigor of the methods used to conduct the monitoring program, but also by the ability of future investigators to decipher the variable codes, measurement units, and other details affecting their understanding of the database. When a potential user of a database is interested in deciphering the details of that database, they often refer to the metadata. Metadata are "data about data" and are an essential aspect of any database because they serve as a guide provided by the developer of the database (Figure 10.3). Metadata facilitate information sharing among current users and are crucial for maintaining the value of the data for future investigations. Indeed, if monitoring data are placed online or meant to be shared among participatory stakeholders, then there needs to be a clear description documenting every relevant step of data curation and processing. There are few things more frustrating for potential data users than receiving a database or map with little or no information

# Database Management

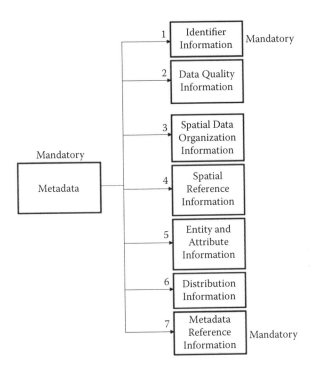

**FIGURE 10.3** Graphical representation of major metadata elements specified by the approved federal standard for geospatial databases (CSDGM Version 2—FGDC-STD-001-1998). (Redrafted from http://biology.usgs.gov/fgdc.metadata/version2.)

on what the variables represent or how the data were collected. The standardized metadata that accompany monitoring databases should be highly valued as one of the principal means for improving transferability of biological monitoring information among different programs, management units, and future data users.

With respect to metadata, the federal government has developed several systems, standards, and templates for database documentation that can be applied to any monitoring database. Since 1995, all federal agencies have adopted a content standard for geospatial data called the Content Standard for Digital Geospatial Metadata (CSDGM; Tsou 2002). This standard was developed by the Federal Geographic Data Committee (FGDC), which is also responsible for reviewing and updating the standard as needed. The currently approved FGDC standard is CSDGM Version 2—FGDC-STD-001-1998 and was developed to be applicable to all geospatial databases. The standard includes seven major elements. Certain GIS packages include software tools that automate a number of these metadata documentation tasks; however, the originator of the database must manually complete most fields. The process of describing data sources, accuracy tests, geoprocessing methods, and organizational information can be tedious and add many hours to the preparation of a data set. This initial cost of the labor and time, however, will ensure that the data can be used for many years into the future, possibly for research or conservation purposes not anticipated by the originator of the data set.

Although prescient, the generic nature of the CSDGM does not provide for standardization of many attributes commonly shared among biological databases. To extend the effectiveness of the CSDGM framework to biological sampling, the FGDC's Biological Data Working Group has standardized the use of terms and definitions in metadata prepared for biological databases with the development of the Biological Metadata Profile (FGDC 1999). The Biological Metadata Profile falls under the broader National Biological Information Structure (NBII) (www.nbii.gov) profile and applies to topics such as taxonomic classification, voucher specimens, environmental attributes, and others not considered in the CSDGM. The Biological Metadata Profile is also applicable to nongeospatial data sets. Considering that most monitoring programs collect biological information, database managers should consult the Biological Metadata Profile before and after data collection.

The FGDC standard and its profiles are widely embraced by the U.S. government and international initiatives (e.g., International Polar Year (IPY)). Nonetheless, several other metadata standards exist, including: (a) Directory Interchange Format (DIF) for a short telephone entry description, which is still widely used by BAS (British Antarctica Service); (b) EML (Ecological Metadata Language) for a rather detailed description of relational databases, which is used by Long Term Ecological Research sites in the United States; and (c) SML (Sensor Metadata Language) for a very powerful and progressive description of high-performance Sensors Networks. There is also a wide array of metadata standards that have local relevance only, and are not compatible with global metadata standards. There is a current movement, however, towards the development and implementation of a more global standardization. Variation among studies in metadata standards impedes global data availability, and deviating from the FGDC NBII standard often results in large information loss. The idea that more local, simpler metadata concepts can simply be mapped, and then cross-walked through automatic parsing software to other standards such as FGDC NBII in order to satisfy delivery needs can prove fatal to data quality, because once an information field is missing, its content can likely never again be filled in a way that maintains the rigor of the database as a whole.

Over 50 collective years of database experience in monitoring populations have led to one major conclusion: the lack of metadata make databases entirely unusable. Thus, metadata and data management needs to occupy a major section of the project budget and should be considered early in the planning process.

## CONSIDER A DATABASE MANAGER

As you may be able to tell from the preceding sections, database management is not a simple task. It requires an understanding of complicated metadata standards and, in the case of many digital databases, a skill set pertaining to computer programming (e.g., SQL language). Unfortunately, perhaps because of the considerable training and effort that are necessary, database management does not receive sufficient attention in terms of time and budgetary allocations in many monitoring efforts. Tasks associated with database management are often given to lower ranking members of the research team, and considered "technician work," not part of "real" science or monitoring. In other settings, databases are simply contracted out to others, and it is hoped that any problems can be fixed that way. Due to these practices, many monitoring databases existing in raw or clumsy formats are published as dead-end PDFs, or are stored in older Excel-type worksheets. The worst-case scenario involves the consignment of years of hardcopy data to a filing cabinet, then to a box, and eventually to the local recycling center (often when the original data manager retires). This situation is particularly discouraging considering that data management has critical implications in the monitoring process and data dissemination and that the proper storage and documentation of years of monitoring data is of utmost relevance for the future use of monitoring data. To overcome this situation, monitoring programs need to make the budgetary commitment necessary to ensure that they have the expertise required for excellent data management. This normally entails hiring experienced database managers.

## AN EXAMPLE OF A DATABASE MANAGEMENT SYSTEM: FAUNA

As an example of a recent database management system, the U.S. Forest Service requires that all monitoring data be stored in the Forest Service's Natural Resource Information System (NRIS) FAUNA database (see Woodbridge and Hargis 2006 for an example). All Forest Service monitoring plans outline several steps in preparing data for entry into the FAUNA database that are intended to be addressed during the development of an inventory or monitoring protocol. Although these basic steps are designed for Forest Service protocols, they can be incorporated into any monitoring initiative:

1. All data collected in the field must be reviewed for completeness and errors before entry into FAUNA. The concerns and techniques specific to the protocol being developed should be discussed.
2. Protocol development teams should become familiar with the major elements of the FGDC-CSDGM, Biological Metadata Profile (FGDC 1999) to better understand metadata standards.
3. Complete descriptions and bibliographic citations for taxonomic, population, or ecological classification systems should be provided, including identification of keywords consistent with the Biological Metadata Profile where appropriate.
4. Sources of maps, geospatial data, and population information that are used to delineate the geographic boundaries of the monitoring program or to locate sampling units should be identified.
5. Units of measurement should be identified.
6. The authors of the protocol and the personnel responsible for maintaining and distributing data resulting from the monitoring programs should be identified (i.e., data steward).
7. The anticipated schedule for data reviews and updates should be described.
8. All data codes, variable names, acronyms, and abbreviations used in the protocol should be defined.
9. An outline or template of the structure of tabular databases in which monitoring program data will be held should be provided.

## SUMMARY

Database management is often an afterthought for many monitoring programs, but proper database management and documentation is critical for the long-term success of a monitoring program. An effective database management details six essential components of data collection including *what, how many, where, when, how,* and *who collected the data.* If possible, data should quickly be incorporated into a digital database either during data collection or shortly thereafter. Digital formats allow for the building of relational database management systems, often automated metadata documentation, and compatibility with current and global metadata standards. Given the increasing sophistication of database management systems, monitoring programs should place a greater emphasis on consulting or hiring a database manager with the necessary skill set needed for maintaining and querying large databases. The failure to develop a strong database management system early in the monitoring program can lead to information loss and an inability to properly analyze and implement the results of a monitoring initiative in the future.

## REFERENCES

Bisby, F.A. 2000. The quiet revolution: Biodiversity informatics and the internet. *Science* 289:2309–2312.

FGDC. 1999. Content standard for digital geospatial metadata, version 2.0. FGDC-STD-001-1998. Federal Geographic Data Committee, Washington, D.C. [http://www.fgdc.gov].

Huettmann, F. 2005. Databases and science-based management in the context of wildlife and habitat: Toward a certified ISO standard for objective decision-making for the global community by using the internet. *Journal of Wildlife Management* 69:466–472.

Jan, L. 2006. Database model for taxonomic and observation data. In *Proceedings of the 2nd IASTED International Conference on Advances in Computer Science and Technology.* ACTA Press, Puerto Vallarta, Mexico.

Traviani, A., J. Bustamante, A. Rodríguez, S. Zapata, D. Procopio, J. Pedrana, and R. Martínez-Peck. 2007. An integrated framework to map animal distributions in large and remote regions. *Diversity and Distributions,* 13.

Tsou, M.H. 2002. An operational metadata framework for searching, indexing, and retrieving distributed geographic information services on the internet. Pages 312–333 in *Geographic Information Science—Second International Conference GIScience 2002.* M. Egenhofer and D. Mark (eds.). Lecture Notes in Computer Science 2478, Springer-Verlag, Berlin.

Woodbridge, B., and C.D. Hargis. 2006. Northern goshawk inventory and monitoring technical guide. USDA Forest Service General Technical Report WO-71. 80 pp.

# 11 Data Analysis in Monitoring

Plant and animal data come in many forms including indices, counts, and occurrences. The diversity of these data types can present special challenges during analysis because they can follow different distributions and may be more or less appropriate for different statistical approaches. As an additional complication, monitoring data are often taken from sites that are in close proximity or surveys are repeated in time, and thus special care must be taken regarding assumptions of independency. In this chapter we will discuss some essential features of these data types, data visualization, different modeling approaches, and paradigms of inference.

This chapter is designed to help you gain biostatistical literacy and build an effective framework for the analysis of monitoring data. These issues are rather complex, and as such, only key elements and concepts of effective data analysis are discussed. Many books have been written on the topic of analyzing ecological data. Thus, it would be impossible to effectively cover the full range of the related topics in a single chapter. The purpose, therefore, of the data analysis segment of this book is to serve as an introduction to some of the classical and contemporary techniques for analyzing the data collected by monitoring programs. After reading this chapter, if you wish to broaden your understanding of data analysis and learn to apply it with confidence in ecological research and monitoring, we recommend the following texts that cover many of the classical approaches: Cochran (1977), Underwood (1997), Thompson et al. (1998), Zar (1999), Scheiner and Gurevitch (2001), Quinn and Keough (2002), Gotelli and Ellison (2004), Crawley (2005, 2007), and Bolker (2008). For a more in-depth and analytical coverage of some of the contemporary approaches we recommend Williams et al. (2002), MacKenzie et al. (2006), Royle and Dorazio (2008), and Thomson et al. (2009).

The field of data analysis in ecology is a rapidly growing enterprise, as is data management (Chapter 10), and it is difficult to keep abreast of its many developments. Consequently, one of the first steps in developing a statistically sound approach to data analysis is to consult with a biometrician in the early phases of development of your monitoring plan. One of the most common (and oldest) lamentations of many statisticians is that people with questions of data analysis seek advice *after* the data have been collected. This has been likened to a postmortem examination; there is only so much a biometrician can do and/or suggest after the data have been collected. Consulting a biometrician is imperative in almost every phase of the monitoring program, but a strong understanding of analytical approaches from the start will help ensure a more comprehensive and rigorous scheme for data collection and analysis.

## DATA VISUALIZATION I: GETTING TO KNOW YOUR DATA

The initial phase of every data analysis should include exploratory data evaluation (Tukey 1977). Once data are collected, they can exhibit a number of different distributions. Plotting your data and reporting various summary statistics (e.g., mean, median, quantiles, standard error, minimums, and maximums) allows you to identify the general form of the data and possibly identify erroneous entries or sampling errors. Anscombe (1973) advocates making data examination an iterative process by utilizing several types of graphical displays and summary statistics to reveal unique features prior to data analysis. The most commonly used displays include normal probability plots, density plots (histograms, dit plots), box plots, scatter plots, bar charts, point and line charts, and Cleveland dotplots (Cleveland 1985; Elzinga et al. 1998; Gotelli and Ellison 2004; Zuur et al. 2007). Effective graphical displays show the essence of the collected data and should (Tufte 2001):

1. Show the data.
2. Induce the viewer to think about the substance of the data rather than about methodology, graphic design, or the technology of graphic production.
3. Avoid distorting what the data have to say.
4. Present many numbers in a small space.
5. Make large data sets coherent and visually informative.
6. Encourage the eye to compare different pieces of data and possibly different strata.
7. Reveal the data at several levels of detail, from a broad overview to the fine structure.
8. Serve a reasonably clear purpose: description, exploration, or tabulation.
9. Be closely integrated with the numerical descriptions (i.e., summary statistics) of a data set.

In some cases, exploring different graphical displays and comparing the visual patterns of the data will actually guide the selection of the statistical model (Anscombe 1973; Hilborn and Mangel 1997; Bolker 2008). For example, refer to the four graphs in Figure 11.1. They all display relationships that produce identical outputs if analyzed using an ordinary least squares (OLS) regression analysis (Table 11.1). Yet, whereas a simple regression model may reasonably well describe the trend in case A, its use in the remaining three cases is not appropriate, at least not without an adequate examination and transformation of the data. Case B could be best described using a logarithmic rather than a linear model and the relationship in case D is spurious, resulting from connecting a single line to the rest of the data cluster. Cases C and D also reveal the presence of outliers (i.e., extreme values that may have been missed without a careful examination of the data). In these cases, the researcher should investigate these outliers to see if their values were true samples or an error in data

Data Analysis in Monitoring

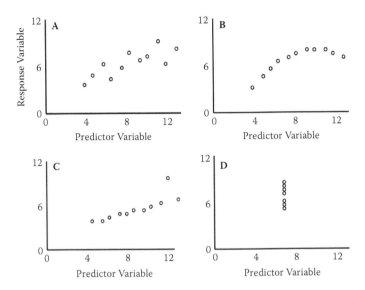

**FIGURE 11.1** Relationships between the four sets of X-Y pairs. (Redrafted from Anscombe, F.J. 1973. *American Statistician* 27:17–21.)

**TABLE 11.1**
**Four Hypothetical Data Sets of X-Y Variable Pairs**

| A | | B | | C | | D | | Analysis Output |
|---|---|---|---|---|---|---|---|---|
| X | Y | X | Y | X | Y | X | Y | |
| 10.0 | 8.04 | 10.0 | 9.14 | 10.0 | 7.46 | 8.0 | 6.58 | N = 11 |
| 8.0 | 6.95 | 8.0 | 8.14 | 8.0 | 6.77 | 8.0 | 5.76 | Mean of Xs = 9.0 |
| 13.0 | 7.58 | 13.0 | 8.74 | 13.0 | 12.74 | 8.0 | 7.71 | Mean of Ys = 7.5 |
| 9.0 | 8.81 | 9.0 | 8.77 | 9.0 | 7.11 | 8.0 | 8.84 | Regression line: |
| 11.0 | 8.33 | 11.0 | 9.26 | 11.0 | 7.81 | 8.0 | 8.47 | Y = 3 + 0.5X |
| 14.0 | 9.96 | 14.0 | 8.10 | 14.0 | 8.84 | 8.0 | 7.04 | Regression SS = 27.50 |
| 6.0 | 7.24 | 6.0 | 6.13 | 6.0 | 6.08 | 8.0 | 5.25 | r = 0.82 |
| 4.0 | 4.26 | 4.0 | 3.10 | 4.0 | 5.39 | 19.0 | 12.50 | $R^2 = 0.67$ |
| 12.0 | 10.84 | 12.0 | 9.13 | 12.0 | 8.15 | 8.0 | 5.56 | |
| 7.0 | 4.82 | 7.0 | 7.26 | 7.0 | 6.42 | 8.0 | 7.91 | |
| 5.0 | 5.68 | 5.0 | 4.74 | 5.0 | 5.73 | 8.0 | 6.89 | |

*Source:* Modified from Anscombe, F.J. 1973. *American Statistician* 27:17–21.

collection and/or entry. This simple example illustrates the value of a visual scrutiny of data prior to data analysis.

The fact that, under some circumstances, visual displays alone can provide an adequate assessment of the data underscores the value of visual analysis to an even greater extent. A strictly visual (or graphical) approach may even be superior to formal data analyses in situations with large quantities of data (e.g., detailed measurements of demographics or vegetation cover) or if data sets are sparse (e.g., in the case of inadequate sampling or pilot investigations). For example, maps can be effectively used to present a great volume of information. Tufte (2001) argues that maps are actually the only means to display large quantities of data in a relatively small amount of space and still allow a meaningful interpretation of the information. In addition, maps allow a visual analysis of data at different levels of temporal and spatial resolution and an assessment of spatial relationships among variables that can help identify potential causes of the detected pattern.

Other data visualization techniques also provide practical information. A simple assessment of the species richness of a community can be accomplished by presenting the total number of species detected during the survey. This is made more informative by plotting the cumulative number of species detected against an indicator of sampling effort such as the time spent sampling or the number of samples taken (i.e., detection curve or empirical cumulative distribution functions). These sampling effort curves can give a quick and preliminary assessment of how well the species richness of the investigated community has been sampled. A steep slope of the resulting curve would suggest the presence of additional, unknown species whereas a flattening of the curve would indicate that most species have been accounted for (Magurran 1988; Southwood 1992). Early on in the sampling, these types of sampling curves are recommended because they can provide some rough estimates of the minimum amount of sampling effort needed.

Constructing species abundance models such as log normal distribution, log series, McArthur's broken stick, or geometric series model can provide a visual profile of particular research areas (Southwood 1992). Indeed, the different species abundance models describe communities with distinct characteristics. For example, mature undisturbed systems characterized by higher species richness typically display a log normal relationship between the number of species and their respective abundances. On the other hand, early successional sites or environmentally stressed communities (e.g., pollution) are characterized by geometric or log series species distribution models (Southwood 1992).

The use of confidence intervals presents another attractive approach to exploratory data analysis. Some even argue that confidence intervals represent a more meaningful and powerful alternative to statistical hypothesis testing since they give an estimate of the magnitude of an effect under investigation (Steidl et al. 1997; Johnson 1999; Stephens et al. 2006). In other words, determining the confidence intervals is generally much more informative than simply determining the $P$-value (Stephens et al. 2006). Confidence intervals are widely applicable and can be placed on estimates of population density, observed effects of population change in samples taken over time, or treatment effects in perturbation experiments. They are also commonly used in calculations of effect size in power or meta-analysis (Hedges and Olkin 1985; Gurevitch et al. 2000; Stephens et al. 2006).

Despite its historical popularity and attractive simplicity, however, visual analysis of the data does carry some caveats and potential pitfalls for hypothesis testing. For example, Hilborn and Mangel (1997) recommended plotting the data in different ways to uncover "plausible relationships." At first glance, this appears to be a reasonable and innocuous recommendation, but one should be wary of letting the data uncover plausible relationships. That is, it is entirely possible to create multiple plots between a dependent variable (Y) and multiple explanatory variables ($X_1$, $X_2$, $X_3$, ..., $X_N$) and discover an unexpected effect or pattern that is an artifact of that single data set, not of the more important biological process that generated the sample data (i.e., spurious effects) (Anderson et al. 2001). These types of spurious effects are most likely when dealing with small or limited sample sizes and many explanatory variables. Plausible relationships and hypotheses should be developed in an *a priori* fashion with significant input from a monitoring program's conceptual system, stakeholder input, and well-developed objectives.

## DATA VISUALIZATION II: GETTING TO KNOW YOUR MODEL

Most statistical models are based on a set of assumptions that are necessary for models to properly fit and describe the data. If assumptions are violated, statistical analyses may produce erroneous results (Sokal and Rohlf 1994; Krebs 1999). Traditionally, researchers are most concerned with the assumptions associated with parametric tests (e.g., ANOVA and regression analysis) since these are the most commonly used analyses. The description below may be used as a basic framework for assessing whether or not data conform to parametric assumptions.

### INDEPENDENCE OF DATA POINTS

The essential condition of most statistical tests is the independence and random selection of data points in space and time. In many ecological settings, however, data points can be counts of individuals or replicates of treatment units in manipulative studies, and one must think about spatial and temporal dependency among sampling units. Dependent data are more alike than would be expected by random selection alone. Intuitively, if two observations are not independent then there is less information content between them. Krebs (1999) argued that if the assumption of independence is violated, the chosen probability for Type I error ($\alpha$) cannot be achieved. ANOVA and linear regression techniques are generally sensitive to this violation (Sokal and Rohlf 1994; Krebs 1999). Autocorrelation plots can be used to visualize the correlation of points across space (e.g., spatial correlograms) (Fortin and Dale 2005) or as a time series (Crawley 2007). Plots should be developed using the actual response points as well as the residuals of the model to look for patterns of autocorrelation.

### HOMOGENEITY OF VARIANCES

Parametric models assume that sampled populations have similar variances even if their means are different. This assumption becomes critical in studies comparing different groups of organisms, treatments, or sampling intervals, and it is the responsibility of the researcher to make these considerations abundantly clear in the protocols of such studies. If the sample sizes are equal, then parametric tests are fairly robust to the departure from homoscedasticity (i.e., equal variance of errors across the data) (Day and Quinn 1989; Sokal and Rohlf 1994). In fact, an equal sample size among different treatments or areas should be ensured whenever possible since most

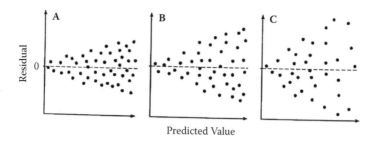

**FIGURE 11.2** Three hypothetical residual scatters. In Case A, the variance is proportional to predicted values, which suggests a Poisson distribution. In Case B, the variance increases with the square of expected values and the data approximate a log-normal distribution. The severe funnel shape in Case C indicates that the variance is proportional to the fourth power of predicted values. (From Sabin, T.E., and S.G. Stafford. 1990. *Assessing the Need for Transformation of Response Variables.* Forest Research Laboratory, Oregon State University, Corvallis, OR. With permission.)

parametric tests are overly sensitive to violations of assumptions in situations with unequal sample sizes (Day and Quinn 1989). The best approach for detecting a violation in homoscedasticity, however, is plotting the residuals of the analysis against predicted values (i.e., residual analysis) (Crawley 2007). This plot can reveal the nature and severity of the potential disagreement/discord between variances (Figure 11.2), and is a standard feature in many statistical packages. Indeed, such visual inspection of the model residuals, in addition to the benefits outlined above, can help determine not only if there is a need for data transformation, but also the type of the distribution. Although several formal tests exist to determine the heterogeneity of variances (e.g., Bartlett's test, Levine's test), these techniques assume a normal data distribution, which reduces their utility in most ecological studies (Sokal and Rohlf 1994).

## NORMALITY

Although parametric statistics are fairly robust to violations of the assumption of normality, highly skewed distributions can significantly affect the results. Unfortunately, nonnormality appears to be the norm in ecology; in other words, ecological data only rarely follow a normal distribution (Potvin and Roff 1993; White and Bennetts 1996; Hayek and Buzas 1997; Zar 1999). Moreover, the normal distribution primarily describes continuous variables whereas count data, often the type of information gathered during monitoring programs, are discrete (Thompson et al. 1998; Krebs 1999). Thus, it is important to be vigilant for large departures from normality in your data. This can be done with a number of tests if your data meet certain specifications. For instance, if the sample size is equal among groups and sufficiently large (e.g., $n > 20$) you can implement tests to assess for normality or to determine the significance of nonnormality. The latter is commonly done with several techniques, including the W-test and the Kolmogorov–Smirnov D-test for larger sample sizes. The applicability of both tests, however, is limited due to the fact that they exhibit a low power if the sample size is small and excessive sensitivity when the sample size is large. To overcome these

# Data Analysis in Monitoring

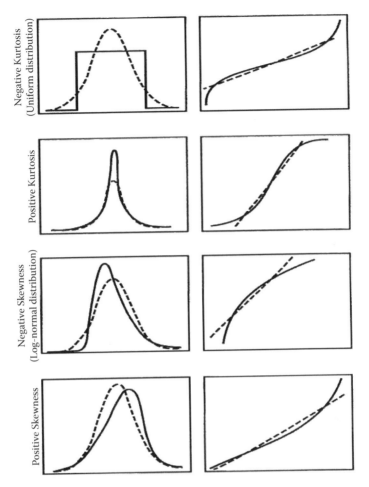

**FIGURE 11.3** Plots of four hypothetical distributions (left column) with their respective normal-probability plots (right column). Solid and broken lines show the observed and normal (expected) distributions, respectively. (From Sabin, T.E., and S.G. Stafford. 1990. *Assessing the Need for Transformation of Response Variables*. Forest Research Laboratory, Oregon State University, Corvallis, OR. With permission.)

complications, visual examinations of the data are also undertaken. Visual examinations are generally more appropriate than formal tests since they allow one to detect the extent and the type of problem. Keep in mind, however, that when working with linear models, the aim is to "normalize" the data/residuals. Plotting the residuals with normal-probability plots (Figure 11.3), stem-and-leaf diagrams, or histograms can help understand the nature of the nonnormal data (Day and Quinn 1989).

## POSSIBLE REMEDIES IF PARAMETRIC ASSUMPTIONS ARE VIOLATED

Data transformations or nonparametric tests are often recommended as appropriate solutions if the data do not meet parametric assumptions (Sabin and Stafford 1990;

Thompson et al. 1998). Those developing monitoring plans, however, should repeatedly (and loudly) advocate sound experimental design as the only effective prevention of many statistical problems. This will keep unorganized data collection and questionable data transformations to a minimum. There is no transformation or magical statistical button for data that are improperly collected. Nonetheless, a properly designed monitoring program is likewise not a panacea; ecosystems are complex. Thus, even proper designs can produce data that confound analysis, are messy, and require some remedies if parametric models are to be used.

## DATA TRANSFORMATION

If significant violations of parametric assumptions occur, it is customary to implement an appropriate data transformation to try to resolve the violations. During a transformation, data will be converted and analyzed at a different scale. In general, researchers should be aware of the need to back-transform the results after analysis to present parameter values on the original data scale or be clear that their results are being presented on a transformed scale. Examples of common types of transformations that a biometrician may recommend for use are presented in Table 11.2. A wisely chosen transformation can often improve homogeneity of variances and produce an approximation of a normal distribution.

## NONPARAMETRIC ALTERNATIVES

If the data violate basic parametric assumptions, and transformations fail to remedy the problem, then you may wish to use nonparametric methods (Sokal and Rohlf 1994; Thompson et al. 1998; Conover 1999). Nonparametric techniques have less stringent assumptions about the data, are less sensitive to the presence of outliers, and are often more intuitive and easier to compute (Hollander and Wolfe 1999).

### TABLE 11.2
### The Most Common Types of Data Transformation in Biological Studies

| Transformation Type | When Appropriate to Use |
|---|---|
| Logarithmic | Use with count data and when means positively correlate with variances. A rule of thumb suggests its use when the largest value of the response variable is at least 10x the smallest value. |
| Square-root | Use with count data following a Poisson distribution. |
| Inverse | Use when data residuals exhibit a severe funnel-shaped pattern, often the case in data sets with many near-zero values. |
| Arcsine square root | Good for proportional or binomial data. |
| Box-Cox objective approach | If it is difficult to decide on what transformation to use, this procedure finds an optimal model for the data. |

*Source:* Modified from Sabin, T.E., and S.G. Stafford. 1990. *Assessing the Need for Transformation of Response Variables.* Forest Research Laboratory, Oregon State University, Corvallis, OR.

Since nonparametric models are less powerful than their parametric counterparts, parametric tests are preferred if the assumptions are met or data transformations are successful (Day and Quinn 1989; Johnson 1995).

## STATISTICAL DISTRIBUTION OF THE DATA

As mentioned above, plant and animal data are often in the form of counts of organisms, and this can present special challenges during analyses. The probability that organisms occur in a particular habitat has a direct bearing on the selection of appropriate sampling protocols and statistical models (Southwood 1992). Monitoring data will most likely approximate random or clumped distributions, yet this should not be blindly assumed. Fitting the data to the Poisson or negative binomial models are common ways to test if they do (Southwood 1992; Zuur 2009). These models are also particularly appropriate for describing count data, which are, once again, examples of discrete variables (e.g., quadrat counts, sex ratios, ratios of juveniles to adults). The following subsections briefly describe how to identify whether or not the data follow either a Poisson or negative binomial distribution model.

### POISSON DISTRIBUTION

Poisson distributions are common among species where the probability of detecting an individual in any sample is rather low (Southwood 1992). The Poisson model gives a good fit to data if the mean count (e.g., number of amphibians per sampling quadrat) is in the range of 1–5. As the mean number of individuals in the sample increases, and exceeds 10, then the random distribution begins to approach the normal distribution (Krebs 1999; Zar 1999).

During sampling, the key assumption of the Poisson (random) distribution is that the expected number of organisms in a sample is the same and that it equals $\mu$, the population mean (Krebs 1999; Zuur 2009). One intriguing property of the random distribution is that it can be described by its mean, and that the mean equals the variance ($s^2$). The probability (frequency) of detecting a given number of individuals in a sample collected from a population with mean = $\mu$ is

$$P_\mu = e^{-\mu}(\mu^\mu/\mu!).$$

Whether or not the data follow a random distribution can be tested with a simple Chi-square goodness-of-fit test or with an index of dispersion ($I$), which is expected to be 1.0 if the assumption of randomness is satisfied:

$$I = s^2/\bar{x},$$

where $\bar{x}$ and $s^2$ are the observed sample mean and variance, respectively.

Zuur et al. (2009) provided excellent examples of tests for goodness of fit for Poisson distributions. In practice, the presence of a Poisson distribution in data can also be assessed visually by examining the scatter pattern of residuals during analysis. If we reject the null hypothesis that samples came from a random distribution, $s^2 < \mu$, and

$s^2/\mu < 1.0$, then the sampled organisms are distributed either uniformly or regularly (underdispersed). If we reject the null hypothesis but the values of $s^2$ and $s^2/\mu$ do not fall within those bounds, then the sampled organisms are clumped (overdispersed).

## NEGATIVE BINOMIAL DISTRIBUTION

An alternative approach to the Poisson distribution, and one of the mathematical distributions that describe clumped or aggregated spatial patterns, is the negative binomial (Pascal) distribution (Anscombe 1973; Krebs 1999; Hilbe 2007; Zuur 2009). White and Bennetts (1996) suggested that this distribution is a better approximation to count data than the Poisson or normal distributions. The negative binomial distribution is described by the mean and the dispersion parameter k, which expresses the extent of clumping. As a result of aggregation, it always follows that $s^2 > \mu$ and the index of dispersion $(I) > 1.0$. Several techniques exist to evaluate the goodness-of-fit of data to the negative binomial distribution. As an example, White and Bennetts (1996) give an example of fitting the negative binomial distribution to point-count data for orange-crowned warblers to compare their relative abundance among forest sites. Zero-inflated Poisson models (ZIP models) are recommended for analysis of count data with frequent zero values (e.g., rare species studies) or where data transformations are not feasible or appropriate (e.g., Heilbron 1994; Welsh et al. 1996; Hall and Berenhaut 2002; Zuur 2009). Good descriptions and examples of their use can be found in Southwood (1992), Krebs (1999), Faraway (2006), and Zuur et al. (2009). Since the variety of possible clumping patterns in nature is practically infinite, it is possible that the Poisson and negative binomial distributions may not always adequately fit the data at hand.

## ABUNDANCE AND COUNTS

### ABSOLUTE DENSITY OR POPULATION SIZE

National policy on threatened and endangered species is ultimately directed toward efforts to increase or maintain the total number of individuals of the species within their natural geographic range (Carroll et al. 1996). Total population size and effective population size (i.e., the number of breeding individuals in a population; Harris and Allendorf 1989) are the two parameters that most directly indicate the degree of species endangerment and/or effectiveness of conservation policies and practices. Population density is informative for assessing population status and trends because the parameter is sensitive to changes in natural mortality, exploitation, and habitat quality. In some circumstances, it may be feasible to conduct a census of all individuals of a particular species in an area to determine the total population size or density parameters. Typically, however, population size and density parameters are estimated using statistical analyses based on only a sample of population members (Yoccoz et al. 2001; Pollock et al. 2002). Population densities of plants and sessile animals can be estimated from counts taken on plots or data describing the spacing between individuals (i.e., distance methods) and are relatively straightforward. Population analyses for many animal species must account for animal response to capture or observation, observer biases, and different detection probabilities among subpopulations (Kéry and Schmid 2004). For instance, hiring multiple technicians for fieldwork and monitoring a species whose behavior or preferred habitat change seasonally are two factors that would need to be addressed in the analysis. Pilot studies are usually required to collect the data necessary to do this. Furthermore, the more common techniques used for animal species, such as mark-recapture studies, catch-per-unit effort monitoring programs, and occupancy studies, require repeated visits to sampling units. This, along with the need for pilot studies, increases the complexity and cost of monitoring to estimate population parameters relative to monitoring of sessile organisms.

For animal species, mark-recapture models are often worth the extra investment in terms of the data generated as they may be used to estimate absolute densities of populations and provide additional information on such vital statistics as animal movement, geographic distribution, and survivorship (Lebreton et al. 1992; Nichols 1992; Nichols and Kendall 1995; Thomson et al. 2009). Open mark-recapture models (e.g., Jolly-Seber) assume natural changes in the population size of the species of interest during sampling. In contrast, closed models assume a constant population size. The program MARK (White et al. 2006) performs sophisticated maximum-likelihood-based mark-recapture analyses, and can test and account for many of the assumptions such as closed populations and heterogeneity.

## Relative Abundance Indices

It is sometimes the case that data analyses for biological inventories and monitoring studies can be accomplished based on indices of population density or abundance, rather than population estimators (Pollock et al. 2002). The difference between estimators and indices is that the former yield absolute values of population density while the latter provide relative measures of density that can be used to assess population differences in space or time. Caughley (1977) advocated the use of indices after determining that many studies that used estimates of absolute density could have used density indices without losing information. He suggested that use of indices often results in much more efficient use of time and resources, and produces results with higher precision (Caughley 1977; Caughley and Sinclair 1994). Engeman (2003) also indicated that use of an index may be the most efficient means to address population monitoring objectives and that the concerns associated with use of indices may be addressed with appropriate and thorough experimental design and data analyses. It is important, therefore, to understand these concerns before utilizing an index, even though indices of relative abundance have a wide support among practitioners who often point out their efficiency and higher precision (Caughley 1977; Engeman 2003). First, they are founded on the assumption that index values are closely associated with values of a population parameter. Because the precise relationship between the index and parameter usually is not quantified, the reliability of this assumption is often brought into question (Thompson et al. 1998; Anderson 2001). Also, the opportunity for bias associated with indices of abundance is quite high. For instance, track counts could be related to animal abundance, animal activity levels, or both. Indices often are used because of logistical constraints. Capture rates of animals over space and time may be related to animal abundance or to their vulnerability to capture in areas of differing habitat quality. If either of these techniques are used to generate the index, considerable caution must be exercised when interpreting results. Given these concerns, the utility of a pilot study that will allow determination, with a known level of certainty, of the relationship between the index and the actual population (or fitness) for the species being monitored is clear. Determining this relationship, however, requires an estimate of the population.

Suitability of any technique, including indices, should ultimately be based on how well it addresses the study objective and the reliability of its results (Thompson 2002). It is also important to consider that statistical analyses of relative abundance data require familiarity with the basic assumptions of parametric and nonparametric models. Some examples of the use and analysis of relative density data can be found in James et al. (1996), Rotella et al. (1996), Huff et al. (2000), Knapp and Matthews (2000), and Rosenstock et al. (2002).

Analyses of relative abundance data require familiarity with the basic assumptions of parametric models. Since the focus is on count data, alternative statistical methods can be employed to fit the distribution of the data (e.g., Poisson or negative binomial). Although absolute abundance techniques are independent of parametric assumptions, they nevertheless do have their own stringent requirements.

When a researcher decides to use a monitoring index, it is important to remember that statistical power negatively correlates with the variability of the monitoring index. This truly underscores the need to choose an appropriate indicator of abundance and accurately estimate its confidence interval (Harris 1986; Gerrodette 1987; Gibbs et al. 1999). An excellent overview of a variety of groups of animals and plants for which the variability in estimating their population is known is given in Gibbs et al. (1998). It is also important to keep in mind that relative measures of density can be less robust to changes in habitat than absolute measures. For instance, forest practices may significantly affect indices that rely on visual observations of organisms. Although these factors may confound absolute measures as well, modern distance and mark-recapture analysis methods can account for variations in sightability and trapability. See Caughley (1977), Thompson et al. (1998), Yoccoz et al. (2001), Pollock et al. (2002), and Rosenstock et al. (2002) and for in-depth discussions of the merits and limitations of estimating relative versus absolute density in population monitoring studies.

## GENERALIZED LINEAR MODELS AND MIXED EFFECTS

Recently, generalized linear models (GLMs) have become increasingly popular and take advantage of the data's true distribution without trying to normalize it (Faraway 2006; Bolker 2008; McCulloch et al. 2008). Oftentimes, the standard linear model cannot handle nonnormal responses, such as counts or proportions, whereas generalized linear models were developed to handle categorical, binary, and other response types (Faraway 2006; McCulloch et al. 2008). In practice, most data have nonnormal errors, and so GLMs allow the user to specify a variety of error distributions. This can be particularly useful with count data (e.g., Poisson errors), binary data (e.g., binomial errors), proportion data (e.g., binomial errors), data showing a constant coefficient of variation (e.g., gamma errors), and survival analysis (e.g., exponential errors; Crawley 2007).

An extension of the GLM is the generalized linear mixed model (GLMM) approach. GLMMs are examples of hierarchal models and are most appropriate when dealing with nested data. What are nested data? As an example, monitoring programs may collect data on species abundances or occurrences from multiple sites on different sampling occasions within each site. Alternatively, researchers might also sample from a single site in different years. In both cases, the data are "nested" within a site or a year, and to analyze the data generated from these surveys without considering the "site" or "year" effect would be considered pseudoreplication (Hurlbert 1984). That is, there is a false assumption of independency of the sampling occasions within a single site or across the sampling period. Traditionally, researchers might avoid this problem by averaging the results of those sampling occasions across the sites or years and focus on the means, or they may simply just focus their analysis within an individual site or sampling period. The more standard statistical approaches, however, attempt to quantify the exact effect of the predictor variables (e.g., forest area, forb density), but ecological problems often involve random effects that are a result of the variation among sites or sampling periods (Bolker et al. 2009). Random effects that come from the same group (e.g., site or time period) will often be correlated, thus violating the standard assumption of independence of errors in most statistical models.

Hierarchal models offer an excellent way of dealing with these problems, but when using GLMMs, researchers should be able to correctly identify the difference between a fixed effect and a random effect. In their most basic form, fixed effects have "informative" factor levels, while random effects often have "uninformative" factor levels (Crawley 2007). That is, random effects have factor levels that can be considered random samples from a larger population (e.g., blocks, sites, years). In this case, it is more appropriate to model the added variation caused by the *differences* between the levels of the random effects and the *variation* in the response variables (as opposed to differences in the mean). In most applied situations, random effect variables often include site names or years. In other cases, when multiple responses are measured on an individual (e.g., survival), random effects can include individuals, genotypes, or species. In contrast, fixed effects then only model differences in the mean of the response variable, as opposed to the variance of the response variable across the levels of the random effect, and can include predictor environmental variables that are measured at a site or within a year. In practice, these distinctions are at times difficult to make and mixed effects models can be challenging to apply. For example, in his review of 537 ecological studies that used GLMM analyses, Bolker (2009) found that 58% used this tool inappropriately. Consequently, as is the case with many of these procedures, it is important to consult with a statistician when developing and implementing your analysis. There are several excellent reviews and books on the subject of mixed effects modeling (Gelman and Hill 2007; Bolker et al. 2009; Zuur 2009).

## ANALYSIS OF SPECIES OCCURRENCES AND DISTRIBUTION

Does a species occur or not occur with reasonable certainty in an area under consideration for management? Where is the species likely to occur? These types of questions have been and continue to be of interest for many monitoring programs (MacKenzie 2005; MacKenzie et al. 2006). Data on species occurrences are often more cost effective to collect than data on species abundances or demographic data. Traditionally, information on species occurrences has been used to:

1. Identify habitats that support the lowest or highest number of species
2. Shed light on the species distribution
3. Point out relationships between habitat attributes (e.g., vegetation types, habitat structural features) and species occurrence or community species richness

For many monitoring programs, species occurrence data are often considered preliminary data only collected during the initial phase of an inventory project and often to gather background information for the project area. In recent years, however, occupancy modeling and estimation (MacKenzie et al. 2002, 2005; MacKenzie and Royle 2005; MacKenzie et al. 2006) has become a critical aspect of monitoring animal and plant populations. These types of categorical data have represented an important source of data for many monitoring programs that have targeted rare or elusive species, or where resources are unavailable to collect data required for parameter estimation models (see Hayek and Buzas 1997; Thompson 2004).

For some population studies, simply determining whether a species is present in an area may be sufficient for conducting the planned data analysis. For example, biologists attempting to conserve a threatened wetland orchid may need to monitor the extent of the species range and proportion of occupied area (POA) in a national forest. One hypothetical approach is to map all wetlands in which the orchid is known to be present, as well as additional wetlands that may qualify as the habitat type for the species within the forest. To monitor changes in orchid distribution at a coarse scale, data collection could consist of a semiannual monitoring program conducted along transects at each of the mapped wetlands to determine if at least one individual orchid (or some alternative criterion to establish occupancy) is present. Using only a list that includes the wetland label (i.e., the unique identifier), the monitoring year, and an occupancy indicator variable, the biologists could prepare a time series of maps displaying all of the wetlands by monitoring year and distinguish the subset of wetlands that were found to be occupied by the orchid.

Monitoring programs to determine the presence of a species typically require less sampling intensity than fieldwork necessary to collect other population statistics. It is far easier to determine if there is at least one individual of the target species on a sampling unit than it is to count all of the individuals. Conversely, to determine with confidence that a species is not present on a sampling unit requires more intensive sampling than collecting count or frequency data because it is so difficult to dismiss the possibility that an individual missed detection (i.e., a failure to detect does not necessarily equate to absence). Traditionally, the use of occurrence data was considered a qualitative assessment of changes in the species distribution pattern and served as an important first step to formulating new hypotheses as to the cause of the observed changes. More recently, however, repeated sampling and the use of occupancy modeling estimation has increased the applicability of occurrence data in ecological monitoring.

## POSSIBLE ANALYSIS MODELS FOR OCCURRENCE DATA

### Species Diversity

The number of species per sample (e.g., in a 1–m² quadrat) can give a simple assessment of local ($\alpha$ diversity) or these data may be used to compare species composition among several locations ($\beta$ diversity) using simple binary formulas such as the Jaccard's index or Sorensen coefficient (Magurran 1988). For example, the Sorensen qualitative index may be calculated as

$$C_S = 2j/(a+b),$$

where $a$ and $b$ are numbers of species in locations A and B, respectively, and $j$ is the number of species found at both locations. If species abundance is known (number individuals/species), species diversity can be analyzed with a greater variety of descriptors such as numerical species richness (e.g., number species/number individuals), quantitative similarity indices (e.g., Sorensen quantitative index, Morista–Horn index), proportional abundance indices (e.g., Shannon index, Brillouin index), or species abundance models (Magurran 1988; Hayek and Buzas 1997).

### Binary Analyses

Since detected/not-detected data are categorical, the relationship between species occurrence and explanatory variables can be modeled with a logistic regression if values of either 1 (species detected) or 0 (species not detected) are ascribed to the data (Trexler and Travis 1993; Hosmer and Lemeshow 2000; Agresti 2002). Logistic regression necessitates a dichotomous (0 or 1) or a proportional (ranging from 0 to 1) response variable. Yet in many cases, logistic regression is used in combination with a set of variables to predict the detection or non-detection a species. For example, a logistic regression can be enriched with such predictors as the percentage of vegetation cover, forest patch area, or presence of snags to create a more informative model of the occurrence of a forest-dwelling bird species. The resulting logistic function provides an index of probability with respect to species occurrence. There are a number of cross-validation functions that allow the user to identify the probability value that best separates sites where a species was found from where it was not found based on the existing data (Freeman and Moisen 2008). In some cases, data points are withheld from formal analysis (e.g., validation data) and used to test the relationships after the predictive relationships are developed using the rest of data (e.g., training data) (Harrell 2001). Logistic regression, however, is a parametric test. If the data do not meet or approximate the parametric assumption, alternatives to standard logistic regression can be used including general additive models (GAMs) and variations of classification tree (CART) analyses.

## Prediction of Species Density

In some cases, occurrence data have been used to predict organism density if the relationship between species occurrence and density is known and the model's predictive power is reasonably high (Hayek and Buzas 1997). For example, one can record plant abundance and species richness in sampling quadrats. The species proportional abundance, or constancy of its frequency of occurrence ($P_o$), can then be calculated as

$P_o$ = Number of species occurrences (+ or 0)/number of samples (quadrats).

Consequently, the average species density is plotted against its proportional abundance to derive a model to predict species abundance in other locations with only occurrence data. Note, however, that the model may function reasonably well only in similar and geographically related types of plant communities (Hayek and Buzas 1997).

## Occupancy Modeling

Note that without a proper design, detected/not-detected data cannot be reliably used to measure or describe species distributions (Kéry and Schmid 2004; MacKenzie et al. 2006; Kéry et al. 2008). Although traditional methods using logistic regression and other techniques may be used to develop a biologically based model that can predict the probability of occurrence of a site over a landscape, occupancy modeling has developed rapidly over the past few years. As with mark-recapture analysis, changes in occupancy over time can be parameterized in terms of local extinction ($\varepsilon$) and colonization ($\gamma$) processes, analogous to the population demographic processes of mortality and recruitment (Figure 11.4; MacKenzie et al. 2003, 2006; Royle and Dorazio 2008). In this case, sampling must be done in a repeated fashion within separate primary sampling periods (Figure 11.4). Occupancy models are robust to missing observations and can effectively model the variation in detection probabilities between species. Of greatest importance, occupancy ($\psi$), colonization ($\gamma$), and local extinction ($\varepsilon$) probabilities can be modeled as functions of environmental covariate variables that can be site specific or count specific or change between the primary periods (MacKenzie et al. 2003, 2009). In addition, detection probabilities can also be functions of season-specific covariates and may change with each survey of a site. More recently, the program PRESENCE has been made available as a sophisticated likelihood-based family of models as estimates of species occurrence and has become increasingly popular in monitoring (www.mbr-pwrc.usgs.gov/software/presence.html). Donovan and Hines (2007) also present an explanation of occupancy models and several online exercises (www.uvm.edu/envnr/vtcfwru/spreadsheets/occupancy/occupancy.htm).

### Assumptions, Data Interpretation, and Limitations

It is crucial to remember that failure to detect a species in a habitat does not mean that the species was truly absent (Kéry and Schmid 2004; MacKenzie et al. 2006; Kéry et al. 2008). Cryptic or rare species, such as amphibians, are especially prone to underdetection and false absences (Thompson 2004). Keep in mind that occasional confirmations

# Data Analysis in Monitoring

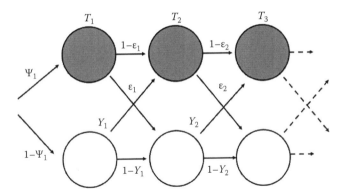

**FIGURE 11.4** Representation of how the occupancy states at a site might change between primary sampling periods ($T_1$, $T_2$, $T_3$). Shaded circles indicate that the site is occupied (species present at some point during a count) during the primary period (T), while empty circles indicate the site is unoccupied (species absent from the count). Processes such as occupancy ($\psi$), colonization ($\gamma$), and local extinction ($\epsilon$) can be modeled independently and account for heterogeneous detection probabilities because of the repeated sampling within each time period. (Redrafted from MacKenzie, D.I. et al. 2006. *Occupancy Estimation and Modeling: Inferring Patterns and Dynamics of Species Occurrence.* Elsevier Academic Press, Burlingame, MA.)

of species presence provide only limited data. For example, the use of a habitat by a predator may reflect prey availability, which may fluctuate daily, weekly, seasonally, or annually or even during one year. A more systematic approach with repeated visits is necessary to generate more meaningful data (MacKenzie and Royle 2005).

Extrapolating density without understanding the species requirements is also likely to produce meaningless results since organisms depend on many factors that we typically do not understand. Furthermore, limitations of species diversity measures should be recognized, especially in conservation projects. For example, replacement of a rare or keystone species by a common or exotic species would not affect species richness of the community and could actually "improve" diversity metrics. Also, the informative value of qualitative indices is rather low since they disregard species abundance and are sensitive to differences in sample size (Magurran 1988). Rare and common species are weighted equally in community comparisons. Often this may be an erroneous assumption since the effect of a species on the community is expected to be proportional to its abundance; keystone species are rare exceptions (Power and Mills 1995). In addition, analyses that focus on species co-occurrences without effectively modeling or taking into account the varying detection probabilities of the species can be prone to error, although new occupancy models are beginning to incorporate detectability in models of species richness (MacKenzie et al. 2004; Royle et al. 2007; Kéry et al. 2009).

## ANALYSIS OF TREND DATA

Trend models should be used if the objective of a monitoring plan is to detect a change in a population parameter over time. Most commonly, population size is repeatedly estimated at set time intervals. Trend monitoring is crucial in the management of species since it may help

1. Recognize population decline and focus attention on affected species
2. Identify environmental variables correlated with the observed trend and thus help formulate hypotheses for cause-and-effect studies
3. Evaluate the effectiveness of management decisions (Thomas 1996; Thomas and Martin 1996)

The status of a population can be assessed by comparing observed estimates of the population size at some time interval against management-relevant threshold values (Gibbs et al. 1999; Elzinga et al. 2001). All monitoring plans, but particularly those designed to generate trend data, should emphasize that the selection of reliable indicators or estimators of population change is a key requirement of effective monitoring efforts. Indices of relative abundance are often used in lieu of measures of actual population size, sometimes because of the relatively reduced cost and effort needed to collect the necessary data on an iterative basis. For example, counts of frog egg masses may be performed annually in ponds (Gibbs et al. 1998) or bird point-counts may be taken along sampling routes (Böhning-Gaese et al. 1993; Link and Sauer 1997a,b, 2007). Analysis of distribution maps, checklists, and volunteer-collected data may also provide estimates of population trends (Robbins 1990; Temple and Cary 1990; Cunningham and Olsen 2009; Zuckerberg et al. 2009). To minimize the bias in detecting a trend, such as in studies of sensitive species, the same population may be monitored using different methods (e.g., a series of different indices; Temple and Cary 1990). Data may also be screened prior to analysis. For example, only monitoring programs that meet agreed-upon criteria may be included, or species with too few observations may be excluded from the analysis (Thomas 1996; Thomas and Martin 1996).

## Possible Analysis Models

Trends over space and time present many challenges for analysis to the extent that consensus does not exist on the most appropriate method to analyze the related data. This is a significant constraint since model selection may have a considerable impact on interpretation of the results of analysis (Thomas 1996; Thomas and Martin 1996).

Poisson regression is independent of parametric assumptions and is especially appropriate for count data. Classic linear regression models in which estimates of population size are plotted against biologically relevant sampling periods have been historically used in population studies since they are easy to calculate and interpret. However, these models are subject to parametric assumptions, which are often violated in count data (Krebs 1999; Zar 1999). Linear regressions also assume a constant linear trend in data, and expect independent and equally spaced data points. Since individual measurements in trend data are autocorrelated, classic regression can give skewed estimates of standard errors and confidence intervals, and inflate the coefficient of determination (Edwards and Coull 1987; Gerrodette 1987). Edwards and Coull (1987) suggested that correct errors in linear regression analysis can be modeled using an autoregressive process model (ARIMA model). Linear route-regression models represent a more robust form of linear regression and are popular with bird ecologists in analyses of roadside monitoring programs (Geissler and Sauer 1990; Sauer et al. 1996; Thomas 1996). They can handle unbalanced data by performing analysis on weighted averages of trends from individual routes (Geissler and Sauer 1990) but may be sensitive to nonlinear trends (Thomas 1996). Harmonic or periodic regressions do not require regularly spaced data points and are valuable in analyzing data on organisms that display significant daily periodic trends in abundance or activity (Lorda and Saila 1986).

For some data where large sample sizes are not possible, or where variance structure cannot be estimated reliably, alternative analytical approaches may be necessary. This is especially true when the risk of concluding that a trend cannot be detected is caused by large variance or small sample sizes, the species is rare, and the failure to detect a trend could be catastrophic for the species. Wade (2000) provides an excellent overview of the use of Bayesian analysis to address these types of problems. Thomas (1996) gives a thorough review of the most popular models fit to trend data and assumptions associated with their use.

### ASSUMPTIONS, DATA INTERPRETATION, AND LIMITATIONS

The underlying assumption of trend monitoring projects is that a population parameter is measured at the same sampling points (e.g., quadrats, routes) using identical or similar procedures (e.g., equipment, observers, time period) at regularly spaced intervals. If these requirements are violated, data may contain excessive noise, which may complicate their interpretation. Thomas (1996) identified four sources of variation in trend data:

1. Prevailing trend—population tendency of interest (e.g., population decline)
2. Irregular disturbances—disruptions from stochastic events (e.g., drought mortality)
3. Partial autocorrelation—dependence of the current state of the population on its previous levels
4. Measurement error—added data noise from deficient sampling procedures

Although trend analyses are useful in identifying population change, the results are correlative and tell us little about the underlying mechanisms. Ultimately, only well-designed cause-and-effect studies can validate causation and facilitate management decisions.

## ANALYSIS OF CAUSE-AND-EFFECT MONITORING DATA

The strength of trend studies lies in their capacity to detect changes in population size. To understand the reason for population fluctuations, however, the causal mechanism behind the population change must be determined. Cause-and-effect studies represent one of the strongest approaches to test cause-and-effect relationships and are often used to assess effects of management decisions on populations. Similar to trend analyses, cause-and-effect analyses may be performed on indices of relative abundance or absolute abundance data.

## POSSIBLE ANALYSIS MODELS

Parametric and distribution free (nonparametric) models provide countless alternatives to fitting cause-and-effect data (Sokal and Rohlf 1994; Zar 1999). Excellent introductory material to the design and analysis of ecological experiments, specifically for ANOVA models, can be found in Underwood (1997) and Scheiner and Gurevitch (2001).

A unique design is recommended for situations where a disturbance (treatment) is applied and its effects are assessed by taking a series of measurements before and after the perturbation (Before–After Control-Impact, BACI; Stewart-Oaten et al. 1986; see Figure 11.5). This model was originally developed to study pollution effects (Green 1979), but it has found suitable applications in other areas of ecology as well (Wardell-Johnson and Williams 2000; Schratzberger et al. 2002; Stanley and Knopf 2002). In its original design, an impact site would have a parallel control site. Further, variables deemed at the beginning of the study to be relevant to the management actions would be planned to be periodically monitored over time. Then, any differences between the trends of those measured variables in the impact site with those from the control site (treatment effect) would be demonstrated as a significant time*location interaction (Green 1979). This approach has been criticized since the design was originally limited to unreplicated impact and control sites (Figure 11.5), but it can be improved by replicating and randomly assigning sites (Hurlbert 1984; Underwood 1994).

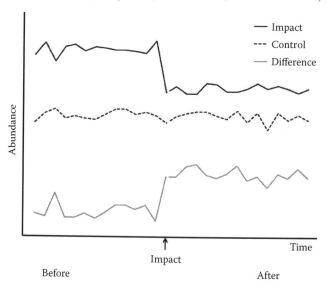

**FIGURE 11.5** A hypothetical example of a BACI analysis where abundance samples are taken at control impact sites before and after the impact and compared to a control site. (Redrafted from Stewart-Oaten, A., W.W. Murdoch, and K.R. Parker. 1986. *Ecology* 67:929–940.)

## Assumptions, Data Interpretation, and Limitations

Since cause-and-effect data are frequently analyzed with ANOVA models (parametric models), one must pay attention to parametric assumptions. Alternative means of assessing manipulative studies may also be employed. For example, biologically significant effect size with confidence intervals may be used in lieu of classic statistical hypothesis testing. An excellent overview of arguments in support of this approach with examples may be found in Hayes and Steidl (1997), Steidl et al. (1997), Johnson (1999), and Steidl and Thomas (2001).

# PARADIGMS OF INFERENCE: SAYING SOMETHING WITH YOUR DATA AND MODELS

## Randomization Tests

These tests are not alternatives to parametric tests, but rather are unique means of estimating statistical significance. They are extremely versatile and can be used to estimate test statistics for a wide range of models, and are especially valuable in analyzing nonrandomly selected data points. It is important to keep in mind, however, that randomization tests are computationally difficult even with small sample sizes (Edgington and Onghena 2007). A statistician needs to be involved in choosing to use and implement these techniques. More information on randomization tests and other computation-intensive techniques can be found in Crowley (1992), Potvin and Roff (1993), and Petraitis et al. (2001).

## Information Theoretic Approaches: Akaike's Information Criterion

Akaike's information criterion (AIC), derived from information theory, may be used to select the best-fitting model among a number of a priori alternatives. This approach is more robust and less arbitrary than hypothesis-testing methods since the $P$-value is often predominantly a function of sample size. AIC can be easily calculated for any maximum-likelihood-based statistical model, including linear regression, ANOVA, and general linear models. The model hypothesis with the lowest AIC value is generally identified as the "best" model with the greatest support (given the data) (Burnham and Anderson 2002). Once the best model has been identified, the results can then be interpreted based on the changes in the explanatory variables over time. For instance, if the amount of mature forest near streams were associated with the probability of occurrence of tailed frogs, then a map generated over the scope of inference could be used to identify current and likely future areas where tailed frogs could be vulnerable to management actions. In addition, one of the more useful advantages of using information theoretic approaches is that identifying a single, best model is not necessary. Using the AIC metric (or any other information criteria), one can rank models and can average across models to calculate weighted parameter estimates or predictions (i.e., model averaging). A more in-depth discussion of practical uses of AIC may be found in Burnham and Anderson (2002) and Anderson (2008).

## BAYESIAN INFERENCE

Bayesian statistics refers to a distinct approach to making inference in the face of uncertainty. In general, Bayesian statistics share much with the traditional frequentist statistics with which most ecologists are familiar. In particular, there is a similar reliance on likelihood models which are routinely applied by most statisticians and biometricians. Bayesian inference can also be used in a variety of statistical tasks, including parameter estimation and hypothesis testing, post hoc multiple comparison tests, trend analysis, ANOVA, and sensitivity analysis (Ellison 1996). Bayesian methods, however, test hypotheses not by rejecting or accepting them, but by calculating their probabilities of being true. Thus, $P$-values, significance levels, and confidence intervals are moot points (Dennis 1996). Based on existing knowledge, investigators assign *a priori* probabilities to alternative hypotheses and then use data to calculate ("verify") posterior probabilities of the hypotheses with a likelihood function (Bayes theorem). The highest probability identifies the hypothesis that is the most likely to be true given the experimental data (Dennis 1996; Ellison 1996). Bayesian statistics has several key features that differ from classical frequentist statistics:

- Bayes is based on an explicit mathematical mechanism for updating and propagating uncertainty (Bayes theorem).
- Bayesian analyses quantify inferences in a simpler, more intuitive manner. This is especially true in management settings that require making decisions under uncertainty.
- Bayesian statistics take advantage of preexisting data, and may be used with small sample sizes.

For example, conclusions of a monitoring analysis could be framed as: "There is a 65% chance that clearcutting will negatively affect this species," or "The probability that this population is declining at a rate of 3% per year is 85%." A more in-depth coverage of the use of Bayesian inference in ecology can be found in Dennis (1996), Ellison (1996), Taylor et al. (1996), Wade (2000), and O'Hara et al. (2002). Even though Bayesian inference is easy to grasp and perform, it is still relatively rare in natural resources applications (although that is quickly changing) and sufficient support resources for these types of tests may not be readily available. It is recommended that it only be implemented with the assistance of a consulting statistician.

## RETROSPECTIVE POWER ANALYSIS

Does the outcome of a statistical test suggest that no real biological change took place at the study site? Did the change actually occur but was not detected due to a low power of the statistical test used, in other words, was a Type II (missed-change) error committed in the process? It is recommended that those undertaking inventory studies should routinely evaluate the validity of results of statistical tests by performing a post hoc, or retrospective power analysis for two important reasons:

# Data Analysis in Monitoring

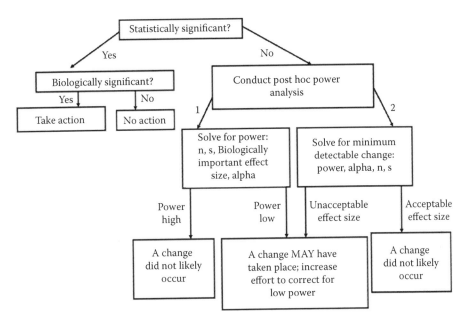

**FIGURE 11.6** A decision process to evaluate the significance of a statistical test. (Redrafted from Elzinga, C.L. et al. 2001. *Monitoring Plant and Animal Populations*. Blackwell Science, Malden, MA.)

1. The possibility of falsely accepting the null hypothesis is quite real in ecological studies, and
2. *A priori* calculations of statistical power are only rarely performed in practice, but are critical to data interpretation and extrapolation (Fowler 1990).

A power analysis is imperative whenever a statistical test turns out to be nonsignificant and fails to reject the null hypothesis ($H_0$); for example, if $P > 0.05$ at 95% significance level. There are a number of techniques to carry out a retrospective power analysis well. For instance, they should be performed only using an effect size other than the effect size observed in the study (Hayes and Steidl 1997; Steidl et al. 1997). In other words, post hoc power analyses can only answer whether or not the performed study in its original design would have allowed detecting the newly selected effect size.

Elzinga et al. (2001) recommend the following approach to conducting a post hoc power analysis assessment (Figure 11.6). If a statistical test was declared nonsignificant, one could calculate a power value to detect a biologically significant effect of interest, usually a trigger point tied to a management action. If the resulting power is low, one must take precautionary measures in the monitoring program. Alternatively, one can calculate a minimum detectable effect size at a selected power level. An acceptable power level in wildlife studies is often set at about 0.80 (Hayes and Steidl 1997). If the selected power can only detect a change that is larger than the trigger point value, the outcome of the study should again be viewed with caution.

Monitoring plans may also encourage the use of confidence intervals as an alternative approach to performing a post hoc power analysis. This method is actually superior to power analysis since confidence intervals not only suggest whether or not the effect was different from zero, but they also provide an estimate of the likely magnitude of the true effect size and its biological significance. Ultimately, for scientific endeavors these are rules of thumb. In management contexts, however, decision making under uncertainty where the outcomes have costs, power calculations, and other estimates for acceptable amounts of uncertainty should be approached more rigorously.

## SUMMARY

Even before the collection of data, researchers must consider which analytical techniques will likely be appropriate to interpret their data. Techniques will be highly dependent on the design of the monitoring program, so a monitoring plan should clearly articulate the expected analytical approaches after consulting with a biometrician. After data collection but before statistical analyses are conducted, it is often helpful to view the data graphically to understand data structure. Assumptions upon which certain techniques are based (e.g., normality, independence of observations, and uniformity of variances for parametric analyses) should be tested. Some violations of assumptions may be addressed with transformations, while others may need different approaches. Detected/nondetected, count data, time series, and before–after control–impact designs all have different data structures and will need to be analyzed in quite different ways. Given the considerable room for spurious analysis and subsequent erroneous interpretation, if possible, a biometrician/statistician should be consulted throughout the entire process of data analysis.

## REFERENCES

Agresti, A. 2002. *Categorical Data Analysis*. 2nd ed. John Wiley & Sons, New York.

Anderson, D.R. 2001. The need to get the basics right in wildlife field studies. *Wildlife Society Bulletin* 29:1294–1297.

Anderson, D.R. 2008. *Model Based Inference in the Life Sciences: A Primer on Evidence*. Springer, New York.

Anderson, D.R., K.P. Burnham, W.R. Gould, and S. Cherry. 2001. Concerns about finding effects that are actually spurious. *Wildlife Society Bulletin* 29:311–316.

Anscombe, F.J. 1973. Graphs in statistical-analysis. *American Statistician* 27:17–21.

Böhning-Gaese, K., M.L. Taper, and J.H. Brown. 1993. Are declines in North American insectivorous songbirds due to causes on the breeding range? *Conservation Biology* 7:76–86.

Bolker, B.M. 2008. *Ecological Models and Data in R*. Princeton University Press, Princeton, NJ. 408 pp.

Bolker, B.M., M.E. Brooks, C.J. Clark, S.W. Geange, J.R. Poulsen, M.H.H. Stevens, and J.S.S. White. 2009. Generalized linear mixed models: A practical guide for ecology and evolution. *Trends in Ecology and Evolution* 24:127–135.

Burnham, K.P., and D.R. Anderson. 2002. *Model Selection and Inference: A Practical Information-Theoretic Approach*. Springer-Verlag, New York. 454 pp.

Carroll, R., C. Augspurger, A. Dobson, J. Franklin, G. Orians, W. Reid, R. Tracy, D. Wilcove, and J. Wilson. 1996. Strengthening the use of science in achieving the goals of the endangered species act: An assessment by the Ecological Society of America. *Ecological Applications* 6:1–11.

Caughley, G. 1977. *Analysis of Vertebrate Populations*. John Wiley & Sons, New York.

Caughley, G., and A.R.E. Sinclair. 1994. *Wildlife Management and Ecology*. Blackwell Publishing, Malden, MA.

Cleveland, W.S. 1985. *The Elements of Graphing Data*. Wadsworth Advanced Books and Software, Monterey, CA.

Cochran, W.G. 1977. *Sampling Techniques*. 3rd ed. John Wiley & Sons, New York.

Conover, W.J. 1999. *Practical Nonparametric Statistics*. 3rd ed. John Wiley & Sons, New York.

Crawley, M.J. 2005. *Statistics: An Introduction Using R*. John Wiley & Sons, West Sussex, England.

Crawley, M.J. 2007. *The R Book*. John Wiley & Sons, West Sussex, England.

Crowley, P.H. 1992. Resampling methods for computation-intensive data-analysis in ecology and evolution. *Annual Review of Ecology and Systematics* 23:405–447.

Cunningham, R.B., and P. Olsen. 2009. A statistical methodology for tracking long-term change in reporting rates of birds from volunteer-collected presence-absence data. *Biodiversity and Conservation* 18:1305–1327.

Day, R.W., and G.P. Quinn. 1989. Comparisons of treatments after an analysis of variance in ecology. *Ecological Monographs* 59:433–463.

Dennis, B. 1996. Discussion: Should ecologists become Bayesians? *Ecological Applications* 6:1095–1103.

Donovan, T.M., and J. Hines. 2007. Exercises in occupancy modeling and estimation. http://www.uvm.edu/envnr/vtcfwru/spreadsheets/occupancy/occupancy.htm

Edgington, E.S., and P. Onghena. 2007. *Randomization Tests*. 4th ed. Chapman & Hall/CRC, Boca Raton, FL.

Edwards, D., and B.C. Coull. 1987. Autoregressive trend analysis—An example using long-term ecological data. *Oikos* 50:95–102.

Ellison, A.M. 1996. An introduction to Bayesian inference for ecological research and environmental decision-making. *Ecological Applications* 6:1036–1046.

Elzinga, C.L., D.W. Salzer, and J.W. Willoughby. 1998. Measuring and monitoring plant plant populations. Technical Reference 1730-1. Bureau of Land Management, National Business Center, Denver, CO.

Elzinga, C.L., D.W. Salzer, J.W. Willoughby, and J.P. Gibbs. 2001. *Monitoring Plant and Animal Populations*. Blackwell Science, Malden, MA.

Engeman, R.M. 2003. More on the need to get the basics right: Population indices. *Wildlife Society Bulletin* 31:286–287.

Faraway, J.J. 2006. Extending the linear model with R: generalized linear, mixed effects and nonparametric regression models. Chapman & Hall/CRC, Boca Raton, FL.

Fortin, M.-J., and M.R.T. Dale. 2005. Spatial analysis: A guide for ecologists. Cambridge University Press, Cambridge, U.K.

Fowler, N. 1990. The 10 most common statistical errors. *Bulletin of the Ecological Society of America* 71:161–164.

Freeman, E.A., and G.G. Moisen. 2008. A comparison of the performance of threshold criteria for binary classification in terms of predicted prevalence and kappa. *Ecological Modelling* 217:48–58.

Geissler, P.H., and J.R. Sauer. 1990. Topics in route-regression analysis. Pages 54–57 in J.R. Sauer and S.Droege, eds. *Survey Designs and Statistical Methods for the Estimation of Avian Population Trends*. USDI Fish and Wildlife Service, Washington, D.C.

Gelman, A., and J. Hill. 2007. *Data Analysis Using Regression and Multilevel/Hierarchical Models*. Cambridge University Press, Cambridge, U.K.; New York.

Gerrodette, T. 1987. A power analysis for detecting trends. *Ecology* 68:1364–1372.

Gibbs, J.P., S. Droege, and P. Eagle. 1998. Monitoring populations of plants and animals. *Bioscience* 48:935–940.

Gibbs, J.P., H.L. Snell, and C.E. Causton. 1999. Effective monitoring for adaptive wildlife management: Lessons from the Galapagos Islands. *Journal of Wildlife Management* 63:1055–1065.

Gotelli, N.J., and A.M. Ellison. 2004. *A Primer of Ecological Statistics*. Sinaeur, Sunderland, MA.

Green, R.H. 1979. *Sampling Design and Statistical Methods for Environmental Biologists*. John Wiley & Sons, New York.

Gurevitch, J., J.A. Morrison, and L.V. Hedges. 2000. The interaction between competition and predation: A meta-analysis of field experiments. *American Naturalist* 155:435–453.

Hall, D.B., and K.S. Berenhaut. 2002. Score test for heterogeneity and overdispersion in zero-inflated Poisson and Binomial regression models. *The Canadian Journal of Statistics* 30:1–16.

Harrell, F.E. 2001. *Regression Modeling Strategies with Applications to Linear Models, Logistic Regression, and Survival Analysis*. Springer, New York.

Harris, R.B. 1986. Reliability of trend lines obtained from variable counts. *Journal of Wildlife Management* 50:165–171.

Harris, R.B., and F.W. Allendorf. 1989. Genetically effective population size of large mammals—An assessment of estimators. *Conservation Biology* 3:181–191.

Hayek, L.-A.C., and M.A. Buzas. 1997. *Surveying Natural Populations*. Columbia University Press, New York.

Hayes, J.P., and R.J. Steidl. 1997. Statistical power analysis and amphibian population trends. *Conservation Biology* 11:273–275.

Hedges, L.V., and I. Olkin. 1985. *Statistical Methods for Meta-Analysis*. Academic Press, Orlando, FL.

Heilbron, D. 1994. Zero-altered and other regression models for count data with added zeros. *Biometrical Journal* 36:531–547.

Hilbe, J. 2007. *Negative Binomial Regression*. Cambridge University Press, Cambridge, U.K.; New York.

Hilborn, R., and M. Mangel. 1997. *The Ecological Detective: Confronting Models with Data*. Princeton University Press, Princeton, NJ.

Hollander, M., and D.A. Wolfe. 1999. *Nonparametric Statistical Methods*. 2nd ed. John Wiley & Sons, New York.

Hosmer, D.W., and S. Lemeshow. 2000. *Applied Logistic Regression*. 2nd ed. John Wiley & Sons, New York.

Huff, M.H., K.A. Bettinger, H.L. Ferguson, M.J. Brown, and B. Altman. 2000. A habitat-based point-count protocol for terrestrial birds, emphasizing Washington and Oregon. USDA Forst Service General Technical Report, PNW-GTR-501.

Hurlbert, S.H. 1984. Pseudoreplication and the design of ecological field experiments. *Ecological Monographs* 54:187–211.

James, F.C., C.E. McCullogh, and D.A. Wiedenfeld. 1996. New approaches to the analysis of population trends in land birds. *Ecology* 77:13–27.

Johnson, D.H. 1995. Statistical sirens—The allure of nonparametrics. *Ecology* 76:1998–2000.

Johnson, D.H. 1999. The insignificance of statistical significance testing. *Journal of Wildlife Management* 63:763–772.

Kéry, M., J.A. Royle, M. Plattner, and R.M. Dorazio. 2009. Species richness and occupancy estimation in communities subject to temporary emigration. *Ecology* 90:1279–1290.

Kéry, M., J.A. Royle, and H. Schmid. 2008. Importance of sampling design and analysis in animal population studies: A comment on Sergio et al. *Journal of Applied Ecology* 45:981–986.

Kéry, M., and H. Schmid. 2004. Monitoring programs need to take into account imperfect species detectability. *Basic and Applied Ecology* 5:65–73.

Knapp, R.A., and K.R. Matthews. 2000. Non-native fish introductions and the decline of the mountain yellow-legged frog from within protected areas. *Conservation Biology* 14:428–438.

Krebs, C.J. 1999. *Ecological Methodology*. 2nd ed. Benjamin/Cummings, Menlo Park, CA.

Lebreton, J.D., K.P. Burnham, J. Clobert, and D.R. Anderson. 1992. Modeling survival and testing biological hypotheses using marked animals—A unified approach with case-studies. *Ecological Monographs* 62:67–118.

Link, W.A., and J.R. Sauer. 1997a. Estimation of population trajectories from count data. *Biometrics* 53:488–497.

Link, W.A., and J.R. Sauer. 1997b. New approaches to the analysis of population trends in land birds: Comment. *Ecology* 78:2632–2634.

Link, W.A., and J.R. Sauer. 2007. Seasonal components of avian population change: Joint analysis of two large-scale monitoring programs. *Ecology* 88:49–55.

Lorda, E., and S.B. Saila. 1986. A statistical technique for analysis of environmental data containing periodic variance components. *Ecological Modelling* 32:59–69.

MacKenzie, D.I. 2005. What are the issues with presence-absence data for wildlife managers? *Journal of Wildlife Management* 69:849–860.

MacKenzie, D.I., L.L. Bailey, and J.D. Nichols. 2004. Investigating species co-occurrence patterns when species are detected imperfectly. *Journal of Animal Ecology* 73:546–555.

MacKenzie, D.I., J.D. Nichols, J.E. Hines, M.G. Knutson, and A.B. Franklin. 2003. Estimating site occupancy, colonization, and local extinction when a species is detected imperfectly. *Ecology* 84:2200–2207.

MacKenzie, D.I., J.D. Nichols, G.B. Lachman, S. Droege, J.A. Royle, and C.A. Langtimm. 2002. Estimating site occupancy rates when detection probabilities are less than one. *Ecology* 83:2248–2255.

MacKenzie, D.I., J.D. Nichols, J.A. Royle, K.H. Pollock, L.L. Bailey, and J.E. Hines. 2006. *Occupancy Estimation and Modeling: Inferring Patterns and Dynamics of Species Occurrence*. Elsevier Academic Press, Burlingame, MA.

MacKenzie, D.I., J.D. Nichols, M.E. Seamans, and R.J. Gutierrez. 2009. Modeling species occurence dynamics with multiple states and imperfect detection. *Ecology* 90:823–835.

MacKenzie, D.I., J.D. Nichols, N. Sutton, K. Kawanishi, and L.L. Bailey. 2005. Improving inferences in popoulation studies of rare species that are detected imperfectly. *Ecology* 86:1101–1113.

MacKenzie, D.I., and J.A. Royle. 2005. Designing occupancy studies: General advice and allocating survey effort. *Journal of Applied Ecology* 42:1105–1114.

Magurran, A.E. 1988. *Ecological Diversity and Its Measurement*. Princeton University Press, Princeton, NJ.

McCulloch, C.E., S.R. Searle, and J.M. Neuhaus. 2008. *Generalized, Linear, and Mixed Models*. 2nd ed. Wiley, Hoboken, NJ.

Nichols, J.D. 1992. Capture-recapture models. *Bioscience* 42:94–102.

Nichols, J.D., and W.L. Kendall. 1995. The use of multi-state capture-recapture models to address questions in evolutionary ecology. *Journal of Applied Statistics* 22:835–846.

O'Hara, R.B., E. Arjas, H. Toivonen, and I. Hanski. 2002. Bayesian analysis of metapopulation data. *Ecology* 83:2408–2415.

Petraitis, P.S., S.J. Beaupre, and A.E. Dunham. 2001. ANCOVA: Nonparametric and randomization approaches. Pages 116–133 in *Design and Analysis of Ecological Experiments*. S.M. Scheiner and J. Gurevitch (eds.). Oxford University Press, Oxford, U.K.; New York.

Pollock, K.H., J.D. Nichols, T.R. Simons, G.L. Farnsworth, L.L. Bailey, and J.R. Sauer. 2002. Large scale wildlife monitoring studies: Statistical methods for design and analysis. *Environmetrics* 13:105–119.

Potvin, C., and D.A. Roff. 1993. Distribution-free and robust statistical methods—Viable alternatives to parametric statistics. *Ecology* 74:1617–1628.

Power, M.E., and L.S. Mills. 1995. The keystone cops meet in Hilo. *Trends in Ecology and Evolution* 10:182–184.

Quinn, G.P., and M.J. Keough. 2002. *Experimental Design and Data Analysis for Biologists.* Cambridge University Press, Cambridge, U.K.

Robbins, C.S. 1990. Use of breeding bird atlases to monitor population change. Pages 18–22 in *Survey Designs and Statistical Methods for the Estimation of Avian Population Trends.* J.R. Sauer and S. Droege (eds.). USDI Fish and Wildlife Service, Washington, D.C.

Rosenstock, S.S., D.R. Anderson, K.M. Giesen, T. Leukering, and M.F. Carter. 2002. Landbird counting techniques: Current practices and an alternative. *Auk* 119:46–53.

Rotella, J.J., J.T. Ratti, K.P. Reese, M.L. Taper, and B. Dennis. 1996. Long-term population analysis of gray partridge in eastern Washington. *Journal of Wildlife Management* 60:817–825.

Royle, J.A., and R.M. Dorazio. 2008. *Hierarchal Modeling and Inference in Ecology: The Analysis of Data from Populations, Metapopulations, and Communities.* Academic Press, Boston, MA.

Royle, J.A., M. Kéry, R. Gautier, and H. Schmid. 2007. Hierarchical spatial models of abundance and occurrence from imperfect survey data. *Ecological Monographs* 77:465–481.

Sabin, T.E., and S.G. Stafford. 1990. *Assessing the Need for Transformation of Response Variables.* Forest Research Laboratory, Oregon State University, Corvallis, OR.

Sauer, J.R., G.W. Pendleton, and B G. Peterjohn. 1996. Evaluating causes of population change in North American insectivorous songbirds. *Conservation Biology* 10:465–478.

Scheiner, S.M., and J. Gurevitch. 2001. *Design and Analysis of Ecological Experiments.* 2nd ed. Oxford University Press, Oxford, U.K.

Schratzberger, M., T.A. Dinmore, and S. Jennings. 2002. Impacts of trawling on the diversity, biomass and structure of meiofauna assemblages. *Marine Biology* 140:83–93.

Sokal, R.R., and F.J. Rohlf. 1994. *Biometry: The Principles and Practice of Statistics in Biological Research.* 3rd ed. W.H. Freeman, San Francisco, CA.

Southwood, R. 1992. *Ecological Methods: With Particular Reference to the Study of Insect Populations.* Methuen, London, U.K.

Stanley, T.R., and F.L. Knopf. 2002. Avian responses to late-season grazing in a shrub-willow floodplain. *Conservation Biology* 16:225–231.

Steidl, R.J., J.P. Hayes, and E. Schauber. 1997. Statistical power analysis in wildlife research. *Journal of Wildlife Management* 61:270–279.

Steidl, R.J., and L. Thomas. 2001. Power analysis and experimental design. Pages 14–36 in *Design and Analysis of Ecological Experiments.* S.M. Scheiner and J. Gurevitch (eds.). Oxford University Press, Oxford; U.K.

Stephens, P.A., S.W. Buskirk, and C. Martinez del Rio. 2006. Inference in ecology and evolution. *Trends in Ecology and Evolution* 22:192–197.

Stewart-Oaten, A., W.W. Murdoch, and K.R. Parker. 1986. Environmental-impact assessment—Pseudoreplication in time. *Ecology* 67:929–940.

Taylor, B.L., P.R. Wade, R.A. Stehn, and J.F. Cochrane. 1996. A Bayesian approach to classification criteria for spectacled eiders. *Ecological Applications* 6:1077–1089.

Temple, S., and J.R. Cary. 1990. Using checklist records to reveal trends in bird populations. Pages 98–104 in J.R. Sauer and S. Droege, eds. *Survey Designs and Statistical Methods for the Estimation of Avian Population Trends.* USDI Fish and Wildlife Service, Washington, D.C.

Thomas, L. 1996. Monitoring long-term population change: Why are there so many analysis methods? *Ecology* 77:49–58.

Thomas, L., and K. Martin. 1996. The importance of analysis method for breeding bird survey population trend estimates. *Conservation Biology* 10:479–490.

Thompson, S.K. 2002. *Sampling*. 2nd ed. John Wiley & Sons, New York.

Thompson, W.L. 2004. *Sampling Rare or Elusive Species: Concepts, Designs, and Techniques for Estimating Population Parameters*. Island Press, Washington.

Thompson, W.L., G.C. White, and C. Gowan. 1998. *Monitoring Vertebrate Populations*. Academic Press, San Diego, CA.

Thomson, D.L., M.J. Conroy, and E.G. Cooch, eds. 2009. *Modeling Demographic Processes in Marked Populations*. Springer, New York.

Trexler, J.C., and J. Travis. 1993. Nontraditional regression analyses. *Ecology* 74:1629–1637.

Tufte, E.R. 2001. *The Visual Display of Quantitative Information*. 2nd ed. Graphics Press, Cheshire, CT.

Tukey, J.W. 1977. *Exploratory Data Analysis*. Addison-Wesley, Reading, MA.

Underwood, A.J. 1994. On beyond BACI: Sampling designs that might reliably detect environmental disturbances. *Ecological Applications* 4:3–15.

Underwood, A.J. 1997. *Experiments in Ecology: Their Logical Design and Interpretation Using Analysis of Variance*. Cambridge University Press, New York.

Wade, P.R. 2000. Bayesian methods in conservation biology. *Conservation Biology* 14:1308–1316.

Wardell-Johnson, G., and M. Williams. 2000. Edges and gaps in mature karri forest, southwestern Australia: Logging effects on bird species abundance and diversity. *Forest Ecology and Management* 131:1–21.

Welsh, A.H., R.B. Cunningham, C.F. Donnelly, and D.B. Lindenmayer. 1996. Modelling the abundance of rare species: Statistical models for counts with extra zeros. *Ecological Modelling* 88:297–308.

White, G.C., and R.E. Bennetts. 1996. Analysis of frequency count data using the negative binomial distribution. *Ecology* 77:2549–2557.

White, G.C., W.L. Kendall, and R.J. Barker. 2006. Multistate survival models and their extensions in Program MARK. *Journal of Wildlife Management* 70:1521–1529.

Williams, B.K., J.D. Nichols, and M.J. Conroy. 2002. *Analysis and Management of Animal Populations: Modeling, Estimation, and Decision Making*. Academic Press, San Diego, CA.

Yoccoz, N.G., J.D. Nichols, and T. Boulinier. 2001. Monitoring of biological diversity in space and time. *Trends in Ecology and Evolution* 16:446–453.

Zar, J.H. 1999. *Biostatistical Analysis*. 4th ed. Prentice Hall, Upper Saddle River, NJ.

Zuckerberg, B., W.F. Porter, and K. Corwin. 2009. The consistency and stability of abundance-occupancy relationships in large-scale population dynamics. *Journal of Animal Ecology* 78:172–181.

Zuur, A. 2009. *Mixed Effects Models and Extensions in Ecology with R*. 1st ed. Springer, New York.

Zuur, A.F., E.N. Ieno, and G.M. Smith. 2007. *Analysing Ecological Data*. Springer, New York; London.

# 12 Reporting

The information that monitoring generates can only be put to use if it is made available in a timely way. Shortly after the monitoring program is terminated, therefore, a formal final report must be developed. Yet this should ideally be the final step in a continual process of communication. Interim monitoring reports should also be provided frequently throughout the duration of monitoring. This may occur annually (e.g., U.S. Geological Survey [USGS] Breeding Bird Survey data; Sauer et al. 2008), or periodically (Forest Inventory and Analysis data; Smith et al. 2004), depending on the program, but must be done often enough for the reporting of monitoring data to allow for rapid response within an adaptive management framework. Frequent communication of monitoring data is also important because it helps inform research approaches. In a sense, the data represent a middle ground between research and monitoring (Figure 12.1).

Aside from the temporal aspects, the successful reporting of data has two components. First, all potential users of the data must be given a means to readily access the report. Web-based dissemination certainly is the most likely way

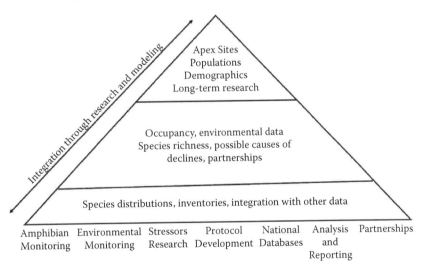

**FIGURE 12.1** The USGS Amphibian Research and Monitoring Initiative (ARMI) conceptual pyramid. Extensive analyses are conducted at the national level (the base of the pyramid), while intensive research occurs at a smaller number of sites (the apex of the pyramid). The middle of the pyramid is where most of the analysis and reporting occurs. (Redrafted from Muths, E. et al. 2006. The Amphibian Research and Monitoring Initiative (ARMI): 5-Year Report: U.S. Geological Survey Scientific Investigations Report 2006–5224. 77 pp.)

of getting information out, especially with powerful search engines now available, but providing copies directly to stakeholders is also necessary. If a report is made available, but some stakeholders are not aware of its availability, then the information is not of use, and worse, the stakeholder may feel marginalized. Proprietary restrictions (if they exist) can hinder the communication process, thus it is best to avoid them provided that doing so does not decrease the quality of the report.

Second, and perhaps most important, the report and the data within them must be presented in a well-organized and visually appealing (i.e., a picture is worth a thousand words) format that is easy for stakeholders to understand. The format is an important consideration, given that even if an effective means of disseminating the reports is chosen, an unclear or otherwise poor format may make the data inaccessible.

## FORMAT OF A MONITORING REPORT

Some Web-based reports are simply interpretations of data in text form for the general public. Others use Web interfaces to provide summaries in a variety of ways for various time periods over various areas, all specified by the user (e.g., annual summaries from the Audubon's Christmas Bird Count). Most, however, are pdf files and have a standard format to allow the users to find the information that they need quickly while still understanding any potential biases, limitations, or interpretations of the data.

Both interim and final reports and other products such as predictive or conceptual models should be designed in the way that best meets the particular information requirements for which the project was intended to address. Hence, there is no single format that should be followed. Within this chapter, we provide an annotated list of elements that together comprise a generalized, commonly used format, but these pieces should be adapted to meet the needs of a particular client or set of stakeholders.

### TITLE

The title should concisely state what, when, and where the monitoring data were collected. Avoid long titles. "Southwestern Willow Flycatcher 2002 Survey and Nest Monitoring Report" is a perfectly acceptable title. Or is it? Albeit succinct and to the point, we do not know where the monitoring occurred. Throughout the range of the species? In one state? In part of one state? Making such a distinction could make the difference between a potential user reading or not reading the report. In this instance, the results are synthesized rangewide, so simply adding that term would clarify the scope of the monitoring effort and report (Sogge et al. 2003).

### ABSTRACT OR EXECUTIVE SUMMARY

Although it may be very important to do so, most readers of a monitoring report will not dig into the details of the Methods and Results. Providing a brief informative

Abstract or Executive Summary is therefore essential. An Executive Summary is a very concise version of the report that includes brief descriptions of data, analysis, and interpretations. An Executive Summary should provide an understanding of:

- The goals and objectives of the monitoring work
- What important decisions or actions these data could help inform
- How the data can (or cannot) answer these questions
- Limitations of the data, including scope of inference both in space and time
- Implications of the trends and recommendations, if appropriate
- Future needs

An Abstract is similar to an Executive Summary but is usually more of a condensed overview of just the objectives and the findings and includes little in the way of interpretation. Generally an Executive Summary for a monitoring report is 2–3 pages, while an Abstract is about a page or less.

## INTRODUCTION

The Introduction should outline the reasons for inventorying and/or monitoring the population, species, community, or ecosystem of interest. It is useful to include ample contextual information that communicates to the reader precisely why monitoring or inventorying was ecologically, economically, culturally, or otherwise justifiable. Suggestions for this section include

1. A statement of the management problem or policy that prompted the inventory and/or monitoring project
2. A summary of current knowledge about the population, species, community, or ecosystem that is relevant to the management problem or policy driving the monitoring plan
3. A statement of the goals and objectives for the monitoring plan
4. Hypotheses and conceptual models that guided sampling, data analysis, and data interpretation

## STUDY AREA

This section should first establish the general spatial and temporal scales of the project and the rationale for the scope. Then, more specifically, the grain, extent, and context for the study should be described in detail. The grain refers to the finest level of detail measured in the study (i.e., patch size), and you should indicate why this grain size was chosen. The extent is the outer bounds of the sampling framework. This may be a species' geographic range, a watershed, or a property boundary. You should make an effort to describe the chosen extent relative to the space used or needed by the populations, species, communities, or ecosystems of interest, especially if the outer bounds are delineated based on anthropogenic criteria. A map should also be included with sufficient detail to allow the reader to understand the context within which the extent is embedded; exactly which components of the

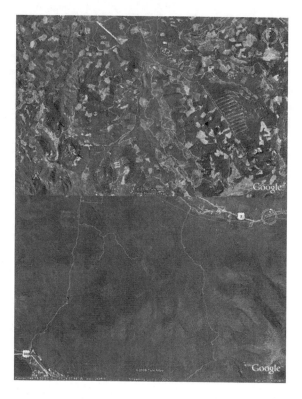

**FIGURE 12.2** Air photos of the Umpqua National Forest, Oregon (top), and the White Mountain National Forest, New Hampshire (bottom), illustrating the level anthropogenic disturbance in each area. Although the lower image seems to show less impact, regenerating patch cuts are scattered throughout the scene. If two similar monitoring programs were implemented in these two locales, the context, both in terms of geographic location and anthropogenic disturbance, would likely result in two significantly different programs on the ground. (Figures captured from Google Earth.)

broader landscape are and are not included in the sampling effort is often highly significant to stakeholders.

Typically the geology, soils, climate, and physiognomy or vegetation is described in this section to reinforce the broad spatial overview. Without describing biophysical factors that could have an influence on the patterns of results seen over space and/or time in sufficient detail, the reader may not understand why patterns were observed. Any pertinent land management actions such as roads, development, timber harvest, or agricultural practices should also be described in sufficient detail so that the reader can understand how these actions may influence results. It is particularly helpful to map how these anthropogenic disturbances relate to the area being monitored and to then provide copies in the report (Figure 12.2). Also, as anthropogenic disturbances typically change in their intensity and location over time, it is important to describe the temporal context of current maps and the implications this history has for the surrounding ecosystem. This can be succinctly done through the use of a chronosequence.

Finally, background information on the population, species, community, or ecosystem of interest should also be provided. Descriptions of the geographic range of the species, its home range, habitat elements, competitors, and predators allow the reader to interpret the results more completely. For example, a report on the monitoring of amphibians in the Mt. Hood National Forest may be particularly important with regards to changes in the abundance or distribution of Larch Mountain salamanders but may be less important with regards to species such as Pacific chorus frogs due to differences in geographic distribution (Figure 12.3). To understand trends in the latter species, coordinated monitoring efforts among many land management agencies and owners would be needed.

## Methods

The gold standard for the Methods section is that a reader be able to use it to repeat exactly what was done and produce comparable results. This necessitates considerable attention to detail in documenting the sampling and analytical procedures. This section should include

1. The rationale for selecting sampling units. Be sure to indicate if sampling was random or systematic, and if not random, then indicate what biases may be inherent in the resulting data. Any irregularities in selection of sampling sites such as failure to acquire permission to enter private lands, access to the sites, or other biases need to be described in detail and interpreted in the Discussion section.
2. A description of the sampling design. Explain the overall sampling design and which state variables were measured (e.g., presence–absence, correlative, before–after control-impact), the rationale for using that design, as well as the analytical model that the data were expected to populate. For instance, if data were collected from random sites among three strata, be clear as to how the strata were defined. Explain if any treatment effects were nested within other "treatments," and what type of statistical approach was appropriate for analysis. Of particular importance is an explanation of the determination of an appropriate sample size used in the monitoring effort. Clearly discuss the sample size from both a statistical power standpoint as well as from a logistical standpoint. Should the sample size be less than optimal based on power analysis, then also note the influence of the reduced sample size on the statistical power. In many cases, a repeated sampling design may have been used to develop detection probability estimates. The report should be clear as to whether the sampling design was of a repeated nature, and if so, in what fashion were the repeated visits standardized.
3. A summary of field methods for locating sampling units and collecting data. Much of this could be included in an appendix containing copies of the field protocols, data sheets, and detailed maps, but give enough detail here to ensure that easily avoidable biases will not be overlooked. Some of the most common biases are a result of differences in observers, or weather and temporal or spatial factors; explicitly addressing these topics is suggested.

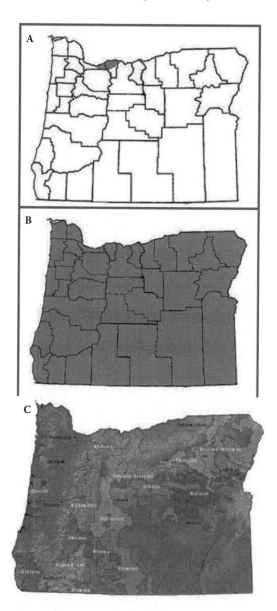

**FIGURE 12.3** Geographic range of the Larch Mountain salamander (A) and the Pacific chorus frog (B), and national forests in Oregon (C). Note that nearly the entire geographic range for the Larch Mountain salamander is included in the Mt. Hood National Forest, but that this national forest represents only a fraction of the geographic range for Pacific tree frogs. (Range maps from Tom Titus, University of Oregon, and used with permission.)

4. Identification of ancillary data used in the analysis. Describe all outside information used, such as satellite imagery or timber inventory data. If remotely sensed data were used, record the time that the images were taken. Similarly, if FIA or other forest inventory data were used, provide the dates of data collection, reference to specific field methods, and the location of the data. If data were acquired or downloaded from an external GIS warehouse and database, then details should be provided regarding the exact URL or appropriate contact information.
5. A description of analytical methods. Provide the specific statistical methods and software platform used to calculate descriptive and comparative statistics. In cases where complex model structures are used, include the programming code used to generate the statistics. Although software changes over time, it is usually far easier to recreate the model structure in another software package from an existing program than from text. If the data are stored in a relational database system, then one should also consider including the exact SQL code for querying these databases and extracting the data. Keep in mind that monitoring programs are often considered legacy projects, and a future user should be able to replicate all steps of data extraction and analysis.
6. A description of measures to ensure data quality. Describe personnel training and sampling activities used to minimize biases in the observation process. The results of observer skill tests, effort data, and detection probability estimates should also be reported so that the reader can understand the degree to which observation biases were included in data collection and analysis. Remember that observation biases, which may increase due to interobserver variability, among other reasons, diminishes the ability to detect trends and estimate state variables. In addition, address data entry, proofing, and cleaning activities in detail.

## RESULTS

This section should describe the results of data collection and analysis with little to no interpretation. The findings should be summarized verbally and statistically, and may also be presented in the form of tables, figures, or maps. Report all relevant aspects of the statistical results (central tendency, variance, drop in deviance, and other parameters) not only so that the information can be clearly interpreted but also so that the information can be used in metaanalysis (Gurevitch and Hedges 1999). Anderson et al. (2001) provided guidance on presenting statistical summaries in scientific papers, particularly with regard to information theoretic approaches to data analyses as well as Bayesian analyses.

It is important to remember that all tables and figures should be able to stand alone. In other words, they should be easily interpretable even if extracted from the report. Hence, the table and figure titles should clearly state what information is being displayed, where the data came from (location), and over what time period (Figure 12.4). Additional explanatory information may be placed in a footnote to the table or figure (Figure 12.5). When maps are used, be sure that legends are provided that include an explanation of the map features, scale, and orientation (e.g., a north arrow).

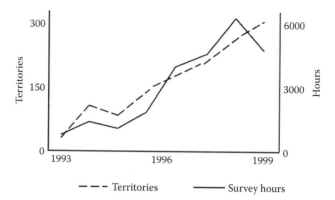

**FIGURE 12.4** Number of survey hours and willow flycatcher territories documented in Arizona, 1993–2000. Note that without also reporting the survey effort on the same chart as number of territories, this result could have been easily be misinterpreted by the reader as an increase in number of flycatchers over time. (Redrafted from Paradzick, C.E. et al. 2001. Southwestern Willow Flycatcher 2000 survey and nest monitoring report. Arizona Game and Fish Department Technical Report 175. Phoenix, AZ.)

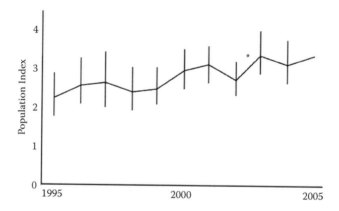

**FIGURE 12.5** Annual population indices (1995–2005) for barred owls, a species monitored in both central and northern Ontario. Data were collected by participants in the Ontario Nocturnal Owl Survey. Asterisks indicate significant differences between pairs of years: * $P < 0.05$, ** $P < 0.01$. (Redrafted from Crewe, T., and D. Badzinski. 2006. Ontario nocturnal owl survey. 2005. Final Report. Ontario Ministry of Natural Resources—Terrestrial Assessment Unit. Port Rowan, ON, Canada.)

## Discussion

Because one set of data can be interpreted a number of ways, depending on the goals and objectives of the party using it, some may argue that it makes sense to allow each individual to interpret the data. It can also be tempting to provide tables and figures that summarize the results without undertaking much interpretation of those results as a means of quickly disseminating monitoring data. Nonetheless, we believe that the individuals best able to interpret the data as objectively as possible are those with

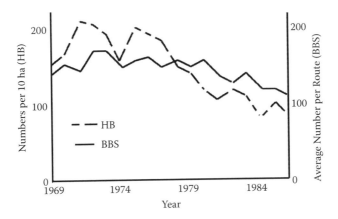

**FIGURE 12.6** Combined abundances of 19 selected species of forest-dwelling birds on the 10-ha study plot at Hubbard Brook (HB) and on the Breeding Bird Surveys (BBS) routes in New Hampshire, 1969–1986. (Redrafted from Holmes, R.T., and T.W. Sherry. 1988. *Auk* 105:756–768.)

the most knowledge about the monitoring program: those responsible for carrying out the actual monitoring work. We suggest that the Discussion, therefore, include an interpretation of the results along with an account of the pertinent knowledge accrued by the parties who undertook the monitoring. Although this may be a comparatively labor-intensive, time-consuming process, it will also make for a fully transparent and more comprehensive presentation and is generally worthwhile.

One important component of such a presentation is to compare the monitoring findings with the results of previous research and monitoring efforts for the species within the study area, as well as comparable efforts elsewhere. You may have read the pertinent literature, but it is unlikely that every user has. For instance, Holmes and Sherry (1988) compared long-term monitoring of a number of bird species on a 10-ha plot to regional patterns of abundance. In this case, generalized trends were remarkably similar (Figure 12.6). But these consistencies among scales are not a rule. In a subsequent study, Holmes and Sherry (2001) identified seven species that exhibited inconsistent trends between local monitoring sites and regional patterns. A brief comparison of these studies within the Discussion section of the latter's report helped place the results in the appropriate context and potentially serves to make the reader more informed than otherwise.

Yet not all monitoring projects have similar precedents. Important questions to address in any Discussion section include

- Did the project satisfy the objectives?
- In what ways did the work extend our knowledge about the species in the study area?
- Do the results support or challenge hypotheses and conceptual models stated in the introduction?
- What should the reader know about these results before using them to make decisions?

The answer to the last question is a key part of a responsible monitoring report. Explicitly informing your readers of the scope and limitations of the results provides a clear framework with which to view interpretations and thereby allows them to interpret the data in a manner that does not overstate the conclusions. Common limitations include scope of inference (both spatial and temporal), potential biases, and unexpected problems encountered during the project.

## Management Recommendations

This section should discuss how the results of the monitoring effort can or cannot be used to improve or otherwise influence resource management, including future monitoring activities. If the trends or differences in populations or habitats have reached or are approaching a threshold or "trigger point" identified in the monitoring plan, corrective management measures should also be proposed (Noon et al. 1999; Moir and Block 2001). With regards to monitoring, it is in this section that recommendations for improvements in methods, changes in parameters monitored, and/or termination of some aspects of monitoring should be proposed.

To decide what to include in a Management Recommendations section, you may have to revisit some of your files from the design and implementation stage of your monitoring program and juxtapose that information with the monitoring results. Information dealing with the program's goals and objectives, and potentially the results of any pilot studies, is particularly helpful. For instance, in a study carried out by McDonnell and Williams (2000), the general goal was to maintain a species-diverse grassland. Early in the research process, they collected a range of diversity values from a number of grassland sites and were able to derive a more specific management objective that defined species-diverse in a way that was pertinent to monitoring: to maintain diversity above 40 species within the grassland (Figure 12.7). From this objective, the researchers determined an appropriate management "trigger" point: "If measured diversity falls below 50 species for two

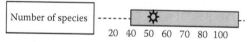

**Trigger:** If measured diversity falls below 50 species for two consecutive years.

**Trigger Response:** Collect data next year; if still < 50 then burn site within the next two years to restore grassland structure and composition.

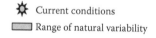

**FIGURE 12.7** A monitoring "scorecard" in which the current species diversity in a hypothetical grassland is assessed against a "desired" range for that parameter. (Redrafted from McDonnell, M.J., and N.S.G. Williams. 2000. Directions in revegetation and regeneration in Victoria. Proceedings of a Forum held at Greening Australia, May 5 and 6, 1999, Heidelberg, Victoria. Australian Research Centre for Urban Ecology. Occasional Publication No. 2, as modified from Hobbs, R., and D.A. Norton. 1996. *Restoration Ecology* 4:93–110.)

consecutive years, then the site should be burned within the next two years to restore the grassland structure and composition" (McDonnell and Williams 2000).

Given this context, if monitoring data reveal that the threshold level of 50 species is reached for two consecutive years, the Management Recommendations section should be used to encourage the development of a burning plan. All of the data that informed the original decisions should be cited to justify the proposal. Conversely, if the data indicate that the threshold has not been reached for two consecutive years, the Management Recommendations section should be used to discourage the development of a burning plan. Once again, all data that informed the original decisions should be cited. This example also underscores the importance of documenting all monitoring decisions and archiving any data or outside sources used in making them.

It is important to realize, however, that monitoring data, especially when they are collected on a long-term basis, may suggest that previously derived objectives and trigger points are unhelpful or unrealistic. Or they may suggest that monitoring itself is ineffective. For instance, if, using the same example, the researchers were to detect only three species in the first two years of sampling, there would likely be an inconsistency between the management plan and the ecosystem, the trigger point and the ecosystem, or the sampling techniques and the species being monitored (provided that nothing has drastically changed since the pilot studies were undertaken). In this scenario, the Management Recommendations section should be used to encourage further research such as a new set of pilot studies to recalibrate management and monitoring.

## List of Preparers

In this section, the authors of the report should identify themselves and other biologists who had supervisory roles in the project by name, title, or position, and provide contact information. Usually a section of acknowledgments lists those who assisted with some aspect of the monitoring design, data collection, or data analysis.

## References

References using standard author-date format must be provided for all statements in the text that are from other sources. Complete citations of all in-text references must be included in the References section. Standard scientific formats should be used such as the one used in this book or in standard scientific writing references such as Huth (1994).

## Appendices

Appendices are especially useful for reporting highly detailed information that may not be necessary for most readers, but which may be critical if other managers or scientists wish to replicate or further interpret the monitoring work. Copies of detailed field data collection protocols, data sheets, programming language used in analyses, detailed statistical summaries, field study site maps, and similar information that may be needed by others in the future can be included in appendices.

## SUMMARY

At times a Summary of the findings may be included if it constitutes more than the reiteration of the Executive Summary. Whereas the Executive Summary precedes the text and provides a brief synopsis of the approach and results, a Summary at the end of the document truly focuses on results and implications of the results. The Summary should be informative with enough detail to allow the reader to walk away knowing the "bottom line" from the monitoring program to date.

## SUMMARY

In summary, timely communication of monitoring results helps ensure that the results will be used and that decisions based on the results can be evaluated by all stakeholders. A well-structured report that allows others to understand how data were collected, what biases might exist, and how reliable inferences from the analyses might be is key to effective use of the data.

## REFERENCES

Anderson, D.R., W.A. Link, D.H. Johnson, and K.P. Burnham. 2001. Suggestions for presenting the results of data analyses. *Journal of Wildlife Management* 65:373–378.

Crewe, T., and D. Badzinski. 2006. Ontario nocturnal owl survey. 2005. Final Report. Ontario Ministry of Natural Resources—Terrestrial Assessment Unit. Port Rowan, ON, Canada.

Gurevitch, J., and L.V. Hedges. 1999. Statistical issues in ecological meta-analysis. *Ecology* 80:1142–1149.

Hobbs, R., and D.A. Norton. 1996. Towards a conceptual framework for restoration ecology. *Restoration Ecology* 4:93–110.

Holmes, R.T., and T.W. Sherry. 1988. Assessing population trends of New Hampshire forest birds: Local vs. regional patterns. *Auk* 105:756–768.

Holmes, R.T., and T.W. Sherry. 2001. Thirty-year bird population trends in an unfragmented temperate deciduous forest: Importance of habitat change. *Auk* 118:589–610.

Huth, E.J. 1994. *Scientific Style and Format: The CBE Manual for Authors, Editors and Publishers.* 6th ed. Cambridge University Press, Cambridge, U.K.

McDonnell, M.J., and N.S.G. Williams. 2000. Directions in revegetation and regeneration in Victoria. Proceedings of a Forum held at Greening Australia, May 5 and 6, 1999, Heidelberg, Victoria. Australian Research Centre for Urban Ecology. Occasional Publication No. 2.

Moir, W.H., and W.M. Block. 2001. Adaptive management on public lands in the United States: Commitment or rhetoric. *Environmental Management* 28:141–148.

Muths, E., A.L. Gallant, E.H. Campbell Grant, W.A. Battaglin, D.E. Green, J.S. Staiger, S.C. Walls, M.S. Gunzburger, and R.F. Kearney. 2006. The Amphibian Research and Monitoring Initiative (ARMI): 5-Year Report: U.S. Geological Survey Scientific Investigations Report 2006–5224. 77 pp.

Noon, B.R., T.A. Spies, and M.R. Raphael. 1999. Chapter 2: Conceptual basis for designing an effectiveness monitoring program. Pages 21–48 in *The Strategy and Design of the Effectiveness Monitoring Program for the Northwest Forest Plan.* USDA Forest Service, Pacific Northwest Research Station, General Technical Report PNW-GTR-437.

Paradzick, C.E., T.D. McCarthey, R.F. Davidson, J.W. Rourke, M.W. Sumner, and A.B. Smith. 2001. Southwestern Willow Flycatcher 2000 survey and nest monitoring report. Arizona Game and Fish Department Technical Report 175. Phoenix, AZ.

Sauer, J.R., J.E. Hines, and J. Fallon. 2008. The North American Breeding Bird Survey, results and analysis 1966–2007. Version 5.15.2008. USGS Patuxent Wildlife Research Center, Laurel, MD.

Smith, W.B., P.D. Miles, J.S. Vissage, and S.A. Pugh. 2004. Forest resources of the United States, 2002. USDA Forest Service General Technical Report NC-241.

Sogge, M.K., P. Dockens, S.O. Williams, B.E. Kus, and S.J. Sferra. 2003. Southwestern Willow Flycatcher Breeding Site and Territory Summary—2002. U.S. Geological Survey. Southwest Biological Science Center, Colorado Plateau Field Station, Flagstaff, AZ.

# 13 Uses of the Data
## *Synthesis, Risk Assessment, and Decision Making*

Imagine the following scenario: You have just finished spending nearly $500,000 over the past 5 years collecting information on changes in the abundance of sharptail snakes in the foothills of the Willamette Valley in Oregon. Data were collected from 30 randomly selected sites on public land managed to restore Oregon oak savannahs, and on another 30 sites on private lands that are grazed. The data are presented in Figure 13.1.

So given this information, what do you do? Continue to monitor? Use the information to make changes to the monitoring protocols or to management? What are the risks of changing versus continuing on with the status quo? Can these data be integrated with monitoring data from other programs to create a broader picture of the state of Oregon's ecosystems? We will follow this example through a few key steps in interpreting monitoring data and see how decisions might be made.

## THRESHOLDS AND TRIGGER POINTS

Clearly there are a number of issues that must be considered not only by managers but also by stakeholders before making any changes. One approach is to agree with

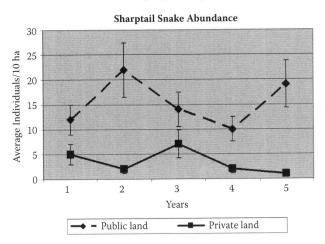

**FIGURE 13.1** Hypothetical patterns of detections of sharptail snakes in the foothills of the Willamette Valley, Oregon.

233

stakeholders at the outset of the monitoring program that if a particular threshold or trigger point is reached, then alternative management actions are to be implemented. Block et al. (2001) differentiated between trigger points that initiate a change to enact recovery, and thresholds that indicate success in a recovery action. In the case of Figure 13.1, a trigger point may be recording <5 snakes/10 ha for 2 or more consecutive years. If such a trigger point is reached, it could be agreed with stakeholders beforehand that a series of steps would be taken by the responsible agency to restore habitat for the species. Or in the case of endangered species, the decision may be made to capture individuals and initiate a captive breeding program. But in our hypothetical case after 5 years, sharptail snake detections meet the trigger point at year 5, so at that point the public management agency biologists may begin meeting with private landowners to explore the following options to restore habitat for the species:

- Provide landowner assistance on habitat restoration.
- Provide incentives to landowners to alter grazing and other land use practices.
- Explore purchase of a conservation easement that allows public biologists to manage land.
- Explore purchase of key properties and begin habitat restoration.

Any one of the above options may be acceptable to one landowner but not to another. As these or other options are implemented, then continued monitoring can allow detection of the point at which a threshold of recovery—say, >10 snakes/10 ha for >2 years—is surpassed and maintained. Monitoring a control area to understand changes in abundance on public conservation land will provide a point of comparison to help ensure that the patterns seen on private lands using the above approaches are more likely caused by management actions than other extraneous effects. For instance, if the abundance on both the public and private lands declined over time despite changes in management practices on the private lands, then declines are more likely due to factors unassociated with management such as changes in climate or disease.

One possible problem with the identification of thresholds is that they are the result of social negotiation, and although they may be based in biology, they may also simply represent an agreed upon, socially acceptable point by managers and stakeholders. Thresholds based on biology may represent population density, probability of occurrence, a change in reproduction or survival (or lambda), genetic heterozygosity, or other population parameters, but the threshold(s) are set jointly by biologists and stakeholders. Use of genetic markers to assess changes in effective population size and other aspects of population ecology has become increasingly popular (Schwartz et al. 2007). Schwartz et al. (2007) described two categories of monitoring using genetic markers: Category I, which can identify individuals, populations, and species; and Category II, which monitors population genetic parameters allowing insights into demographic processes and "retrospective monitoring" to better understand historical changes (Figure 13.2). Thresholds may also, however, constitute more of a reflection of society's tolerance of or desires for a particular species. For instance, the threshold for the number of cougars in a residential area

Uses of the Data

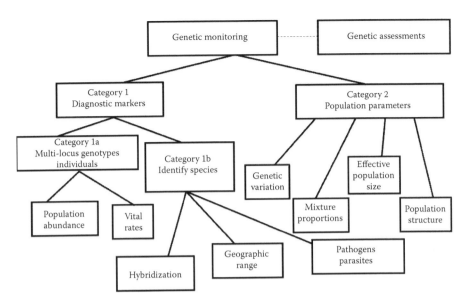

FIGURE 13.2 Examples of genetic monitoring and the types of information that can be garnered from these techniques. (Redrafted from Schwartz, M.K., G. Luikart, and R.S. Waples. 2007. *Trends in Ecology and Evolution* 22:25–33.)

of California may be the level that the public can tolerate rather than what is most significant in terms of the population dynamics of the species.

## FORECASTING TRENDS

With 5 years of data, trends can begin to emerge from the data (Figure 13.3) that provide information to guide management actions. In our hypothetical example, trends on public lands are rather stable, whereas those on private lands are declining. If we forecast the trend from private lands into the future we can see that in 2.5 years the x intercept for the trend will reach 0. The degree of precision in estimating the x intercept decreases dramatically as forecasts are extended further into the future, so forecasting attempts should be viewed as one tool to guide decision making. It is not clear if the x intercept will be reached in 1 year, 2.5 years, or 10 years, or at all, but the trend line does raise concerns about the long-term viability of the species on private lands and may initiate a more rapid response than if the trend line had an x intercept of 15 years. Dunn (2002) used an approach similar to this and categorized over 200 bird species into conservation alert categories.

But these are simply linear trends, and the variability associated with trends, especially for rare species, is often very high. Indeed, the power associated with detecting a significant trend is often very low with rare species, thus statistical trend lines must be interpreted cautiously to avoid making an error of concluding that no trend exists when it actually is in a decline. This is especially problematic when populations have already reached very low levels, and the probability of detecting an additional decline is very low (Staples et al. 2005). In these cases it may be more

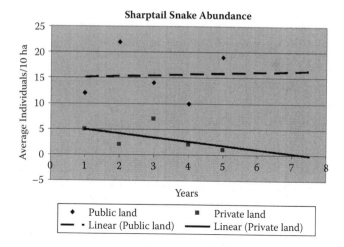

**FIGURE 13.3** Trend lines associated with hypothetical sharptail snake abundance and forecast estimates of abundance 2.5 years into the future. Note that at 7.5 years, the estimate falls to 0 on private land.

useful to employ risk assessments based on population viability analyses (PVA; Lande et al. 1993; Morris et al. 2002). If the data collected in monitoring can be used to aid in parameterizing a PVA model, then at least relative changes in future population abundance or time-to-extinction can be estimated (Dennis et al. 1991; Morris and Doak 2002). Staples et al. (2005) proposed a viable population monitoring approach in which yearly risk predictions are used as the monitoring indicator. Staples et al. (2005) defined "risk" as "the probability of population abundance declining below a lower threshold within a given time frame." Predicting that risk will increase over time could constitute a trigger point and prompt alternative management actions.

## PREDICTING PATTERNS OVER SPACE AND TIME

Clearly managers would like to know where on a landscape species are likely to occur so that management actions can be taken to increase or decrease populations or at least have minimal effects on desired species. Monitoring occurrence of organisms across a landscape can provide information in the spatial distribution of individuals within populations and can provide a better understanding of metapopulation structure and connectivity among subpopulations. If information on reproduction and survival is also included in the monitoring effort, then additional information on the value of subpopulations as sources or sinks can also be gained. And if this information is collected over time, then information on the probabilities of subpopulations becoming locally extinct in patches and subsequent recolonization can also be understood through long-term monitoring.

Although this baseline information on the distribution and fitness of organisms over a planning area is valuable information for understanding the impacts of managing landscapes, issues such as land use and climate change make the information

Uses of the Data

even more valuable. In the face of such changes, the risk of species loss from an area, or even overall extinction, depends on the rate at which a species can adapt to changing conditions. Monitoring information can provide evidence to more fully understand both the rates of change in the biophysical environment and the associated fitness of organisms. In the following sections we use monitoring in the context of climate change as an example to show how environmental stressors can influence the way managers act to attempt to conserve biodiversity, but also the difficulties of confronting such comprehensive ecological changes.

If we continue to pump $CO_2$ into the atmosphere at current rates, then approximately 20%–30% of plant and animal species assessed by the IPCC (2007) are likely to be at increased risk of extinction as global average temperature increases by 1.5°C to 2.5°C or more. Hence, understanding certain aspects of the environment and species responses through monitoring is key to providing opportunities for species to adapt to or recover from climate change. But climate change is probably one of the more difficult environmental stressors to respond to even with good monitoring data because it is global; the opportunity for comparisons between sites affected by climate change and those unaffected by change is rare if they exist at all. Indeed, we are not usually given the opportunity to use BACI or comparative mensurative approaches when designing a monitoring plan affecting regional or global stressors, so we must rely instead on associations over time. To be more specific, effect may be inferred from these data only with care since other factors associated with change may have a greater or lesser effect in any observed trends. Nonetheless, there are a number of potential factors that are often assessed when trying to understand effects of global changes like climate change on loss of biodiversity.

## Geographic Range Changes

If global average temperature increases exceed 1.5°C to 2.5°C, then major changes in ecosystem structure and function, species' ecological interactions, and shifts in species' geographical ranges are anticipated, with predominantly negative consequences for biodiversity (IPCC 2007). Because geographic ranges of species are often dictated by climatic conditions (or by topographic barriers) that influence physiological responses, changes in geographic ranges of species are frequently predicted using bioclimate envelopes (Pearson and Dawson 2003), and observed changes are used as an early indicator of a species' response and ability to adapt to climate change. But bioclimate envelopes are coming under scrutiny and being questioned because biotic interactions, evolutionary change, and dispersal ability also influence the ability or inability of a species to respond to changes in its environment (Pearson and Dawson 2003). One can easily imagine how the impacts of climate change on subpopulations could be exacerbated by land use change that leads to their isolation; indeed, these subpopulations would become more vulnerable to local extinction through inability to disperse, infectious disease, or competition with invasive species as their habitat changes in response to climate change.

Zuckerberg et al. (2009) used the New York State Breeding Bird Atlas surveyed in 1980–1985 and 2000–2005 to test predictions that changes in bird distribution are related to climate change. They found that 129 bird species showed an average northward range shift in their mean latitude of 3.6 km (Zuckerberg et al. 2009) and that the southern range boundaries of some bird species moved northward by 11.4 km. Clearly these monitoring programs can provide evidence for associations between climate change and changes in geographic ranges, yet other factors should not be ruled out. Human population density has changed over that time as have land use patterns, and both could have had similar effects on the geographic range of certain species. Nonetheless the compelling fact is that all of the 129 species that they examined showed a northward shift in distribution; thus, in this case, the data suggest that the driving influence is something more global and consistent in its impact. Similar efforts at using monitoring information over time can elucidate changes for less mobile species such as plants, invertebrates, and amphibians (Walther et al. 2002).

## Home Range Sizes

Resource availability is related to home range size for many species. Climate change quite likely will influence the dispersion or concentration of available food and cover resources for many species (McNab 1963). Therefore, monitoring home range sizes also constitutes a method for assessing ecological effects of climate change on some species.

Documenting the sizes of home ranges can be costly and estimates can suffer from low precision for a number of reasons (Borger et al. 2006). Estimating the effect size that could be detected (a power analysis would help determine this; Zielinski and Stauffer 1996) can allow a better understanding of the actual risks of losing species.

Some effects are obvious in the higher latitudes. As sea ice is lost and shifts in its locations, polar bears must extend their foraging bouts into new locations (Derocher et al. 2004). If the energy that they expend in foraging exceeds the energy they gain from catching prey, then they will die. With polar bears and other species, expanding home range size can be an early warning indicator of decreased or dispersed resource availability and an indication that the species may be facing imminent population declines. Changes in home range sizes hence can be an important aspect of risk analysis. If home range sizes are expanding then risk of population decline is greater than if they are stable to contracting, and changes in home range size may be detectable prior to a decline in abundance.

## Phenological Changes

Another early warning sign of impending impacts of climate change on populations is changes in the phenological patterns of plants and animals. Indeed, events such as the arrival at breeding or wintering sites from migrations, onset of flowering or other reproductive activities, leaf-out, or leaf-fall function are indicators because they tend to be influenced at least in part by temperature (Parmesan 2007). Phenological studies have been conducted for years (e.g., Menzel 2000), but not on the global scale necessary for monitoring global climate change. Schwartz (1994) provided a discussion on the detection of large-scale changes using phenological information.

He commented on how past efforts at recording phenological patterns have often been done on small scales, and then suggested that by integrating ground-based data collection with remotely sensed data, local patterns can be appropriately scaled to ecoregions, continents, and ideally the globe, allowing larger scale patterns to be inferred. White et al. (2005) proposed a global framework for monitoring phenological responses to climate change using remotely sensed data. If White et al.'s (2005) approach can be implemented effectively, then the physical mechanisms responsible for observed patterns can be used to assess the effectiveness of global-scale models in predicting changes in phenological events (Schwartz 1994).

### Habitat Structure and Composition

For some purposes, simply understanding changes in the availability of habitat for a species may be sufficient to infer likely changes in a species' potential abundance or distribution. The devil is in the details, however. For many species, knowledge of abundance and spatial arrangement of fine-scale habitat elements such as large trees, snags, logs, or shrubs is important. However, gathering this knowledge on a large scale can pose problems; satellite imagery will not detect many of these features. LIDAR or other remotely sensed data, however, can often provide information at a fine enough scale to detect habitat components (Hyde et al. 2006). LIDAR in particular can provide information on the fine-scale vertical complexity of a forest including canopy heights and canopy biomass (Hyde et al. 2006). For those species associated with vegetative layers in forests, remotely sensed data may be useful. For species associated with dead wood or other habitat elements that are not detectable using remote techniques, then combining remotely sensed data with ground-plot data becomes the only logical approach. These fine-scale habitat elements can be imputed to pixels from known locations of ground plots using nearest-neighbor techniques (Ohmann and Gregory 2002).

## SYNTHESIS OF MONITORING DATA

Monitoring data can be integrated with other information on terrain, climate, disturbance probabilities, land use, land ownership, and infrastructure to paint a generalized integrated picture of the state of a landscape. These approaches allow managers to monitor not only the individual pieces of the landscape but also the integrated whole over time. For instance, changes in the structure and composition of forest stands in Oregon with and without certain silvicultural practices can be incorporated into maps of forest age classes and habitat types (Spies et al. 2007). These can then be linked to models of forest growth and development (many of which are based on continuous forest inventory monitoring plots), and to transition probabilities associated with land management decisions, allowing projections of possible future conditions for planning purposes and to better understand the implications of possible changes in land use policy (Spies et al. 2007). Other approaches have not explicitly used vegetation growth models but have developed scenarios of past conditions, current conditions, and likely alternative future conditions of landscapes (Baker et al. 2004). It is important to stress that these approaches use monitoring information not only to parameterize many of the spatial and temporal projections, but also to improve our understanding of possible future conditions. Indeed, it is the ability to use data to create models that allows projections of conditions into the future based on interacting stressors such as climate change (IPCC 2007) and land use planning (Kaiser et al. 1995). These model projections not only raise the potential for developing "what if" scenarios to compare alternative policies, but they can identify key parameters that should be monitored into the future to help stakeholders understand if the results of a policy change are being realized as projected. There are so many interacting assumptions that enter into these complex landscape projections that without monitoring data the projections are at best a likely future condition and at worst an artifact of an incorrect assumption. Some practitioners also attempt to integrate ecological monitoring data with economic, social, and institutional information in order to create bodies of data that function as sustainability indicators. This has often been done for agricultural systems and for communities in developing countries but is expanding to include other regions, such as highly developed urban environments (Olewiler 2006; Van Cauwenbergh et al. 2007). Not all of these initiatives necessarily include the monitoring of populations or habitat, but many do. For instance, to assess the sustainability of the terrestrial resource use of communities in tropical ecosystems, several researchers have integrated wildlife monitoring and the mapping of hunting kill-sites with data regarding the use of other terrestrial resources, access to new technologies, and changing local land uses (Koster 2008; Parry et al. 2009). In the Sustainability Assessment of Farming and the Environment (SAFE) framework for developing a set of variables that indicate the sustainability of agroecosystems, variables that measure the retention of biodiversity and the "functional quality of habitats" are considered an integral component of the monitoring framework (Van Cauwenbergh et al. 2007). While monitoring wildlife and habitat is not explicitly discussed within the framework guidelines, it would be difficult to make such assessments without doing so. It is also important to realize that the concept of "sustainability indicators" and previous attempts at deriving them have their share of critics. Scerri and James (2009), for instance, discuss how many practitioners reduce the complex concept of sustainability and the generation of sustainability indicators that is likely context specific to a very technical, quantitative task.

PVA models typically compare the estimated risk of a species or population going extinct among several management alternatives. PVA models are notoriously data hungry, requiring age- or stage-specific estimates of survival, reproduction, and movements with associated ranges of variability for each parameter estimate (Beissinger and Westphal 1998; Reed et al. 2002). As with projections of landscape models, monitoring aspects of PVA projections not only allow an assessment of risk associated with not achieving an expected result but also highlight the weaknesses in the model assumptions. Monitoring programs that inform the validity of assumptions can provide the opportunity for developing more reliable model structures and resulting projections. Deciding which assumptions or parameters to monitor based on a model structure can be problematic, especially with large complex models such as the two described above. Identification of variables to monitor may be based on subjective assessment of the reliability of the underlying data or on more structured sensitivity analyses that identify variables that have an overriding influence on the model results (McCarthy et al. 1995; Fieldings and Bell 1997). Quite often the least reliable parameters in these models are those that are the most difficult to measure. This can create a dilemma for a program manager developing a monitoring program since these data may be the most important to decreasing uncertainty in future predictions, but they may also be the most expensive to acquire. Hence, a benefit: cost assessment will need to be made with stakeholders to develop a priority list of variables.

Despite the ability to develop more reliable estimates of key variables from monitoring data, projections into the future are always faced with the inability to predict unknown threshold events that would not have been foreseen at the outset. For instance, barred owl invasions into spotted owl habitat were not seriously considered as much of a threat as habitat loss when early PVAs for spotted owls were developed (Peterson and Robins 2003). And even when models can consider new or confounding variables, the inter-relationships among the variables can give rise to new states or processes that could not be foreseen.

Climates have always changed on this earth but the rate of change likely to be seen in the next century could be unprecedented. Alterations in vegetative community structure and inter-specific relationships are likely to change, but their ability to adapt to changing climatic conditions is in question. Williams and Jackson (2007) provided an overview of no-analog plant communities associated with historic "novel" climates and future novel climates that are likely to be warmer than any at present. Ecological models such as forest dynamics models and PVA models are at least partially parameterized from relatively recently collected data, so they may not accurately predict responses to novel climates (Williams and Jackson 2007). The uncertainty raised by the potential development of no-analog conditions must be explicitly considered during risk analyses.

## RISK ANALYSIS

Risk analyses have been formally developed with regards to direct and indirect effects of pollutants on wildlife species. The U.S. Environmental Protection Agency defines Ecological Risk Assessment (ERA) as

> ... an evaluation of the potential adverse effects that human activities have on the living organisms that make up ecosystems. The risk assessment process provides a way to develop, organize, and present scientific information so that it is relevant to environmental decisions. When conducted for a particular place such as a watershed, the ERA process can be used to identify vulnerable and valued resources, prioritize data collection activity, and link human activities to their potential effects. ERA results provide a basis for comparing different management options, enabling decision-makers and the public to make better informed decisions about the management of ecological resources. (http://epa.gov/superfund/programs/nrd/era.htm)

The steps used by the EPA are outlined in Figure 13.4, and could be adapted for use in other situations where risks from other environmental stressors or disturbances may be of key importance to managers (e.g., fires, land use, floods, etc.). For instance, Hull and Swanson (2006) provided a stepwise process for assessing risk to

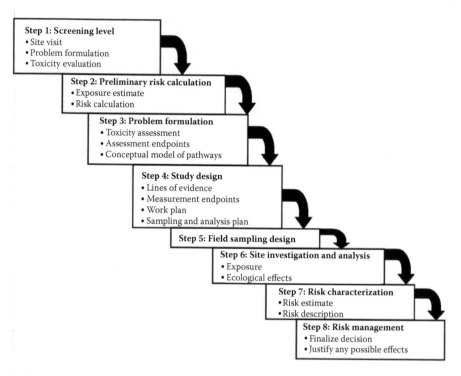

**FIGURE 13.4** Example of a risk-assessment process formulated by the U.S. Environmental Protection Agency to identify and mitigate ecological risks. (From http://www.epa.gov/superfund/programs/nrd/era2.htm, accessed July 6, 2009.)

wildlife species from exposure to pollutants. Similar approaches have been proposed to assess risk to loss of biodiversity. Kerns and Ager (2007) described risk assessment as a procedure to assess threats and understand uncertainty by "... providing (1) an estimation of the likelihood and severity of species, population, or habitat loss or gain, (2) a better understanding of the potential tradeoffs associated with management activities, and (3) tangible socioeconomic integration." They proposed a quantitative and probabilistic risk assessment to provide a bridge between planning and policy that includes stakeholder involvement (Kerns and Ager 2007). Such formal approaches are needed within ecological planning processes if both managers and stakeholders are to understand uncertainty, and the costs associated with the risks of not achieving the intended results.

## DECISION MAKING

From a logical standpoint, decisions should be made using a sequence of steps: characterize the problem or question, identify a full range of alternatives and determine criteria for selecting one, collect information about each option and rate it based on the criteria, then make the final decision based on the rating (Lach and Duncan 2007). But Klein (2001) found that only 5% of all decisions are made using such a logical approach. Individuals often make their decisions using intuition and mental simulations (quickly relating the outcome of a decision to some experience) (Lach and Duncan 2007). Groups may make decisions differently and are better able to make better decisions on complex problems than are individuals (Lach and Duncan 2007). People with different worldviews structure the world around them in different ways and in so doing bring a different perspective to a group decision. Ensuring that a range of worldviews is represented in a group can be particularly useful when trying to reach a balanced decision on a complex issue, though discussions needed to reach that decision may necessarily become protracted.

## SUMMARY

Considerable time and money are invested in many monitoring programs, so not only must the design of these programs be scientifically and statistically rigorous, it must be clear to the managers and stakeholders how the information will be used to make decisions. During the design phase, trigger points or thresholds should be identified to ensure that managers know when changes in management approaches should be considered. In many circumstances it is easy for managers to simply wait for more information without taking an action, not realizing that waiting places greater risk on the achievement of desired outcomes. Using monitoring data as the basis for forecasting trends over space and time can allow managers to understand the implications of waiting too long before taking remedial actions. Factors such as changes in climatic characteristics, phenology, geographic ranges, and home range sizes of some species can be particularly informative in the face of global changes to climate for which the only reference condition is the past. Using monitoring information as a means of parameterizing models of landscape or climate change allows projections over space and time of more complex conditions. Such integrative approaches further allow comparisons among alternative management strategies or policies, and can be an important component of a risk analysis, a formalized approach to identifying uncertainties, and assessing direct and indirect effects of stresses on organisms and ecosystems. The results of monitoring, modeling, and risk analysis are then used to make decisions by individuals or by groups. Although we typically assume that decisions are made in a logical manner, many decisions are made based on intuition or as the result of group discussions among people with various worldviews.

## REFERENCES

Baker, J.P., D.W. Hulse, S.V. Gregory, D. White, J. Van Sickle, P.A. Berger, D. Dole, and N.H. Schumaker. 2004. Alternative futures for the Willamette river basin. *Ecological Applications* 14:313–324.

Beissinger, S.R., and M.I. Westphal. 1998. On the use of demographic models of population viability in endangered species management. *Journal of Wildlife Management* 62:821–841.

Block, W.M., A.B. Franklin, J.P. Ward, Jr., J.L. Ganey, and G.C. White. 2001. Design and implementation of monitoring studies to elucidate the success of ecological restoration on wildlife. *Restoration Ecology* 9:293–303.

Borger, L., N. Franconi, F. Ferretti, F. Meschi, G. De Michele, A. Gantz, A. Manica, S. Lovari, and T. Coulson. 2006. Effects of sampling regime on the mean and variance of home range size estimates. *Journal of Animal Ecology* 75:1393–1405.

Dennis, B., P.L. Munholland, and J.M. Scott. 1991. Estimation of growth and extinction parameters for endangered species. *Ecological Monographs* 61:115–143.

Derocher, A., N.J. Lunn, and I. Stirling. 2004. Polar bears in a warming climate. *Integrative Comparative Biology* 44:163–176.

Dunn, E.H. 2002. Using decline in bird populations to identify needs for conservation action. *Conservation Biology* 16:1632–1637.

Fieldings, A.H., and J.F. Bell. 1997. A review of methods for the assessment of prediction errors in conservation presence: Absence models. *Environmental Conservation* 24:38–49.

Hull, R.N., and S. Swanson. 2006. Sequential analysis of lines of evidence—An advanced weight-of-evidence approach for ecological risk assessment. *Integrated Environmental Assessment and Management* 2:302–311.

Hyde, P., R. Dubayah, W. Walker, J.B. Blair, M. Hofton, and C. Hunsaker. 2006. Mapping forest structure for wildlife habitat analysis using multi-sensor (LiDAR, SAR/InSAR, ETM+, Quickbird) synergy. *Remote Sensing of Environment* 102:63–73.

IPCC. 2007. Climate change 2007: Synthesis report. Contribution of working groups I, II and III to the fourth assessment report of the Intergovernmental Panel on Climate Change. IPCC, Geneva, Switzerland, 104 pp.

Kaiser, E., D. Godschalk, and F.S. Chapin. 1995. *Urban Land Use Planning*. 4th ed. University of Illinois Press, Urbana, IL.

Kerns, B.K., and A. Ager. 2007. Risk assessment for biodiversity conservation planning in Pacific Northwest forests. *Forest Ecology and Management* 246:38–44.

Klein, G. 2001. Understanding and supporting decision making: An interview with Gary Klein. *Information Knowledge Systems Management* 2(4):291–296.

Koster, J. 2008. The impact of hunting with dogs on wildlife harvests in the Bosawas Reserve, Nicaragua. *Environmental Conservation* 35(3):221–220.

Lach, D., and S. Duncan. 2007. How do we make decisions? Chapter 2. Pages 12–20 in Johnson, K.N., S. Gordon, S. Duncan, D. Lach, B. McComb, and K. Reynolds. Conserving creatures of the forest: A guide to decision making and decision models for forest biodiversity. National Commission on Science for Sustainable Forestry Final report. NCSSF, Washington, D.C.

Lande, R. 1993. Risks of population extinction from demographic and environmental stochasticity and random catastrophes. *The American Naturalist* 142:911–927.

McCarthy, M.A., M.A. Burgman, and S. Ferson. 1995. Sensitivity analysis for models of population viability. *Biological Conservation* 73:93–100.

McNab, B.K. 1963. Bioenergetics and the determination of home range size. *The American Naturalist* 97:133–141.

Menzel, A. 2000. Trends in phenological phases in Europe between 1951 and 1996. *International Journal of Biometeorology* 44:76–81.

Morris, W.F., P.L. Bloch, B.R. Hudgens, L.C. Moyle, and J.R. Stinchcombe. 2002. Population viability analysis in endangered species recovery plans: Past use and future improvements. *Ecological Applications* 12:708–712.

Morris, W.F., and D.F. Doak. 2002. *Quantitative Conservation Biology: Theory and Practice of Population Viability Analysis*. Sinauer Associates, Sunderland, MA.

Ohmann, J.L., and Gregory, M.J. 2002. Predictive mapping of forest composition and structure with direct gradient analysis and nearest neighbor imputation in coastal Oregon, USA. *Canadian Journal of Forest Research* 32:725–741.

Olewiler, N. 2006. Environmental sustainability for urban areas: The role of natural capital indicators. *Cities* 23(3):184–195.

Parmesan, C. 2007. Influences of species, latitudes and methodologies on estimates of phenological responses to global warming. *Global Change Biology* 13:1860–1872.

Parry, L., J. Barlow, and C.A. Peres. 2009. Allocation of hunting effort by Amazonian smallholders: Implications for conserving wildlife in mixed-use landscapes. *Biological Conservation* 142:1777–1786.

Pearson, R.G., and T.P. Dawson. 2003. Predicting the impacts of climate change on the distribution of species: Are bioclimate envelope models useful? *Global Ecology and Biogeography* 12:361–371.

Peterson, A.T., and C.R. Robins. 2003. Using ecological-niche modeling to predict barred owl invasions with implications for spotted owl conservation. *Conservation Biology* 17:1161–1165.

Reed, J.M., L.S. Mills, J.B. Dunning, E.S. Menges, K.S. McKelvey, R. Frye, S.R. Beissinger, M.C. Anstett, and P. Miller. 2002. Emerging issues in population viability analysis. *Conservation Biology* 16:7–19.

Scerri, A., and P. James. 2009. Communities of citizens and "indicators" of sustainability. *Community Development Journal*. doi:10.1093/cdj/bsp013.

Schwartz, M.D. 1994. Monitoring global change with phenology: The case of the spring green wave. *International Journal of Biometeorolology* 38:18–22.

Schwartz, M.K., G. Luikart, and R.S. Waples. 2007. Genetic monitoring as a promising tool for conservation and management. *Trends in Ecology and Evolution* 22:25–33.

Spies, T.A., K.N. Johnson, K.M. Burnett, J.L. Ohmann, B.C. McComb, G.H. Reeves, P. Bettinger, J.D. Kline, and B. Garber-Yonts. 2007. Cumulative ecological and socioeconomic effects of forest policies in coastal Oregon. *Ecological Applications* 17:5–17.

Staples, D.F., M.L. Taper, and B.B. Shepard. 2005. Risk-based viable population monitoring. *Conservation Biology* 19:1908–1916.

Van Cauwenbergh, N., K. Biala, C. Bielders, V. Brouckert, L. Franchois, V. Garcia Cidad, M. Hermy, E. Mthij, B. Muys, J. Rejinders, X. Sauvenier, J. Valckx, M. Vancloster, B. Van der Veken, E. Wauters, and A. Peeters. 2007. SAFE—A hierarchical framework for assessing the sustainability of agricultural systems. *Agriculture, Ecosystems and Environment* 120:229–242.

Walther, G.-R., E. Post, P. Convey, A. Menzel, C. Parmesan, T.J.C. Beebee, J.-M. Fromentin, O. Hoegh-Guldberg, and F. Bairlein. 2002. Ecological responses to recent climate change. *Nature* 416:389–395.

White, M.A., F. Hoffman, W.W. Hargrove, and R.R. Nemani. 2005. A global framework for monitoring phenological responses to climate change. *Geophysical Research Letters* 32:L04705.

Williams, J.W., and S.T. Jackson. 2007. Novel climates, no-analog communities and ecological surprises. *Frontiers in Ecology and the Environment* 5:475–482.

Zielinski, W.J., and H.B. Stauffer. 1996. Monitoring *Martes* populations in California: Survey design and power analysis. *Ecological Applications* 6:1254–1267.

Zuckerberg, B., A.M. Woods, and W.F. Porter. 2009. Poleward shifts in breeding bird distributions in New York State. *Global Change Biology* 15:1–18.

# 14 Changing the Monitoring Approach

Despite the best efforts at designing a monitoring plan, it is nearly inevitable that changes will be made to the monitoring strategy sooner or later. Indeed, if monitoring data are used as intended in an adaptive management framework, the information gained *should* be used to refine the monitoring approach and improve the quality and utility of the data that are collected (Vora 1997). Marsh and Trenham (2008) summarized 311 surveys sent to individuals involved with monitoring programs in North America and Europe and estimated that 37% of the programs made changes to the overall design of the monitoring program at least once and that data collection techniques changed in 34% of these programs. Adding novel variables to measure is also often incorporated into monitoring programs as information reveals new, perhaps more highly valued, patterns and processes. Another change, dropping variables, oftentimes must accompany these additions simply due to increased costs associated with measuring more things. And clearly the vagaries of budget cycles can cause variables to be dropped for one or more monitoring cycles and then readded if budgets improve. But changes to monitoring programs can result in some or all of the data collected to date being incompatible with data collected after a change is made.

Increased precision in data collection is always a goal, but if data are collected with two levels of precision over two periods of time, then pooling the data becomes problematic. Similarly, changing the locations or periodicity of sampling can lead to discontinuities in the data; this makes analyses more challenging. Given these concerns, changing a monitoring program should not be done lightly and is a process that necessitates as much preparation as establishing the initial monitoring plan. Despite this, there is remarkably little information available to inform monitoring program managers when they consider changing approaches within their program.

## GENERAL PRECAUTIONS TO CHANGING METHODOLOGY

In light of the paucity of references, it is important to carefully consider those that do exist. For instance, Shapiro and Swain (1983) noticed the enormous impact a change in methodology had on data analysis in a monitoring program involving water quality in Lake Michigan. Indeed, the apparent decline in silica concentration between 1926 and 1962 was an artifact of having changed methods and not a result of increased phosphorus loads as had been reported prior to their work (Shapiro and Swain 1983).

Strayer et al. (1986) described this classic example of the dangers associated with changing methodology during a monitoring program further and used the experience to suggest several general precautions to take when changing methodology in the middle of a long-term monitoring program:

1. Calibrate the new methods against the old methods for a sufficient period of time.
2. Maintain a permanent detailed record of all protocols used.
3. Archive reference samples of materials collected in the field where appropriate. Voucher specimens, soil or water samples, and similar materials should be safely and securely archived for future reference. Consider collecting hair, feather, scales, or other tissues from animals for future DNA analyses.
4. Change methods as infrequently as possible.

## WHEN TO MAKE A CHANGE

Although alterations should certainly be minimized, there will be instances in which a change will make monitoring more valuable than maintaining the original design. There will also be instances in which changes are unavoidable due to data deficiencies or stakeholder desires. Changes made to the design, in the variables measured, in the sampling techniques or locations, in the precision of the samples, or in the frequency of sampling are all potentially helpful. Yet they also all have the potential to detract from a monitoring program in unique and powerful ways. Determining when or if to change your program, therefore, can be a difficult task. So, when should a monitoring program be changed? The following examples describe some potentially appropriate or necessary scenarios, but it is important to keep in mind that all potential changes merit careful consideration.

### CHANGING THE DESIGN

The initial stages of a monitoring program result in a series of data points and associated confidence intervals that describe a trend over space or time. Once that trend is established, new questions often emerge or, as new techniques are developed that are more useful or precise, a change in the design might be warranted. For example, the pattern observed in Figure 14.1 represents a positive trend, but the precision of the estimates may come into question since the data in time periods 4, 5, and 6 reflect a plateau. This realization leads to a number of questions that would undermine the monitoring program if left unanswered: Is this plateau real or a function of imprecise data collection? Are other variables such as fecundity or survival better indicators of population response than simply population sizes? If a change in design can answer these questions and make data more useful, and staying the course cannot, managers and stakeholders may decide that it is time to make some changes even after only 5 years of data collection.

## CHANGING THE VARIABLES THAT ARE MEASURED

Changing the variables measured may be considered for a number of reasons, but should be undertaken prudently because it can set a monitoring program back to the very beginning. One legitimate reason to change the variables, however, is a failure to meet the goals and objectives of the project or a sudden change in the desires of stakeholders. If goals and objectives are not being met by the data being collected, it is obvious that changes must be made. In such a case, data collected to date may still be very useful to decision makers. These data may help inform decisions made regarding how the changes ought to be made (altering variables measured or intensity of sampling). Using data in post hoc power analyses has also become quite popular when trying to understand why a significant trend was not observed, but the logic behind post hoc power analysis has been considered inherently flawed by some authors because the point of power analysis is to ensure during the design phase of a project that if a trend is real then there is an "x" percent chance that it can be detected (Hoenig and Heisey 2001). Data analyses that include confidence intervals on parameter estimates are particularly useful in understanding the deficiencies of the underlying data and informing decisions regarding what changes to make, and how and when to make them. This is especially true when failure to reject a null hypothesis (e.g., unable to detect a trend) could be erroneous and jeopardize a population.

Once the data have been analyzed, and the uncertainty associated with parameter estimates is understood, then all stakeholders can be informed about the changes that could be made in the monitoring plan to better meet their goals and objectives and then ensure that they are involved in the decision-making process.

Returning to our example (Figure 14.1), positive trends may be encouraging, but if the data are needed to describe the degree of recovery and potential delisting under the Endangered Species Act, then stakeholders likely would suggest that population parameters describing reproductive rates and survival rates may also

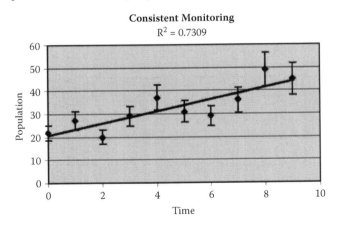

**FIGURE 14.1** Example of an increasing population over time using consistent monitoring techniques in each time period.

need to be measured. In this case, an entirely new monitoring program may be added to the populations monitoring program, or monitoring of demographic parameters may simply be added to the existing protocol. Alternatively, sampling animal abundance may even be dropped and replaced with estimates of animal demographics, and hence truncating the continued understanding of population trends. The extent and form the changes take will depend on budgets, logistics, and stakeholder support and will ultimately be determined by assessing the potential costs associated with change relative to the perceived benefits and opinions of the stakeholders and funders.

## CHANGING THE SAMPLING TECHNIQUES

As new techniques become available that provide more precise or more accurate estimates of animal numbers, habitat availability, or demographic parameters, the tendency is to use the new methods in place of less precise or less accurate methods used to date. This can also be a very appropriate scenario in which to change a monitoring program, but should be undertaken with a controlled approach if at all possible.

What does this entail? Consider the pattern in Figure 14.2A. Changing techniques in year 5 results in a higher $R^2$ and an abrupt change in population estimates. Because changes over time are confounded with changes in techniques, we cannot be sure if the observed trend line is real or an artifact of the changes in techniques.

Using a more controlled approach by making the change to the new technique in year 5 but also continuing to use the old technique as well for years 5–9, however, has the potential to inform managers of the effect due to technique change (Sutherland 1996). This is especially true if a statistical relationship (regression) between the data collected using both techniques allows the managers to standardize the data points for years 0–4 (Figure 14.2B). Such an approach requires extrapolation to years where only one technique was used (not both), and such extrapolations are accompanied by confidence intervals that describe the uncertainty in the data.

This more controlled process has been taken by others and led to informed, helpful changes to already established monitoring programs (Buckland et al. 2005). When the Common Bird Census (CBC) in the United Kingdom was established, the proposed techniques were state-of-the-art. Over time, however, the methods were questioned and increasingly viewed as obsolete. Despite these concerns, the flawed methods were still used due to a fear that any change would undermine the value of the long-time series (Buckland et al. 2005). Eventually, it became obvious that changing the methods had the potential to rectify the problem of misleading and unhelpful data, and was therefore necessary. The British Trust for Ornithology decided to replace the CBC with a Breeding Bird Survey (BBS) in the United Kingdom that was similar to the North American BBS. Yet they also decided to conduct both approaches simultaneously for several years to allow calibration of CBC data to BBS data to provide a bridge in understanding how the results from one technique were associated with another, which then allowed them to move to the BBS approach (Buckland et al. 2005). Strayer et al. (1986) also advised employing new and old techniques simultaneously for a calibration period when changing techniques.

# Changing the Monitoring Approach

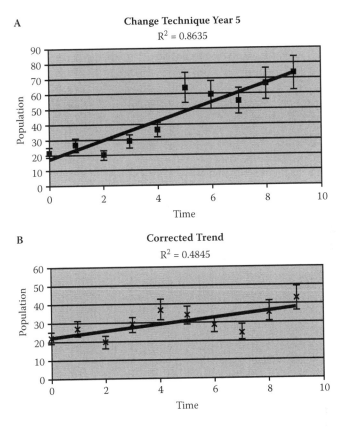

**FIGURE 14.2** Example of a monitoring data set in which an alternative technique was instituted in time period 5 (A). The associated trend line may be due to increased numbers of animals, increased detectability of animals using the new technique, or both. This trend has been standardized to the original technique and shows a much more modest slope (B).

## CHANGING THE SAMPLING LOCATIONS

Changing the sampling locations over time can introduce variability into the data that may make detections of patterns difficult or impossible and many practitioners would therefore be very hesitant to do so. Nonetheless, in several cases it may be required for monitoring to be meaningful. First, not changing sampling location in environments that change more rapidly than the populations that are being monitored will confound data so much that analysis may be impossible unless the location is changed. Consider sampling beach mice on coastal dune environments. If a grid of sample sites are established and animals are trapped and marked year after year, the trap stations may become submerged as the dune location shifts over time. Similar problems arise when sampling riparian systems. Moving the sampling locations is required in these instances. In order to reduce the variance introduced into the sample by continually changing locations, stratification of sampling sites based on topographic features or (if necessary) vegetation structure or composition can help reduce variability. Nonetheless, such stratification will not entirely compensate for increases in variability due to changing sampling locations. It is important to keep this in mind at the outset of a monitoring program designed to sample organisms in dynamic ecosystems. Unless the variability in samples due to changing locations is considered during experimental design, it is possible, and indeed likely, that the statistical power estimated from a pilot study that does not sample new places each year will be inadequate to detect patterns. The pilot study should explicitly consider the variability associated with changing sampling locations from year to year.

The second case in which the sampling location may warrant a change is due to a deficient or misleading pilot study and involves a one-time alteration. While organisms may behave according to certain trends over time or space, one can never be certain that the data from a pilot study embody those trends. To be more specific, if the goal of a monitoring program is to collect data on a particular population of a particular organism, it is likely necessary to choose a sampling location that either constitutes or is embedded within the home range of that population. A pilot survey is therefore necessary to choose the location. Yet, as organisms are normally not physically restricted to their home range, it is possible that survey data from a pilot study could suggest a sampling location that is hardly ever frequented by the focal species. This may be particularly likely for wide-ranging species such as the white-lipped peccary. Given this species' tendency to travel in herds of several hundred animals over an enormous area, the presence of individuals or of sign at one point in time, even if abundant, may not be a good indication of the frequency with which a location is used (Emmons and Feer 1997). If the sampling location chosen inhibits the collection of the desired population parameters, an alteration to the sampling location is necessary.

## CHANGING THE PRECISION OF THE SAMPLES

Despite having collected pilot data and designed a monitoring protocol to detect a given rate of change in a population, unexpected sources of variability may arise (human disturbances, climate, etc.) that reduce your ability to detect a trend. Increasing sample size or reducing sampling error can lead to more precise estimates as illustrated in Figure 14.3 after year 5. After making this change the fit of the line to the points is considerably tighter and the variance about the points is considerably less, leading to greater assurance that the population is indeed increasing. Such changes increase the $R^2$ somewhat (0.74 to 0.78) but the $R^2$ rose from 0.58 for the first 5 years to 0.99 during the second 5 years after improving precision.

But given the risks of altering a monitoring program, is it worth making a change just to increase precision? At what level is an increase in precision worth the risks of making a change? There are no simple answers to questions such as these; the program managers and stakeholders would need to decide if the added costs associated with increasing precision are worth the increased level of certainty associated with the trend estimates.

# Changing the Monitoring Approach

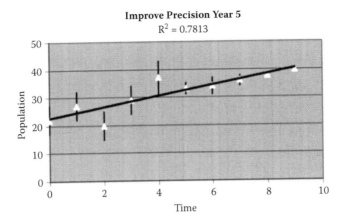

**FIGURE 14.3** Example of a monitoring data set in which sampling intensity was increased and/or sampling error was decreased at year 5 to improve the fit of the trend line. The $R^2$ prior to the change over 5 years was 0.58 and after the change the fit improved to 0.99.

## Changing the Frequency of Sampling

Given constraints on time, money, and people, a decision might have to be made to reduce the level of effort associated with monitoring. Consider the trend in Figure 14.4 where the trend over the first 5 years was positive ($R^2 = 0.58$). In this case, program managers and stakeholders might agree that given the slope of the line, there is little need to be concerned about this population, but that they want to continue some level of monitoring to ensure that the population does not begin to decline in the future. They may decide to reduce the frequency with which monitoring is conducted to every other year with the agreement that should there be more than two consecutive samples showing a decline that they would then revert back to annual sampling.

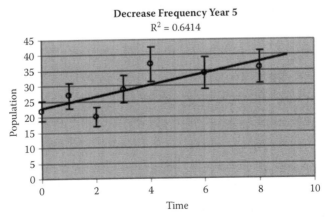

**FIGURE 14.4** Influence of reducing the frequency of monitoring from every year to every other year after 5 years. The decision to move back to annual sampling would be based on a trigger point such as two consecutive time periods showing a decline.

Such an approach should not be taken lightly, however. For instance, in our example, the reduced sampling did continue to demonstrate continual increases in abundance, but the $R^2$ associated with the trend declined. Such a decline in explanatory power may not be particularly important to the stakeholders so long as the population is increasing, but the level of certainty in that trend should also be valuable information to certain stakeholders and therefore carefully considered before making the change. It is also necessary to consider that high precision now may be important for the future. Should the population show declines over time, maintaining the sampling frequency and the high precision of the estimates now may facilitate future management decisions. Changes such as these come at a cost in certainty, money, and time, and truly must be made prudently.

## LOGISTICAL ISSUES WITH ALTERING MONITORING PROGRAMS

One other key component of prudently changing a monitoring program is that the program manager consider the logistical constraints associated with those changes. Training personnel in new techniques requires added time prior to the field sampling season. Where data standardization is necessary, collecting data using both the old and the new techniques adds considerable time and effort to field sampling. Changing locations may mean reestablishing sampling points and recording new GPS locations.

Changing variables that will be measured may require new equipment, additional travel, or different sampling periods. For example, sampling survival of postfledging birds will require different techniques, sampling strategies, and sampling times than estimating abundance of adults from variable circular plot data. These new logistical constraints will need to be evaluated relative to the societal and scientific value of the monitoring program to determine if making the changes is a tenable endeavor.

## ECONOMIC ISSUES WITH ALTERING MONITORING PROGRAMS

Changing monitoring programs in any way usually results in at least an initial expenditure of funds for equipment, travel, or training. It is thus important to address the economic questions of whether the change will result in increased costs and, if so, will the new data be worth the increases in costs. This clearly must be done prior to making the change.

Keep in mind that a modest increase now will tend to compound itself over time, especially in the case of added per diems or salaries for new staff, which must be iteratively increased to reflect cost-of-living trends, or added fuel consumption, which will almost certainly bring increasing costs over time due to rising fuel prices. A simple cost:benefit analysis is a useful way to estimate the marginal increases (or decreases) in costs associated with alternative changes in the monitoring program. The ultimate question that must be addressed is, "Where will you be getting the best information at the least cost and still stay within your budget?"

# Changing the Monitoring Approach

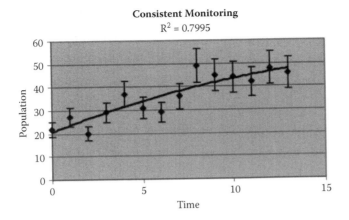

**FIGURE 14.5** Extending the monitoring data portrayed in Figure 14.1 resulted in an asymptote being reached. At this point one could conclude that a carrying capacity has been reached and that further monitoring is unnecessary.

## TERMINATING THE MONITORING PROGRAM

The decision to terminate a monitoring program is probably the most difficult decision that a program manager can make. Obviously, you end a program when you have answered the questions associated with your goals and objectives, correct? But how do you know when that is?

Consider the information in Figure 14.1. It seems quite obvious that the population is increasing. What if we extended the monitoring program another 4–5 years? Perhaps we would see the population approach an asymptote, which we would expect if a carrying capacity was reached (Figure 14.5). At this point, if the monitoring program was established to determine when and if a certain population target was attained, it would be logical to terminate the monitoring program; or if any monitoring was to be continued, it would be appropriate to only maintain it at a minimal level and a low cost.

Alternatively, if the data showed a different trend, such as a decline (Figure 14.6), it would be unwise to assume that the population was increasing or even remaining stable. The decline in population during the last 5 years could simply be due to chance, or to a population cycle, or to some biophysical factor leading to a true long-term decline. Unless comparable data had been collected on reference sites, we would not know if this pattern is likely a cyclic population or if a local event had occurred that would lead to a continued decline. In this case, key questions have not been answered, the goals and objectives of the monitoring program have not been attained, and there is a strong argument for the continuation of monitoring.

## SUMMARY

Changing a monitoring program is done frequently, and has significant consequences with regard to the utility of the data, costs, and logistics. If the data that are collected

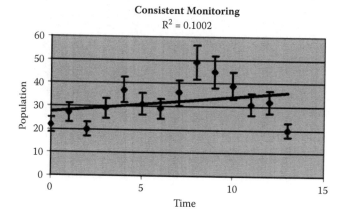

**FIGURE 14.6** An alternative to the pattern in Figure 14.5 is a decline in populations with continued sampling. In this instance monitoring should not be terminated until the cause for the decline is determined.

and analyzed suggest that the goals and objectives are not being met adequately as determined by program management and stakeholders, then revisions in the protocol will be required. Adding or deleting variables, altering the frequency of data collecting, altering sample sizes, or changing techniques to increase precision can all improve a monitoring program and help meet goals and objectives more comprehensively. Yet all of these changes can also lead to changes in costs, in power to detect trends, or other patterns, thus any potential alteration to a monitoring program must be carefully considered.

## REFERENCES

Buckland, S.T., A.E. Magurran, R.E. Green, and R.M. Fewster. 2005. Monitoring change in biodiversity through composite indices. *Philosophical Transactions of the Royal Society Biology*. 360.

Emmons, L., and F. Feer. 1997. *Neotropical Rainforest Mammals: A Field Guide*. The University of Chicago Press, Chicago, IL. 307 pp.

Hoenig, J.M., and D.M. Heisey. 2001. The abuse of power: The pervasive fallacy of power calculations for data analysis. *The American Statistician* 55(1):19–24.

Marsh, D.M., and P.C. Trenham. 2008. Current trends in plant and animal population monitoring. *Conservation Biology* 22:647–655.

Shapiro, J., and E.B. Swain. 1983. Lessons from the silica decline in Lake Michigan. *Science* 221(4609):457–459.

Strayer, D., J.S. Glitzenstein, C.G. Jones, J. Kolas, G. Likens, M.J. McDonnell, G.G. Parker, and S.T.A. Pickett. 1986. Long-term ecological studies: An illustrated account of their design, operation, and importance to ecology. Occasional Paper 2. Institute for Ecosystem Studies, Millbrook, NY.

Sutherland, W.J. 1996. *Ecological Census Techniques*. Cambridge University Press, UK.

Vora, R.S. 1997. Developing programs to monitor ecosystem health and effectiveness of management practices on Lakes States National Forests, USA. *Biological Conservation* 80:289–302.

# 15 The Future of Monitoring

Our world is a dynamic place. This constant—change—has myriad manifestations, some of which we view as negatively impacting us and others as positively impacting us. It should come as no surprise, therefore, that monitoring these changes to understand and prepare for their implications is not a novel endeavor, but a human enterprise that has undergone a long evolutionary process.

Perhaps the earliest form of what we would consider statistically based monitoring arose around the turn of the 17th century in the midst of the worst years of the plague. During this time, the lord mayor of London mandated that parish clerks compile "bills of mortality" to keep track of the ravages of the disease (Mlodinow 2008). From this monitoring data, a man by the name of John Graunt not only created the first life table, but also reached several groundbreaking conclusions about the prevalence and utility of the normal distribution (Mlodinow 2008). Over time, as our values have shifted and we have been forced to confront distinct changes and challenges in our environment, the targets of monitoring have expanded. We still monitor human health, of course, but now also monitor the economy, our education systems, technological advances, and wildlife and their habitat.

The latter has been the topic of this book and we have attempted to provide a fairly consummate and practical overview of the current state of monitoring wildlife and their habitat. We are not the first group to have tackled this topic, neither was the author before us, nor the author before that publication, nor the one before that; indeed, every few years a new monitoring book is published to concisely update practitioners, researchers, and interested parties of the developments in the field. In other words, just like monitoring in a general sense, the monitoring of wildlife and habitat has undergone a long process of evolution that continues unabated even today. In fact, we are currently living in a time during which climate change, the state of international politics, and rapid scientific and technological advancement make the number of changes in our environment and the rate at which they occur astonishing. Thus, over the past 30 years, ecological monitoring and ecology in a more general sense have developed at a particularly impressive pace and incorporated a number of novel monitoring methods and mathematical ways of thinking (Moore et al. 2009).

So given these historical precedents and the volatility of our reality, where is the field headed? What techniques, technology, and mathematics are going to usurp those popular among today's scientists? Will there be any changes in how monitoring data are applied? In this final chapter, we attempt to tackle some of these difficult questions and provide our interpretation of what the indicators of the present might mean for the future of monitoring.

## EMERGING TECHNOLOGIES

### GENETIC MONITORING

Monitoring populations over time through the use of genetic analyses is not necessarily a cutting-edge idea, but the increasing affordability and precision of testing DNA, along with the increasing prevalence of fully sequenced genomes, are slowly creating a more practical, defensible, and widely used system for monitoring animal populations (Schwartz et al. 2006). Increased use frequently begets increased innovation, and this certainly appears to be the case with genetic monitoring.

Schwartz et al. (2006) discuss the field as having two distinct approaches. The first is to undertake diagnostic assays in order to identify "individuals, populations, species and other taxonomic levels" (Schwartz et al. 2006). Data generated from iteratively sampling DNA and undertaking such assays for a population can be used in traditional population models estimating abundance or vital rates. In comparison with many traditional capture–mark–recapture (CMR) techniques, this can be done in a relatively noninvasive manner (i.e., through using hair from hair snares or fecal samples) and may help both reduce biases associated with capturing animals and resolve the controversy and difficulty of capturing rare and elusive species. Diagnostic assays may also prove increasingly helpful in the monitoring of species' range shifts and rates of hybridization as alterations in habitat and forced migration due to anthropogenic changes such as urban sprawl and climate change become more acute (Schwartz et al. 2006).

The second approach uses the monitoring of population genetic metrics such as effective population size, changes in allele frequencies, or estimates of changes to genetic diversity based on expected genetic heterozygosity, as indicators of more traditional population metrics (Schwartz et al. 2006). This approach will be particularly helpful if evolutionary principles can be reliably correlated to population dynamics such that inferences can be made about wild populations. For instance, think of the implications of being able to defensibly compare characteristics of DNA extracted from museum specimens with the DNA of wild specimens; this would allow retrospective monitoring and, potentially retrospective BACI experimental designs.

There are many other exciting potential applications of genetic monitoring. As a recent example, researchers used genetic analysis to estimate the population density and distribution of grizzly bears in and around Glacier National Park (Kendall et al. 2008, 2009). Hair samples were collected through two sampling methods including systematically distributed, baited barbed-wire hair traps and unbaited bear rub trees found along trails. The researchers estimated there was an average number of over 240 bears in the study area resulting in a density of 30 bears/1,000 km$^2$. These noninvasive genetic methods provided critical baseline information for managing one of the few remaining populations of grizzlies in the contiguous United States, and hold promise for monitoring other large mammals through similar methods (Kendall and McKelvey 2008). Genetic information could provide state wildlife agencies with an abundance of new information on how hunters are affecting game populations over time. This could allow them to more carefully regulate hunting in a way that maintains more genetically diverse and economically desirable populations. Using DNA analyses to monitor mixed-species fish stocks (i.e., some species of salmon), bird flocks (i.e., black ducks versus mallard ducks), or mammal populations (i.e., New England cottontail versus Eastern cottontail) that include rare and common species that are difficult to differentiate from one another but are nonetheless harvested due to their economic value could also lead to improved hunting regulations. Indeed, DNA analyses could provide insight into temporal or spatial patterns that are unique to each species, which could serve as a basis for more specific harvest regulations that effectively conserve the rare species. A similar approach has been effective with sockeye salmon in British Columbia (Beacham et al. 2004).

Kilpatrick et al. (2006) used DNA analyses to monitor the blood inside mosquitoes and were able to correlate a shift in feeding behavior from birds to mammals with patterns in West Nile virus outbreaks in North America. Using a genetic monitoring approach to other zoonotic diseases has enormous potential, especially if recent upward trends in urban wildlife populations and the transmission of their diseases to urban citizens are substantiated (Tsukada et al. 2000). Finally, the application of evolutionary principles to genetic monitoring in a general sense will almost certainly provide invaluable insights into how we manage and conserve populations and their habitat.

Despite the enormous potential, there are still a number of limitations to the use of DNA in monitoring. These range from the additional expense of iteratively undertaking DNA assays, the ease with which fraudulent samples can be inserted into the collected data, the prevalence and implications of genotyping errors on any inferences derived from monitoring, and a lack of powerful statistical tools to assess genetic metrics (Schwartz et al. 2006). Yet as additional research is undertaken and more sophisticated simulation software that models these metrics is derived, genetic monitoring will almost certainly help us carry out wildlife and habitat monitoring more comprehensively.

## Monitoring Environmental Change with Remote Sensing

Regardless of one's personal opinions or conclusions concerning climate change, its current and potential impact on our world as an increasing societal, political, and economic concern is undeniable. Further, the science linking climate change to inflated atmospheric levels of greenhouse gases is incontrovertible (IPCC 2007). In light of this, many governments and environmental organizations have either mandated or proposed tighter restrictions on society's $CO_2$ emissions via cap and trade systems, carbon taxes, stricter automobile regulations, or international treaties (Stavins 2008). As discussed in the Introduction to this textbook, these governments and organizations are going to want to know if the money spent designing and implementing these strategies to curb emissions is attaining their objectives. An increase in the monitoring of $CO_2$ emissions in terms of prevalence and strategies, therefore, can only be expected.

One of the most recent innovations involves the use of satellite technology designed specifically for this purpose. In 2009, the Japanese government launched the Greenhouse Gases Observing Satellite (GOSAT), which is expected to collect useful data on global patterns of greenhouse gas emissions (GOSAT Project

2008). A similar U.S. initiative ended in failure (the satellite went into the ocean rather than space), but it seems likely that further efforts will be undertaken (Morales 2009).

Directly related to $CO_2$ emissions is the carbon sequestration capacity of forests. This has historically been monitored to assess the impacts of deforestation on atmospheric $CO_2$ levels, but also represents a way to monitor a locale's contributions to mitigating carbon emissions through conservation and a means to generate data to justify programs to pay for this ecological service. The traditional approach is to undertake limited destructive harvesting in order to measure the capacity of individual trees to store carbon, carry out a ground-based forest inventory, and then use these two sets of data in conjunction with one another to make inferences about an entire forest's, region's, or country's carbon sequestration capacity (Gibbs et al. 2007). Yet given that forest inventories are almost always local in extent by necessity, and that small changes in a tree's characteristics may translate into large changes in carbon sequestration capacity, this approach is rife with potential biases. Although techniques to reduce them based on empirical studies of soils, topography, or climate have been advanced, extrapolating these local data across larger scales can still be tenuous (Gibbs et al. 2007). This has resulted in efforts to more effectively utilize remotely sensed data, which allows for the collection of information specific to each individual habitat type across a region. The typical approach to derive estimates for a forest's capacity to sequester carbon with remotely sensed data is to measure proxies, such as all individual tree heights and crown diameters, and then apply allometric relationships between these proxies and carbon sequestration generated from ground-based studies (Gibbs et al. 2007). Nonetheless, this approach is also vulnerable to several biases, and the reliability of such remotely sensed data in dense forests, such as many of those in the tropics, is questionable (Gibbs et al. 2007). Thus, despite significant advances, especially with the use of radar sensors and light detection and ranging (LiDAR) systems, significant work remains to be done before a comprehensively reliable system is created.

Such advances regarding the monitoring of carbon sequestration, as well as more general advances regarding the monitoring of greenhouse gas emissions in terms of sampling techniques and methods of analyzing data that are global in their extent, should be expected. If the monitoring of climate change continues to reveal the enormous importance of the oceans in mitigating the impacts of this global phenomenon, we may also see the design of a rigorous system to measure and monitor the oceans' capabilities as carbon sinks.

## ADVANCES IN COMMUNITY MONITORING AND THE INTERNET

If, as indicated in Chapter 3, community monitoring becomes even more prevalent than it currently is, it is likely that monitoring techniques designed to attain a high degree of scientific rigor in the hands of the public as well as capture and keep the attention of nonscientists will become even more common. Many such community-oriented innovations to date have simply been variations of time-tested monitoring approaches such as avian surveys (e.g., atlases) or simplified tools to

# The Future of Monitoring

measure a river's nitrate and phosphate levels. There are also a number of efforts underway to increase the rigor of techniques historically popular among citizens, yet frowned upon by scientists. For instance, track-based monitoring has become increasingly adapted in recent years as sign, such as black bear bites and claw marks on trees, has been incorporated into designs previously based solely on track and scat counts (S. Morse, pers. comm.). Such efforts to include indicators that can withstand precipitation and are not as strongly impacted by variations in substrate reduce the potential for certain biases that have always plagued tracking techniques.

Several entirely novel innovations have come about with increases in the public availability of satellite imagery, increasing Internet access, and the ease with which many citizens can now undertake sophisticated mapping exercises. For instance, the Green Map System enables citizens to create maps of their hometowns and insert data indicating the area's most sustainable options for visitors and citizens alike (Green Map 2008). Open source style maps on the system's Website allow users to monitor changes in these locales on the ground and update maps when needed. These projects are akin to the monitoring *of* communities *by* communities, and researchers have only begun to scratch the surface of using social networking Internet sites (e.g., Facebook) for community-based monitoring projects.

The Internet will continue to allow communities to monitor their own natural resources and local animal populations in novel and exciting ways. As Google Maps, Google Earth, and Google Oceans are refined, more of these interactive, democratic community-monitoring and mapping projects will likely evolve in unexpected ways. The participatory monitoring of wildlife populations and their habitats in a public forum is one way for humankind to conceptualize and rigorously keep abreast of our impacts on the ecosystems in which we are embedded and the enormous scale on which they act.

## A NEW CONCEPTUAL FRAMEWORK FOR MONITORING

Although the monitoring of wildlife and their habitat, and the particular scientific theories that inform it, draw heavily from current ideas in ecology, there is also a clear disconnect between ecologists in academia and many who design and implement monitoring programs. Indeed, there tends to be a time lag between the implementation of new ideas in strict ecological research and the subsequent implementation of those ideas in ecological monitoring and management. This is likely due to the conflict between the inevitable uncertainty and theoretical basis of many new concepts in ecology and the need for land managers and practitioners carrying out monitoring to be confident in their protocols and to accomplish specific objectives that are planned far in advance (Moore et al. 2009). To put it simply, land managers and those given a mandate to monitor often have to minimize the risk involved with their projects to maximize their job security. Given this relationship between ecological thinking and monitoring and management, will the key concepts in contemporary ecological thinking manifest themselves in the world of monitoring wildlife and their habitat? If so, how?

## A Reflection on Ecological Thinking

Moore et al. (2009) undertook a Delphi study using a panel of professional ecologists to determine where those concepts currently stand. The most unanimously agreed upon ideas were the following:

1. Disturbances are extremely prevalent and historically contingent phenomena that can impact ecosystems.
2. Considering multiple levels and how each impacts on the ecosystem and on one another is integral to understanding an ecosystem.
3. Simple biodiversity is a poor measure and functional diversity is largely what determines the future characteristics of the ecosystem (Moore et al. 2009).

These ideas—particularly the concept of ecosystems changing via disturbance and consisting of multiple levels—have supported a strong movement toward conceptualizing ecosystems as complex, dynamic, open systems rather than the more parochial views of the past (Moore et al. 2009).

There is, of course, no means of telling how or if these concepts will be involved in future monitoring programs. However, if not only ecological thinking but also contemporary societal values are taken into account, it seems likely that many of these ideas will certainly be integrated into ecological monitoring. In a general sense, contemporary societal values relative to wildlife and their habitats are increasing in complexity due to the global scale on which our current environmental crises and issues act. Global climate change, the globalization of our economy, and an increasing desire to buy "green" products have citizens more carefully monitoring how their everyday behaviors impact on the global environment. In other words, citizens are beginning to incorporate many of the ecological science ideas discussed by Moore et al. (2009) in their lives. For instance, citizens have begun to display the belief that the actions that cause disturbances today will partially determine the state of the ecosystem in the future. Indeed, many strive to emit less $CO_2$ under the assumption that it will make for a more agreeable global climate with fewer health complications in the future (Fay Cortez and Morales 2009; Terrapass 2009). Also, citizens are behaving in ways that display a belief that an ecosystem is impacted by several different interacting levels; there is a strong movement, for example, to buy "green" products that advertise how their producers support conservation elsewhere (UPFRONT 2009). There is a consumer movement based on the idea that environmentally friendly producers in a particular locale can be supported by broader economic activity and that the interaction of and activity on both levels impact the environment. There is also a strong movement to eradicate invasive species based on the assumption that native species are more ecologically healthy and support higher diversity than invasive species (Ruiz and Carlton 2003). Finally, the breadth of these conservation activities, which occur in the grocery market, the gas station, and the local farm, indicates that citizens are implementing, whether consciously or not, a more systems-based approach to the global environment. Conservation and preservation are no longer defined by fencing off protected areas and strictly regulating access and use, but by a variety of consumer behaviors, personal decisions, and lifestyle changes.

# The Future of Monitoring

Given that both ecologists and citizen behaviors and beliefs, which are driven by their values, are largely congruous, and that these are important determinants of the current state of monitoring wildlife and their habitats, it seems highly likely that monitoring will also begin to exhibit similar trends. This means that a more systems-based approach to monitoring may become prevalent. Such approaches monitor indicators at different scales and levels and seek to integrate not simply components of the local ecosystem, but a broader ecosystem that involves the impacts of human beings on several levels. The typical scale of monitoring may become even larger, given widespread concern about global warming and the shifts in flora and fauna that it will cause.

To be sure, monitoring has already exhibited some of these trends. The Breeding Bird Atlas and TRANSECT programs, for instance, have very large geographical scopes. Project BudBurst, albeit designed for younger students, enlists citizens across the United States to monitor the phenophases of their local plants over time, which they hope will create an informative series of maps that describe trends in plant growth that can be compared with climate data to look for correlations (Project BudBurst 2009). Further, as indicated earlier in this chapter, the indicators that wildlife and habitat monitoring programs utilize and the manner in which they do so are expanding. This includes looking at novel indicators on a small scale (DNA) to those on a larger scale ($CO_2$ emission). If a systems-based approach becomes the norm, monitoring programs that include a variety of both new and old techniques that all address different levels impacting a locale may become the norm.

## Dealing with Complexity and Uncertainty

The likely transition of ecological monitoring to a more holistic endeavor that seeks to track changes in multiple animal populations and ecosystems will undoubtedly have a number of analytical challenges in the future. As has been discussed throughout this book, natural systems and populations are dynamic and complex. This complexity changes over time, and in an unpredictable fashion, yet scientists are expected to make predictions of these systems based on the data they collect and their management actions. This is the ultimate moving target in a system full of uncertainty.

Recently, ecologists have turned to advanced mathematics and statistics to aid them in dealing with this dynamic uncertainty. As an example, Chades et al. (2008) proposed partially observable Markov decision processes (POMDP) as an approach for placing resource allocation and monitoring decisions into an objective decision-making process. The authors model three possible scenarios regarding the management of the Sumatran tiger within the Kerinci Seblat region including population management, population surveys to assess whether it is still extant in the region, or the cessation of all conservation efforts to focus resources elsewhere (Chades et al. 2008). The approach identifies which approach should be made each year, for a series of years, given the current belief about the state of the population (extinct or extant). The POMDP approach has several advantages to decision making in monitoring and may have much to offer population monitoring and adaptive management in the future (MacKenzie 2009). What is becoming increasingly clear, however, is that the monitoring of animal populations and their habitats will rely on quantitative and statistical advancements for dealing with uncertainty. This poses a significant problem for many ecologists and managers who are not professionally trained in advanced statistics or computational mathematics, but are responsible for studying and managing natural resources. Consequently, the future of monitoring may involve nontraditional collaborations between ecologists, managers, computational scientists, and statisticians. As an example, computational sustainability is an emerging field that aims to apply techniques from computer science, information science, operations research, applied mathematics, and statistics for balancing environmental, economic, and societal needs for sustainable development. This field promises to have a major influence in ecological research and monitoring in the near future and in the development of computational and mathematical models for decision making in natural resources management. The advantage of these types of collaborations and approaches is that they often involve combinatorial decisions for the management of highly dynamic and uncertain environments. The first annual conference in computation sustainability was held in June 2009 at Cornell University and brought together over 200 computer scientists, applied mathematicians, statisticians, biologists, environmental scientists, biological and environmental engineers, and economists. The future of monitoring and predicting complex ecological systems may very well depend on these types of partnerships.

## SUMMARY

Monitoring is a process of gaining concatenate information and revising approaches to management based on the information gained. This book and others like it have been and will continue to be a part of the monitoring process. Recent approaches that show considerable promise for expansion and proliferation in use among monitoring processes are DNA approaches, community monitoring systems, and systems-based frameworks for collecting and synthesizing monitoring data. Open-source monitoring frameworks allow direct input and utilization of monitoring data that benefit many stakeholders simultaneously and allow many minds to contribute solutions to complex problems based on the data available.

Analytical approaches also will have to adapt to these changing systems to allow rigorous analysis of a steady flow of incoming data so that stakeholders can interpret results to address their goals and objectives. Results will need to directly quantify uncertainty, and they will need to be easily synthesized into systems-based projections of current and likely future conditions. Synthetic approaches must extend beyond biologists and ecologists to economists, social scientists, and mathematicians, among others, to build team approaches to addressing the complex challenges facing wildlife populations and the habitats on which they survive.

## REFERENCES

Beacham, T.D., M. Lapointe, J.R. Candy, K.M. Miller, and R.E. Withler. 2004. DNA in action: Rapid application of DNA variation to sockeye salmon fisheries management. *Conservation Genetics* 5:411–416.

Chades, I., E. McDonald-Madden, M.A. McCarthy, B. Wintle, M. Linkie, and H.P. Possingham. 2008. When to stop managing or surveying cryptic threatened species. *Proceedings of the National Academy of Sciences of the United States of America* 105:13936–13940.

Fay Cortez, M., and A. Morales. 2009. Global warming called top health threat: Medical problems expected with environmental changes. Worcester Telegram and Gazette. /17/2009.

Gibbs, H., S. Brown, J.O. Niles, and J.A. Foley. 2007. Monitoring and estimating tropical forest carbon stocks: Making REDD a reality. *Environmental Research Letters* 2(4):13.

GOSAT Project. 2008. http://www.gosat.nies.go.jp/index_e.html

Green Map. 2008. http://www.greenmap.org/

IPCC (Intergovernmental Panel on Climate Change). 2007. Climate change 2007: The physical science basis. Summary for policymakers. Geneva, Switzerland: Intergovernmental Panel on Climate Change.

Kendall, K.C., and K.S. McKelvey. 2008. *Noninvasive Survey Methods for North American Carnivores*. Island Press, Washington, D.C.

Kendall, K.C., J.B. Stetz, J. Boulanger, A.C. Macleod, D. Paetkau, and G.C. White. 2009. Demography and genetic structure of a recovering grizzly bear population. *Journal of Wildlife Management* 73:3–17.

Kendall, K.C., J.B. Stetz, D.A. Roon, L.P. Waits, J.B. Boulanger, and D. Paetkau. 2008. Grizzly bear density in Glacier National Park, Montana. *Journal of Wildlife Management* 72:1693–1705.

Kilpatrick, A.M., L.D. Kramer, M.J. Jones, P.P. Marra, and P. Daszak. 2006. West Nile virus epidemics in North America are driven by shifts in mosquito feeding behavior. *PLoS Biology* 4:606–610.

MacKenzie, D.I. 2009. Getting the biggest bang for our conservation buck. *Trends in Ecology and Evolution* 24(4):175–177.

Mlodinow, L. 2008. *The Drunkard's Walk: How Randomness Rules Our Lives*. Pantheon Books, New York. 252 pp.

Moore, S.A., T.J. Wallington, R.J. Hobbs, P.R. Ehrlich, C.S. Holling, S.L. Levin, D. Lindenmayer, C. Pahl-Wostl, H. Possingham, M.G. Turner, and M. Westoby. 2009. Diversity in current ecological thinking: Implications for environmental management. *Environmental Management*. 43(1):17–27.

Morales, A. 2009. Satellite to study global-warming gases lost in space. Bloomberg. Accessed online: http://www.bloomberg.com/apps/news?pid=20601082&sid=aj64Vi2YnzMM&refer=Canada

Project BudBurst. 2009. http://www.windows.ucar.edu/citizen_science/budburst/

Ruiz, G.M., and J.T. Carlton. 2003. *Invasive Species: Vectors and Management Strategies*. Island Press, Washington, DC. 528 pp.

Schwartz, M.K., G. Luikart, and R.S. Waples. 2006. Genetic monitoring as a promising tool for conservation and management. *Trends in Ecology and Evolution*. 22:25–33.

Stavins, R.N. 2008. Addressing climate change with a comprehensive U.S. cap-and-trade system. *Oxford Review of Economic Policy*. 24(2):298–321.

Terrapass. 2009. http://www.terrapass.com/

Tsukada, H., Y. Morishima, N. Nonaka, Y. Oku, and M. Kamiya. 2000. Preliminary study of the role of red foxes in *Echinococcus multilocularis* transmission in the urban area of Sapporo, Japan. *Parasitology* 120:423–428.

UPFRONT 2009. People, projects, and programs, news from the field. *Sustainability: The Journal of Record* 2(2):69–78.

# Appendix

## Scientific Names of Species Mentioned in the Text

| Common Name | Scientific Name |
|---|---|
| **Plants** | |
| Common reed | *Phragmites australis* |
| Oak–pine | *Quercus–Pinus* |
| Oregon oak | *Quercus garryana* |
| San Diego ambrosia | *Ambrosia pumila* |
| **Invertebrates** | |
| Giant clam | *Tridacna gigas* |
| **Fish** | |
| Arapaima | *Arapaima* sp. |
| Brown trout | *Salmo trutta* |
| Coho salmon | *Oncorhynchus kisutch* |
| Sockeye salmon | *Oncorhynchus nerka* |
| Whitefish | *Coregonus lavaretus* |
| **Amphibians** | |
| Ensatina salamander | *Ensatina eschscholtzii* |
| Larch Mountain salamander | *Plethodon larselli* |
| Pacific chorus frog | *Hyla regilla* |
| Spring salamander | *Gyrinophilus porphyriticus* |
| Tailed frog | *Ascaphus truei* |
| Torrent salamanders | *Rhyacotriton* spp. |
| Weller's salamander | *Plethodon welleri* |
| **Reptiles** | |
| Desert tortoise | *Gopherus agassizii* |
| Gopher snake | *Pituophis catenifer* |
| Rattlesnake | *Crotalus* spp. |
| Sharp-tailed snake | *Contia tenuis* |
| **Birds** | |
| American robin | *Turdus migratorius* |
| American woodcock | *Scolopax minor* |
| Band-tailed pigeon | *Columba fasciata* |
| Barred owl | *Strix varia* |
| Black duck | *Anas rubripes* |
| Black-and-white warbler | *Mniotilta varia* |
| Black-capped chickadee | *Poecile atricapilla* |
| Blue grouse | *Dendragapus obscurus* |

## Scientific Names of Species Mentioned in the Text (continued)

| Common Name | Scientific Name |
|---|---|
| **Birds (continued)** | |
| Bluebird | *Sialia* spp. |
| Blue-winged warbler | *Vermivora pinus* |
| Brown creeper | *Certhia americana* |
| Brown-headed cowbird | *Molothrus ater* |
| Canada goose | *Branta canadensis* |
| Carolina wren | *Thryothorus ludovicianus* |
| Cooper's hawk | *Accipiter cooperii* |
| Downy woodpecker | *Picoides pubescens* |
| Eastern meadowlark | *Sturnella magna* |
| Eastern towhee | *Pipilo erythrophthalmus* |
| Grasshopper sparrow | *Ammodramus savannarum* |
| Gray jay | *Perisoreus canadensis* |
| Hermit warbler | *Dendroica occidentalis* |
| Hummingbirds | *Archilochus* spp. |
| Mallard | *Anas platyrhinchos* |
| Marbled murrelet | *Brachyramphus marmoratus* |
| Marsh wren | *Cistothorus palustris* |
| Northern goshawk | *Accipiter gentilis* |
| Northern spotted owl | *Strix occidentalis caurina* |
| Olive-sided flycatcher | *Contopus cooperi* |
| Orange-crowned warbler | *Vermivora celata* |
| Pigeon | *Columba livia* |
| Pileated woodpecker | *Dryocopus pileatus* |
| Red-cockaded woodpecker | *Picoides borealis* |
| Song sparrow | *Melospiza melodia* |
| Varied thrush | *Ixoreus naevius* |
| Western bluebird | *Sialia mexicana* |
| White-crowned sparrow | *Zonotrichia leucophrys* |
| Willow flycatcher | *Empidonax traillii* |
| Winter wren | *Troglodytes troglodytes* |
| Wood duck | *Aix sponsa* |
| Wood thrush | *Hylocichla mustelina* |
| Yellow-billed cuckoo | *Coccycus americanus* |
| **Mammals** | |
| African elephant | *Loxodonta africana* |
| American marten | *Martes americana* |
| Beaver | *Castor canadensis* |
| Black bear | *Ursus americanus* |
| Bowhead whale | *Balaena mysticetus* |
| Caribou | *Rangifer tarandus* |
| Cheetah | *Acinonyx jubatus* |

## Scientific Names of Species Mentioned in the Text (continued)

| Common Name | Scientific Name |
|---|---|
| **Mammals (continued)** | |
| Eastern chipmunk | *Tamias striatus* |
| Eastern cottontail | *Sylvilagus floridanus* |
| Fisher | *Martes pennanti* |
| Gopher | *Thomomys* spp. |
| Gray squirrel | *Sciurus carolinensis* |
| Grizzly bear | *Ursus arctos* |
| Ground squirrel | *Spermophilus* spp. |
| Mule deer | *Odocoileus hemionus* |
| New England cottontail | *Sylvilagus transitionalis* |
| Northern flying squirrel | *Glaucomys sabrinus* |
| Northern raccoon | *Procyon lotor* |
| Polar bear | *Ursus maritimus* |
| Red tree vole | *Arborimus longicaudus* |
| Snowshoe hare | *Lepus americanus* |
| Sumatran tiger | *Panthera tigris sumatrae* |
| Virginia opossum | *Didelphis virginiana* |
| White-footed mouse | *Peromyscus leucopus* |
| White-lipped peccary | *Tayassu pecari* |
| White-tailed deer | *Odocoileus virginanus* |

# Index

## A

Absolute density, 198–199
Abstracts, report, 220–221
Abundance
    absolute density or population size, 198–199
    indices, 133, 199–200
    relative, 199–200
Accountability, economic, 3–5
Accuracy assessment, photography, 162
Active adaptive management, 12
Adaptive cluster sampling, 111–112
Adaptive management, 11–13
    Experimental Design for, 97
Adaptive sampling, 111–112
Aerial photography, 160–161
Akaike's information criterion (AIC), 209
Amphibian Research and Monitoring Initiative (ARMI), *219*
Amphibians and reptiles, 141–142
Animal sampling
    aquatic, 135–136
    techniques, 134–141
    terrestrial and semi-aquatic, 136–141
ANOVA designs, 94–95, 210
Aquatic organisms sampling, 135–136
Associations, 80
Assumption of normality, 194–195
Assurance, quality, 126
Atlases, bird, 31–33
Audio recordings, 139, 142
Availability, habitat, 170

## B

Basal areas, 168–169
Bayesian inference, 210
Before-After Control Versus Impact (BACI) designs, 95–97, 100, 208
Belt transects, 116
Binary analyses, 203
Binomial distribution, negative, 198
Biological Metadata Profile, 186
Biological study ethics, 122
Biomass, 169
Biomonitoring of Environmental Status and Trends (BEST), 18–20
Biotic integrity, 147
Birds
    Christmas Bird Count (CBC), 30–33, 220
    Common Bird Census (CBC), 250
    community-based monitoring of, 40–41
    life history and population characteristics, 142
    New York State Breeding Bird Atlas (BBA), 31–33, 46, 237–238
    North American Breeding Bird Survey (BBS), 3, 20–23
Budgets, 110, 127–128, 254
    constraints, 147–148

## C

Carbon dioxide emissions, 260
Cause and effect relationships, 73, 80
    Before-After Control Versus Impact (BACI) designs and, 95–97
    data analysis, 207–209
    monitoring designs, 94
Centric systematic area sampling, 118
Changes
    climate and habitat, 236–239
    design, 248
    economic issues with, 254
    logistical issues with, 254
    methodology, 247–248
    sample precision, 252, *253*
    sampling frequency, 253–254
    sampling location, 251–252
    sampling technique, 250, *251*
    terminating monitoring programs and, 255
    variable, 249–250
    when to make, 248–254
Christmas Bird Count (CBC), 30–33, 220
Citizen-based monitoring, 30–33
Classification schemes, vegetation, 162–163
Clothing, observers', 127
Cluster sampling, adaptive, 111–112
Collaborative approach to community-based monitoring, 48–50
Common Bird Census (CBC), 250
Communication with stakeholders, 63
Community-based monitoring programs (CBMP)
    collaborative approach to, 48–50
    conflict over benefits, 38–44
    defined, 37
    designing and implementing, 44–50, 54
    economic constraints, 38–39
    education and community enrichment through, 40–43
    effectiveness of, 43–44
    ethical considerations in, 39–40
    Internet access and, 260–261

271

participatory action research and, 51–52
prescriptive approach to, 45–47
scientists and, 50–54
systems thinking and, 52–54
Community structure, estimating, 145–147
Complexity and uncertainty of monitoring, 263–264
Comprehensive Conservation Plans (CCPs), 5
Confidence intervals, 192
Contaminants monitoring, 18–20
Content Standard for Digital Geospatial Metadata (CSDGM), 185–186
Coordination and schedule plans, 122–123
Cornell Lab of Ornithology, 40–41
Costs, 110, 127–128, 147–148, 254
Crises, monitoring in response to, 7–10
Cultural value of monitoring, 3
CyberTracker, 180

## D

Data. *See also* Sample(s); Sampling
becoming familiar with, 190–193
collected to meet objectives, 70–74
determining relevant, 62
fitness, 133–134
forecasting trends using, 235–236
forms, 182–184, 189
identifying information needs and, 63–64
inventory, 172–174
meta-, 184–186
models, 193–195
normality assumption of, 194–195
occurrence and distribution, 131–132, 202–205
point independence, 193
Poisson distribution of, 197–198
population size and density, 132–133
predicting patterns over space and time using, 236–239
previously collected, 82–87
remotely sensed, 159–163
requirements, 131–134
statistical distribution of, 197–198
storage, 184
synthesis of, 239–241
thresholds and trigger points, 233–235
transformation, 196
use of existing, 103–110
visualization, 190–195
Data analysis
abundance and counts in, 198–201
Akaike's information criterion (AIC), 209
assumptions, data interpretation, and limitations to, 204–205
Bayesian inference, 210
becoming familiar with data in, 190–193
binary, 203

cause and effect monitoring, 207–209
data transformation in, 196
decision making using, 242–243
generalized linear models (GLM) in, 200–201
homogeneity of variances in, 193–194
independence of data points in, 193
inference, 209–210
models in, 193–195
negative binomial distribution in, 198
nonparametric analysis in, 196–197
normality assumption of, 194–195
occupancy modeling, 204
occurrence and distribution, 202–205
Poisson distribution of data in, 197–198
possible remedies if parametric assumptions are violated in, 195–197
randomization tests, 209
retrospective power analysis, 210–212
risk analysis, 241–242
species density prediction, 204
trend, 205–207
visualization in, 190–195
Databases
digital, 180–182
FAUNA management system, 187
general structure of monitoring, 180
management basics, 179
managers, 186–187
Dead wood sampling, 169
Decision making, 242–243
Density
absolute, 198–199
extrapolating, 205
habitat element, 166
population size and, 132–133
species, 204
Design. *See also* Implementation
ANOVA, 94–95, 210
articulating questions to be answered and, 80–82
Before-After Control Versus Impact (BACI), 95–97, 100, 208
beginning monitoring plan and, 97–101
cause and effect monitoring, 94
changes, 248
detecting the desired effect size in, 100
Experimental Design for Adaptive Management (EDAM), 97
and implementation of community-based monitoring programs, 44–50, 54
incidental observations, 88
inventory, 82, 88–89
previously collected data and, 82–87
process, 79
proposed statistical analyses and, 100
sample, 98
scope of inference and, 101

# Index

selection of specific indicators in, 98–99
types of monitoring, 87–97
use of existing data to inform sampling, 103–110
Detection, 80, 82, *83*, 94–95
  desired effect size, 100
  distances estimation, 105
  effects of terrain and vegetation on, 143–144
  estimating variance associated with indicators using, 105
  existing data to inform sampling design, 104
Digital databases, 180–182
Discussion sections, report, 226–228
Distribution
  analysis of species occurrences and, 202–205
  data, statistical, 197–198
  habitat, 172
  negative binomial, 198
  and occurrence data, 131–132
  Poisson, 197–198
  spatial, 107–109
Diurnal variability, 127
Diversity, species, 203
Documentation
  of field monitoring plans, 125–126
  sample sites, 163

## E

Ecological Risk Assessment (ERA), 241–242
Ecological thinking, 262–263
Economic accountability, 3–5
Economic constraints on community-based monitoring, 38–39
Economic value of monitoring, 2–3
Education
  and community enrichment through community-based monitoring, 40–43
  value of monitoring, 3
Effectiveness
  community-based monitoring, 43–44
  monitoring objective, 64–66
Elephant monitoring, 26–30
Elevation and season considerations, 126–127
Environmental change, 236–239, 259–260
Environmental Monitoring and Assessment Program (EMAP), 23–26, 99
Estimators, 144–145
  biotic integrity, 147
  community structure, 145–147
  habitat elements, 170–172
Ethics
  biological study, 122
  considerations in community-based monitoring, 39–40
Executive summaries, report, 220–221
Existing data, use of, 103–110

Experimental Design for Adaptive Management (EDAM), 97
Extent, spatial, 134

## F

FAUNA database, 187
Federal monitoring
  Biomonitoring of Environmental Status and Trends (BEST), 18–20
  Environmental Monitoring and Assessment Program (EMAP), 23–26
  North American Breeding Bird Survey (BBS), 3, 20–23
Field monitoring plans, documenting, 125–126
Fitness data, 133–134
Forecasting, trend, 235–236
Forest Inventory and Analysis program, U. S. Forest Service, 2–3
Forms, data, 182–184, 189
Frequency of sampling, changing, 253–254
Future of monitoring
  emerging technologies and, 258–261
  new conceptual frameworks and, 261–264

## G

Generalized linear models (GLM), 200–201
Genetic analyses, 141
Genetic monitoring, 141, 258–259
Geographic range changes, 237–238
Global positioning systems (GPS), 124, 162, 163, 180
Gradient nearest neighbor (GNN) approach, 172–173
Greenhouse Gases Observing Satellite (GOSAT), 259–260
Ground-based sampling of habitat elements, 165–169
Ground measurements of habitat elements, 163–165
Ground-truthing, 162

## H

Habitats
  accuracy assessment and ground-truthing of images of, 162
  aerial photography of, 160–161
  availability, 170
  consistent documentation of sample sites for, 163–165
  defined, 155
  distribution across landscapes, 172
  ground-based sampling of, 165–169
  ground measurements of, 163–165
  hierarchical selection of, 157–159

measuring landscape structure and change in, 174–175
predicting patterns over space and time in, 236–239
random sampling, 165
remotely sensed data on, 159–163
resource availability, 155–156
satellite imagery of, 161–162, 261
selection, 156–159
structure and composition changes, 239
suitability, 170–172
vegetation classification schemes and, 162–163
vegetation sampling and, 166–169
Heights, tree, 167
Hierarchical selection of habitat, 157–159
Home range sizes, 238
Homogeneity of variances, 193–194

## I

Identification of information needs, 63–64
Implementation. *See also* Design
biological study ethics and, 122
budgets, 127–128
creating a standardized sampling scheme for, 115–117
logistics, 120–122
permits and, 121–122
qualifications for personnel and, 123
resources needed and, 121
safety plans and, 120–121
sampling unit marking and monuments, 123–124
schedule and coordination plan in, 122–123
selection of sample sites for, 117–120
voucher specimens and, 122
Incidental observations, 88
Independence, data point, 193
Indicators
estimating variance associated with, 105
selection of specific, 98–99
Indices, 144–145
relative abundance, 199–200
Inference
Bayesian, 210
paradigms of, 209–210
scope of, 101, 134
Information needs, identifying, 63–64
Institutional Animal Care and Use Committee (IACUC), 122
Internet, 260–261
Introductions, report, 221
Inventory data, 172–174
Inventory designs, 82, 88–89

## J

Journals, scientific, 51

## K

Kesterson National Wildlife Refuge, 18
Kolmogorov-Smirnov D-test, 194

## L

Landsat Thematic Mapper (TM) imagery, 173
Landscapes
distribution of habitat across, 172
scale, 67–68
structure and change measurements, 174–175
Large River Monitoring Network (LRMN), 20
Legal challenges, monitoring in response to, 10, *11*
Life history, 141–143
Line transect sampling, 138–139
Logistics
issues with altering monitoring programs, 254
resources needed, 121
safety plan, 120–121
tradeoff scenarios, 106
Long Term Ecological Research (LTER), 3

## M

Mammal life history and population characteristics, 142–143
Management
adaptive, 11–13, 97
database, 179, 186–187
Experimental Design for Adaptive, 97
objectives, 65
recommendations in monitoring reports, 228–229
Managers, database, 186–187
Manta tows, 135–136
Mapping, spot, 136, 138
Maps, aerial, 160–161
Marking, sampling unit, 123–124
McLean Game Refuge, 10, *11*
Metadata, 184–186
Microsoft Access and Excel, 180
Mixed effects, GLM, 200–201
Models, data, 193–195
cause and effect, 207–209
inference, 209–210
trend analysis, 205–207
Monitoring. *See also* Community-based monitoring programs (CBMP); Population monitoring
adaptive management in, 11–13
citizen-based, 30–33
complexity and uncertainty of, 263–264
design types, 87–97
ecological thinking in, 262–264
federal, 18–26
future of, 257–264
genetic, 258–259

# Index

incorporating stakeholder objectives in, 61–63
intended users of plans for, 75–76
methodology changes, 247–248
nongovernmental organizations and initiatives, 26–30
objectives, 64–66
as part of resource planning, 5–6, *7*
plan implementation, 97–101
reasons for establishing, 1, 13–14
in response to crises, 7–10
in response to legal challenges, 10, *11*
targeted versus surveillance, 59–61
types of programs for, 17
value of, 2–5
Monitoring Avian Productivity and Survivorship (MAPS) Program, 3
Monitoring the Illegal Killing of Elephants (MIKE), 26–30
Monuments, sampling unit, 123–124

## N

National Ecological Observatory Network (NEON), 3
Natural resources monitoring, 23–26
Negative binomial distribution, 198
New American Amphibian Monitoring Program, 120
New York State Breeding Bird Atlas (BBA), 31–33, 46, 237–238
Nongovernmental organization (NGOs), 4–5
    Monitoring the Illegal Killing of Elephants (MIKE), 26–30
Nonparametric analysis, 196–197
Normality assumption, 194–195
North American Amphibian Monitoring Program, 3
North American Breeding Bird Survey (BBS), 3, 20–23
Northwest Forest Plan (NWFP), 7–8, 12–13

## O

Objectives
    articulating scales of population monitoring, 67–70
    data collected to meet, 70–74
    effective monitoring, 64–66
    management, 65
    sampling, 65
    scientific, 64–65
    stakeholder, 61–63
    what, where, when, who objectives, 65–66
Observation, 80
    aquatic organisms, 135–136
    incidental, 88
    observers' clothing and, 127
    terrestrial and semi-aquatic organisms, 136–141
Occupancy modeling, 204
Occurrence and distribution data, 131–132, 202–205
Organism-centered perspective, 70

## P

Parametric analyses, 93, 195–197
Partially observable Markov decision processes (POMDP), 263
Participatory action research (PAR), 51–52
Passive adaptive management, 11
Peer review, 112
Percent cover, 166–167
Permits, 121–122
Personal Digital Assistants (PDAs), 180
Personnel
    database management, 186–187
    qualifications, 123
Phenological changes, 238–239
Photography, aerial, 160–161
Pin flags, 123
Plans
    budget, 127–128
    documenting field monitoring, 125–126
    resource, 5–6, *7*
    safety, 120–121
    schedule and coordination, 122–123
Plastic flagging, 123, 124
Plausible relationships, 192–193
Plotless sampling, 108–109, 155–156
Point counts, 136
Point transect sampling, 138–139
Poisson distribution, 197–198
Poisson regression, 206–207
Population monitoring, 8–10
    absolute density or population size in, 198–199
    articulating scales of, 67–70
    bird, 3, 20–23
    data collected to meet objectives of, 70–74
    data requirements, 131–134
    effects of terrain and vegetation on, 143–144
    elephant, 26–30
    estimating community structure in, 145–147
    life history and population characteristics in, 141–143
    merits and limitations of indices compared to estimators in, 144–145
    plotless sampling and, 108–109
    selecting appropriate scales of, 156–159
    size and density, 132–133
    spatial distributions, 107–109
    temporal variation, 109–110
Power analysis, retrospective, 210–212

Precision, sample, 252, *253*
Predications, pattern, 236–239
Prescriptive approach to community-based monitoring, 45–47
Project scale, 67
Proportion of occupied area (POA), 202
Protocol review and standardization, 147
PVA models, 240–241

## Q

Qualifications, personnel, 123
Quality control, 126
Questions to be answered, articulating, 80–82, 85

## R

Radio-telemetry, 140–141
Randomization tests, 209
Random sampling
  habitat elements, 165
  simple, 117
  stratified, 118–120
Random site selection, 111
Rangewide scale, 68–70
Recommendations, management, 228–229
Regression, Poisson, 206–207
Relational database management system (RDMS), 181–182
Relative abundance indices, 199–200
Remote sensing, 140, 159–160, 259–260
Reports, monitoring
  abstract or executive summaries, 220–221
  appendices, 229
  components of successful, 219–220
  discussion sections, 226–228
  introductions, 221
  lists of preparers, 229
  management recommendations, 228–229
  methods section, 223–225
  references, 229
  results summaries, 225, *226*
  study area sections, 221–223
  summaries, 230
  titles, 220
Reptiles and amphibians, 141–142
Research, 82
  participatory action, 51–52
  studies, 134
Resource planning, monitoring as part of, 5–6, *7*
Results summaries, 225, *226*
Retrospective analyses, 94–95, 210–212
Review, peer, 112
Risk analysis, 241–242
Road counts, 139

## S

Safety plans, 120–121
Sample(s)
  design, 98
  precision changes, 252, *253*
  sites selection, 100, 117–118
  size estimation, 105
  stratification of, 111
Sampling
  adaptive, 111–112
  aquatic organisms, 135–136
  centric systematic area, 118
  changing frequency of, 253–254
  creating a standardized scheme for, 115–117
  dead wood, 169
  design and cost effectiveness, 110
  design and use of existing data, 103–110
  design peer review, 112
  documenting field monitoring plans for, 125–126
  ground-based habitat element, 165–169
  habitat element random, 165
  location changes, 251–252
  objectives, 65
  plotless, 108–109, 155–156
  schedule and coordination plans, 122–123
  simple random, 117
  stratified random, 118–120
  systematic, 117–118
  technique changes, 250, *251*
  techniques for animal, 134–141
  terrestrial and semi-aquatic organisms, 136–141
  unit marking and monuments, 123–124
  unit selection, 115
  unit size and shape, 115–117
  vegetation, 166–169
Satellite imagery, 161–162, 172–174, 261
Scales of population monitoring, 67–70, 156–159
Schedule and coordination plans, 122–123
Scientific objectives, 64–65
Scientists and community-based monitoring, 50–54
Scope of inference, 101, 134
Season and elevation considerations, 126–127
Selection
  habitat, 156–159
  of monitored species, 74–75
  random site, 111
  sample site, 100, 117–120
  sampling units, 115
  of specific indicators, 98–99
Semi-aquatic organisms, 136–141
Simple random sampling, 117
Sites
  changes in sampling, 251–252
  consistent documentation of sample, 163

# Index

scale, 67
selection of sample, 117–120
Social systems, 53–54
Social value of monitoring, 3
Southern Alliance for Indigenous Resources (SAFIRE), 48
Spatial extent, 134
Spatial patterns, 107–109
Spatial scale of monitoring programs, 44, 92
Species
   assumptions, data interpretation, and limitations, 204–205
   density, 204
   diversity, 203
   occupancy modeling, 204
   occurrences and distribution analysis, 202–205
   selection for monitoring, 74–75
Spot mapping, 136, 138
Square plots, 116
Stabilization, variance, 106–107
Stakeholder objectives, 61–63
Standardization
   critical areas for, 126–127
   protocol review and, 147
   sampling scheme, 115–117
   season and elevation considerations in, 126–127
Status and trends, 73
   designs, 90–93
Storage, data, 184
Stratification of samples, 111
Stratified random sampling, 118–120
Strip transects, 116
Study areas, report, 221–223
Suitability, habitat, 170–172
Surveillance versus targeted monitoring, 59–61
Surveys, 71–72
Sustainability Assessment of Farming and the Environment (SAFE), 240
Synthesis of monitoring data, 239–241
Systemic sampling, 117–118
Systems thinking, 52–54

## T

Targeted versus surveillance monitoring, 59–61
Technologies, emerging, 258–261
Telemetry, radio, 140–141
Temporal variation, 109–110
Terminating monitoring programs, 255
Terrain and vegetation effects, 143–144
Thresholds, data, 233–235

Titles, report, 220
Transects, belt, 116
Transformation, data, 196
Tree heights, 167
Trends and patterns
   data analysis, 205–207
   forecasting, 235–236
   predictions over space and time, 236–239
   spatial distribution, 107–109
Trigger points, data, 233–235

## U

Unbiased estimate of abundance, 72
Uncertainty and complexity of monitoring, 263–264
Users of monitoring plans, 75–76

## V

Value, monitoring
   economic, 2–3
   economic accountability and, 3–5
   social, cultural, and educational, 3
Variable changes, 249–250
Variance(s)
   diurnal, 127
   estimations, 105
   homogeneity of, 193–194
   stabilization, 106–107
   temporal, 109–110
Vegetation
   aerial photography of, 160–161
   basal area estimations, 168–169
   biomass estimation, 169
   classification schemes, 162–163
   dead wood sampling, 169
   density, 166
   percent cover, 166–167
   sampling, 166–169
   satellite imagery of, 161–162
   and terrain effects on population monitoring, 143–144
   tree heights, 167
Visualization, data, 190–195
Visual searches, 139
Voucher specimens, 122

## W

Wildlife habitat relationships, 74
W-test, 194

An environmentally friendly book printed and bound in England by www.printondemand-worldwide.com

PEFC Certified

This product is from sustainably managed forests and controlled sources

www.pefc.org

PEFC/16-33-415

MIX
Paper from responsible sources
FSC® C004959
www.fsc.org

This book is made entirely of sustainable materials; FSC paper for the cover and PEFC paper for the text pages.

#0140 - 241014 - C0 - 234/156/20 [22] - CB